PROOFS AND IDEAS
A Prelude to Advanced Mathematics

AMS/MAA | TEXTBOOKS

VOL 68

PROOFS AND IDEAS
A Prelude to Advanced Mathematics

B. Sethuraman

Providence, Rhode Island

2020 *Mathematics Subject Classification*. Primary 00-01, 00A05.

For additional information and updates on this book, visit
www.ams.org/bookpages/text-68

Library of Congress Cataloging-in-Publication Data

Names: Sethuraman, B. A., author.
Title: Proofs and ideas : a prelude to advanced mathematics / B. Sethuraman.
Description: Providence, Rhode Island : MAA Press, an imprint of the American Mathematical Society, 2022. | Series: AMS/MAA textbooks, 2577-1205 ; Volume 68 | Includes index.
Identifiers: LCCN 2021023870 | ISBN 9781470465148 (paperback) | 9781470467616 (ebook)
Subjects: LCSH: Mathematics. | AMS: General and overarching topics; collections – Introductory exposition (textbooks, tutorial papers, etc.) pertaining to mathematics in general. | General and overarching topics; collections – General and miscellaneous specific topics – Mathematics in general.
Classification: LCC QA37.3 .S48 2021 | DDC 510–dc23
LC record available at https://lccn.loc.gov/2021023870

∞ The paper used in this book is acid-free and falls within the guidelines
established to ensure permanence and durability.
Visit the AMS home page at `https://www.ams.org/`

10 9 8 7 6 5 4 3 2 1 27 26 25 24 23 22

This book is dedicated to my mother.
An infinite source of strength and support.

Contents

Preface

Abstract mathematics is a mixture of rigor and intuition, and to develop either of these, we need repeated exposure and keen practice. As students progress through high school and even early college, they are typically made to spend years working exclusively with algorithmic mathematics ("differentiate this," "move this term over there," "plug that value into it,") before they are allowed to see any abstraction. It is therefore natural for students to have trouble in their first course with abstraction. They may not have seen the language of advanced mathematics. They may not yet be familiar with basic concepts. To them, even the simplest abstract ideas with which the subject works may feel overwhelming.

To help students make the change to abstract mathematics from a more algorithmic approach, many universities offer transition courses designed to provide some warm-up to the students. This book was written for one such course at California State University Northridge.

In the competition between developing the language and tools for writing proofs—statements, logical expressions, sets, functions, induction—on the one hand and developing mathematical intuition on the other, the balance in this book tips slightly towards the intuition side of the equation. I feel that many tools are best developed in context. So, while this book definitely has whole chapters on statements and on sets and functions and on induction, complete with exercises, the bulk of the book focuses on mathematical ideas and on developing creativity. Thus, the book has chapters on combinatorics (counting, the pigeonhole principle), on elementary number theory (divisibility, primes, the Unique Prime Factorization theorem and consequences), on analysis (convergence, continuity, the completeness of the real number system and consequences), on graph theory (the Königsberg Bridge problem, Eulerian trails and circuits), and on elementary group theory (permutations, cyclic groups, matrix groups). Each of these chapters is liberally augmented with exercises, parenthetical remarks, comments on mathematical culture and tips on how to study mathematics.

I feel that a strictly logical presentation of topics, Bourbaki style with all terms defined and every assertion carefully proved, can be confusing to a student at this stage of development—contradictory as this view might be in a course designed to expose students to rigorous mathematics! I believe learning is non-linear: sometimes we need to work with concepts for a long time before we can truly understand them, and it is often pedagogically sound to let students see concepts early even if all the background i's have not been dotted nor all the background t's crossed. So, where necessary, I have not hesitated in doing a bit of "hand-waving" to get to the heart of a subject without getting bogged down in technicalities. For instance, in the first chapter ("Introduction") I introduce students informally to the rings $\mathbb{Z}/n\mathbb{Z}$, representing the elements by just the remainders $0, \ldots, n-1$, and ask them to make computations with these remainders and even prove various assertions, without formally indicating why basic algebraic rules like associativity and distributivity of multiplication over addition hold. I believe that when material is presented in this manner, students can start to work with such objects, and as they do so, these objects will appear less and less intimidating, and learners will develop confidence and build intuition. (Of course, associativity and distributivity are considered later, in Chapter 9, after equivalence relations are introduced.)

The prerequisites for this book are simply high school level mathematics (and naturally, a strong desire to learn!). Various chapters of the book can be used to teach a course on mathematical reasoning in a liberal arts program. (And indeed, I have taught precisely such a course at Krea University from this book.) Just as easily, this book could be used by a bright high school student wanting to learn mathematics beyond the traditional school curriculum. While I invoke complex numbers in various examples, the usage of these numbers in these examples is not critical, and one can simply substitute complex numbers with real numbers without any measurable loss. Calculus is not a prerequisite at all, but since many students are likely to have seen it, I make allusions on occasion to examples or ideas from calculus. Once again, these are not critical, and a student who has not seen calculus can simply skip the relevant material and remain unaffected.

Chapters 1 through 9 deal with the basic bricks from which mathematics is constructed: statements, sets and their cardinalities, functions, and equivalence relations, along with elementary tools: principle of induction, the pigeonhole principle, and simple counting techniques. This is followed by an introductory study of four areas of mathematics: elementary number theory involving divisibility and unique prime factorization in the integers (Chapter 10), beginning analysis covering convergence of sequences and continuity of functions and limits (Chapter 11), the least upper bound principle and various consequences, including the monotone convergence theorem, the Bolzano-Weierstrass theorem, and tests for convergence of series (Chapter 12), beginning group theory covering examples of the dihedral, symmetric, and cyclic groups, and also subgroups, leading up to Lagrange's theorem for finite groups (Chapter 13), and finally, beginning graph theory built from the Königsberg Bridge Problem, leading up to a proof of necessary and sufficient conditions for a graph to have an Eulerian circuit (Chapter 14).

Several courses can be fashioned from this book. Really, this is a matter of instructor taste, but here are some suggestions. For a semester-long course intended for students who need to have their foundations built from scratch, a leisurely study of Chapters 1 through 9, with just a toe-dip, if there is time at all, into any of the remaining chapters on unique factorization (Chapter 10), or sequences (Chapter 11), or groups (Chapter 13) or graphs (Chapter 14) will be apt. For a semester-long course intended for students with greater mathematical preparation, to whom the language and culture of mathematics will come a bit more naturally, a relatively fast study of Chapters 1 through 9, followed by a detailed study of elementary number theory (Chapter 10), followed perhaps by the two analysis chapters (Chapters 11 and 12), or perhaps by a combination of the group theory and graph theory chapters (Chapters 13 and 14), would be more suitable. For a course on mathematical reasoning in a general education program, one could first focus on the introductory chapter (Chapter 1), on the pigeonhole principle (Chapter 2), on counting in finite sets (Chapter 4), and induction (Chapter 7), so that students will be exposed to some fun problem solving, and after a brief study of the chapters on statements (Chapter 3) and sets and functions (Chapter 5), move either to the chapter on the Königsberg Bridge Problem (Chapter 14) or to the proof of the unique prime factorization theorem (Chapter 10). As for bright high school students, such students are likely to be reading this book as a self-study project. Once they have absorbed the basic material on statements (Chapter 3) and sets and functions (Chapter 5), they can pick and choose as they please: it will all be useful and hopefully all be fun and beautiful!

I wish to thank Steve Kennedy at AMS Publications, who shepherded this book through the publication process, as well as my editor Suzanne Larson and her team of reviewers. The suggestions they all made were detailed and invaluable. Of course, the usual caveats apply: only I am to blame for mistakes that remain. I also wish to take this opportunity to thank Loretta Bartolini, the gentle and encouraging editor at Springer-Verlag where too this book was accepted. She and her reviewers also helped influence this book, and I am grateful to have been considered by them.

I wish to thank my colleagues in the Mathematics department at California State University Northridge who used this book to teach the transition course there. Katherine Stevenson provided me with many suggestions as she tested this book in her course. Sungjin Kim was an enthusiastic adopter, who also provided me with comments. Jerry Rosen was a firm supporter, teaching out of the earliest version of the book.

I am grateful to the National Science Foundation for their generous research grant CCF 1318260, under whose broad support the core of this book was written.

B. Sethuraman
California State University Northridge
Krea University, India

1

Introduction

Perhaps you are a college student with a calculus sequence under the belt, and are transitioning to higher level mathematics courses, possibly as part of a mathematics or science degree. Or perhaps you are a college student majoring in the arts or humanities, with a solid high school mathematics background, and perhaps you are curious about what advanced mathematics is all about — maybe you are even taking a mathematics appreciation course to answer your question. Or maybe you are a bright high school student who has mastered much mathematics already, and you are fascinated by the subject and are hungry for more. If you are any of these, and you pick up a textbook of advanced mathematics, with inviting titles like "Introduction to Algebra,", or "Elementary Number Theory," or "Basic Real Analysis," or "Undergraduate Topology," you will find a way of doing mathematics that will be quite unfamiliar, even strange. In fact the very objects that you will see will be new, with names that are not particularly enlightening, and you will wonder whether what you are holding in your hands is even a textbook of mathematics!

Example 1.1. Here are three examples of sentences from introductory books in advanced mathematics; the first is the definition of continuity in analysis (or "advanced calculus"), the second the definition of a normal subgroup in abstract algebra, and the third an assertion in topology:

> **Example 1.1.1.** A function $f : \mathbb{R} \to \mathbb{R}$ is said to be continuous at a if given any $\epsilon > 0$, there exists $\delta > 0$ such that $|x-a| < \delta$ implies $|f(x)-f(a)| < \epsilon$.

> **Example 1.1.2.** A subgroup N of a group G is said to be normal if for all $g \in G$ and $n \in N$, $gng^{-1} \in N$.

> **Example 1.1.3.** The fundamental groups of $\mathbb{R}^2$ minus the origin and $\mathbb{R}^3$ minus the origin are different, so $\mathbb{R}^2$ and $\mathbb{R}^3$ cannot be homeomorphic.

Without some warmup, some intellectual stretching, your first reaction on reading these sentences is likely to be one of bewilderment! There is just so much going on here that is different from what you saw in your calculus courses, and certainly different from what you saw in your high school mathematics courses. To begin with, there is the confounding notation: the colons, the right arrows, the strange fonts, the stuff inside braces, and the profusion of Greek letters. Then, there is the disorienting use of some very familiar words (*group, normal*) to mean something obviously very different from their day-to-day usage, as also the presence of a rather strange word: *homeomorphic*. And finally, there are the complicated conditions and properties that have to be absorbed. For instance, in Example 1.1.1 above, the definition requires that to check continuity of a function at a point you need to keep δ and ϵ apart in your mind, you need to learn which comes first—you are given a positive ϵ, and for this ϵ you have to be able to find a δ—, you need to realize that you have to be able to do this *for any positive* ϵ, and finally, you need to fully grasp the roles that ϵ and δ play—if x is within δ of a, then $f(x)$ must be within ϵ of $f(a)$, etc. To add insult to injury, there is not a single calculation in sight. No application of tidy formulas, no multiplying out terms or factoring polynomials or adding fractions, nothing comforting that the mind can latch on to and operate lazily on auto pilot. Instead, every word in each of these sentences above has you sitting up straight, scratching your head, wondering what is going on.

The purpose of this book is to provide you that warmup, that intellectual stretching needed to help you tackle the mathematics that you will see in your advanced courses. There is indeed a transition to be made. In advanced mathematics, you will need to get away from viewing mathematics as a set of computational rules and focus on what the rules are really saying. You will need to focus on what the objects that are being talked about really mean. Importantly—and this is where the mathematician gets truly excited—you will need to focus on "global" and "structural" issues. For instance: How can we construct number systems in which we can solve all polynomial equations? How can we estimate accurately the number of prime numbers below a given integer? How can we quantify symmetry? What do we really mean by geometry? How can we classify shapes? Etc., etc.

The good news is that *advanced mathematics is a lot of fun, even as it is tremendously deep.* As you become more and more comfortable with the depth, you will not only continue to have the fun you had in your earlier courses, but you will in all likelihood have even more fun, as you discover facets of mathematics that you had not known before to exist, let alone consider. So, as you work your way through the book, even if your mind at first resists the new ways of looking at things, focus on the fun, and you will find yourself joyfully transported to higher planes of thought!

One caveat. *To understand mathematics, you need to pay attention to detail. Every word in a mathematical statement carries weight and means something.* If you tell yourself consciously to pay attention to detail, it will soon become a habit, and you will find it easier and easier to decode a mathematical statement.

In this introductory chapter, just to get our toes wet, we will recall how to prove a few elementary facts from school mathematics, using some simple "algebra." (In advanced mathematics, "algebra" refers to a very deep sub-speciality in mathematics, that often goes by the fuller name of "abstract algebra." When we use the word "algebra" in this textbook without the qualifying word "abstract" in front, we will mean the usual body of rules for combining and simplifying expressions that we learn in high school, such as $(a+b)^2 = a^2 + 2ab + b^2$.) We will then show how the ideas behind these elementary facts can be generalized, leading to some very rich mathematics and new number systems!

Recall that integers are any of the numbers $0, \pm 1, \pm 2, \ldots$. We will denote this collection by $\mathbb{Z}$. We will assume at the very outset that you are familiar with the integers. By "familiar," we simply mean that you have the level of knowledge about these numbers that most students would have after finishing high school. Similarly, we will assume that you are familiar with the rational numbers (or reduced fractions with positive denominators): we will denote this collection by $\mathbb{Q}$; the real numbers (representing lengths on the number line): we will denote this collection by $\mathbb{R}$; and the complex numbers: we will denote this collection by $\mathbb{C}$. (Now, it is quite possible that you did not learn about complex numbers in school, but this is no barrier at all. You can simply ignore the examples we study that involve the complex numbers, and you will not be any the worse for doing so. Of course, we encourage you to learn about complex numbers as soon as you can: we will consider them ourselves in Example 5.41 in Chapter 5.)

Positive integers (1, 2, 3, ...) are also known as *natural numbers*. Accordingly, we will denote the collection of positive integers by $\mathbb{N}$. We will denote the collection of nonnegative integers (0, 1, 2, ...) by $\mathbb{Z}_{\geq 0}$. Similarly, we will denote by $\mathbb{Q}_{>0}$ and $\mathbb{Q}_{\geq 0}$, and $\mathbb{R}_{>0}$ and $\mathbb{R}_{\geq 0}$, the corresponding appropriate collections of rational numbers and real numbers.

Example 1.2. Let us start with a very elementary problem: Prove that the sum of the squares of two odd integers is even.

You can try to test the truth of this statement by taking a few pairs of odd integers at random, squaring them, and adding the squares. For instance, $3^2 + 7^2 = 58$, and 58 is even, $5^2 + 1^2 = 26$, and 26 is even, and so on. This is reassuring, of course, but does not constitute a proof: a proof should reveal why this statement must be true for the squares of *any two* odd integers.

You need to invoke the fact that the given integers are odd. Recall that odd integers are precisely those that leave a remainder of 1 when divided by 2: they are therefore precisely those that are expressible as $2x + 1$ for some integer x. (And of course, even integers are precisely those that are expressible as $2y$ for some integer y.) We need to show that the sum of the *squares* of two integers is even, but let us first recall an easier result: the plain sum of two odd integers (that is, without first squaring the integers) is even. Both because its proof will serve as a model for us and because we will need this fact later, here is a proof of this easier result: let us express one integer as $2x + 1$ for some integer x and the other as $2y + 1$ for some integer y (note that we are using y the second time around—we must use a different letter or else we would be assuming that the two integers we start with are equal, an invalid assumption on our part!). Then their sum is $2x + 1 + 2y + 1 = 2(x + y) + 2 = 2(x + y + 1)$, and this is even because it is a multiple of two.

Before proceeding further, try to solve Exercises 1.2.1 and 1.2.2 below. It will strengthen your understanding of the argument above, so that you can use it with ease later on.

Exercise 1.2.1. Modify the argument above and show that the sum of two even integers is even, and the sum of an even integer and an odd integer is odd.

Exercise 1.2.2. Now modify the argument above further and show that the product of two odd integers is odd, the product of two even integers is even, and the product of an even integer and an odd integer is even. (For instance, to show that the product of two odd integers is odd, we let the first integer be $2x+1$ for some integer x, and the other be $2y+1$ for some integer y, just as above. The product of these two integers is $(2x+1)(2y+1) = 4xy+2x+2y+1 = 2(2xy+x+y)+1$. Now $2xy+x+y$ is some other integer, so the product is of the form 2 times an integer plus 1. Thus, the product is odd.)

Now let us prove the assertion at the start of this example. We will give two proofs:

First Proof. Let the first odd integer be $2x+1$ for some integer x, and the second odd integer be $2y+1$ for some integer y. We square them and add: $(2x+1)^2+(2y+1)^2 = (4x^2+4x+1)+(4y^2+4y+1) = \big(2(2x^2+2x)+1\big)+\big(2(2y^2+2y)+1\big) = (2k+1)+(2l+1)$, where we have written k for the integer $2x^2+2x$ and l for the integer $2y^2+2y$. But $(2k+1)+(2l+1) = 2k+2l+2 = 2(k+l+1) = 2m$, where we have written m for the integer $k+l+1$. Since $2m$ is even, we have proved our assertion. □

Second Proof. Call the first odd integer a and the second odd integer b. Then, by Exercise 1.2.2 above, a^2 $(= a \cdot a)$, being the product of two odd integers, is odd. Similarly, by that same exercise, b^2 $(= b \cdot b)$ is also odd. Finally, as we saw earlier in the paragraph above Exercise 1.2.1, the sum of two odd integers is even, so a^2+b^2 is even. □

A mathematician would describe the second proof as "cleaner" than the first proof! There is an elegance to it: it brings into play a hierarchy of ideas. While the first proof repeats calculations already done in Exercise 1.2.2 and in the discussions above that exercise, the second proof simply recalls certain base results that have already been proved, and then shows that the new result to be proved falls out from these earlier results by just organizing them in the correct hierarchy. Nevertheless, the first proof is also very useful. It is always reassuring to be able to repeat arguments and techniques that have been used before. Moreover, repetition is a sound way to learn a subject.

Now here is something *key* to understanding mathematics: you shouldn't stop here and go home! Ask yourself: what other results must be based on similar reasoning? For instance, what about the the sum of the squares of three odd integers, will it be even or odd? Or, going in a different direction, what about the sum of the cubes of two odd integers? The sum of the squares of four odd integers? Etc., etc.! Formulate your own possible generalizations, and try to prove them! And try to find different proofs (even if they are only superficially different) if you can! When you work at mathematics with vigor, not only will your understanding of the subject deepen considerably, but you will find yourself enjoying the subject immensely!

Notice something about our solution in the example above: *we wrote out our proofs in complete and grammatically correct sentences.* It is very critical that you do this when you write out mathematics from now on. Despite the apparent computational nature of your previous courses, mathematics is really about *ideas*. To describe ideas, and in particular to describe the flow of reasoning from one idea to another, you need the full power of language. (This could be any language that you are comfortable in, although we will use English in this book.) But it goes the other way too: when you use full sentences, and when you focus on getting the grammar right, you force yourself to think carefully about the flow of logic, about what your assumptions are and what your conclusions are, and you prevent yourself from making lazy leaps of faith based on incomplete thought processes. Moreover, when you write out your ideas, you should *explain what all your symbols mean.* If you introduce a symbol "x" for instance in your writeup, you must specify what x is. Is it supposed to be an integer? A real number? Can it take on arbitrary values? Etc. Once again, focusing on getting this right will make you that much more precise in your own thinking: you won't be careless in your mind about what x is supposed to be! (Of course, sometimes, it will be very clear from the context what x is, and in such situations, to avoid tedium, it is acceptable to not specify what x is.)

As already exemplified in the last paragraph of the example above, mathematics needs *active* participation from you if it is to reveal all its secrets to you. You don't just stop after you have solved a problem, instead, *you sit back and contemplate your solution.* What exactly made this solution work? What other results will this same solution technique reveal if we were to use it just a bit differently? What will happen if we were to tweak some of the parameters in the question, e.g., change "two odd integers" to "three odd integers" or change "square" to "cube?" Can we find a different solution altogether? This is a habit you want to get into very quickly, and if you do so, you will soon discover very pretty mathematical results for yourself!

We will leave the solutions of these new questions posed at the end of Example 1.2 as exercises (see Exercise 1.8 through Exercise 1.12 ahead). You just need to push the idea of expressing odd integers as $2x + 1$ for a suitable integer x a little further than in the example.

But let us now proceed in a different direction, to illustrate how abstract mathematics evolves from just simple results like the kind we have considered above. We have seen above that the sum of two odd integers is even—and this does not depend on which two odd integers we pick. Thus, this property may be properly ascribed to the underlying "oddness" of the two integers. Similarly, we have seen that the product of an even integer and an odd integer is even—once again, this does not depend on which even integer and which odd integer we pick. Thus, this property may be properly ascribed to the underlying "evenness" of the first integer and "oddness" of the second integer. We will now make a giant leap; such leaps are very typical of advanced mathematics and lead to immense richness of ideas:

Example 1.3. Let us consider *all* even integers together as just *a single number,* and all odd integers as another single number. How would these new numbers behave? To help us think through this, let us invent symbols for these new numbers: let us denote the collection of all even integers as $[0]_2$ and the collection of all odd integers as $[1]_2$. We know that the sum of any two even integers is even (see Example 1.2 and also the exercises within that example). Since this fact is independent of which even integers we pick, we can interpret this as a property of the entire collection of even integers, and we can decree that $[0]_2 + [0]_2 = [0]_2$. Similarly, since the product of any even integer and any odd integer is even, we can interpret this as a property of the entire collections of even integers and odd integers, and we can decree that $[0]_2 \cdot [1]_2 = [0]_2$.

Exercise 1.3.1. Arguing thusly, fill out the following addition and multiplication tables for the number system comprising of the entire collection of even integers and the entire collection of odd integers:

"+"	$[0]_2$	$[1]_2$
$[0]_2$	$[0]_2$	
$[1]_2$		

"•"	$[0]_2$	$[1]_2$
$[0]_2$		$[0]_2$
$[1]_2$		

Do not be fooled by the simplicity: this new "number system" is a vital object in mathematics and is the basis of all computer science! This number system is known as *the integers mod* 2 and is denoted $\mathbb{Z}/2\mathbb{Z}$. (The word "mod" is short for "modulo," an adjective that in this context means "remainder on division by.") The individual "numbers" $[0]_2$ and $[1]_2$ are referred to as "zero mod 2" and "one mod 2." (Sometimes, if the context of $\mathbb{Z}/2\mathbb{Z}$ is clear, $[0]_2$ and $[1]_2$ are loosely called "zero" and "one," but this is strongly dependent on the context being clear!)

You may want to see Remark 1.6 ahead on the term "number system."

Example 1.4. In Example 1.3 we developed a new number system $\mathbb{Z}/2\mathbb{Z}$, from rules of the form "odd integer times odd integer gives you an odd integer," or "an even integer plus an even integer gives you an even integer." This was already a giant conceptual leap, but we will leap even higher in this example!

Recall that odd integers are those integers that leave a remainder of 1 when divided by 2, and even integers are those that leave no remainder when divided by 2. Here is a typical way mathematicians think: they ask "why only consider remainders on dividing by 2?" "Why not consider remainders on dividing by some fixed integer n, where n could be not just 2, but 3, 4, 5, etc.?" As it turns out, considering remainders on dividing by a general integer n ($n \geq 3$) leads to a very rich body of mathematics that lies at the heart of number theory, and which you will study in your future courses. In particular, it leads to the number systems $\mathbb{Z}/n\mathbb{Z}$ ($n = 3, 4, \dots$) analogous to $\mathbb{Z}/2\mathbb{Z}$. In this example, we will introduce these number systems:

Definition 1.4.1. In the same way as in Example 1.3 above, for any $n \geq 3$, we collect all integers that leave a remainder of 0 when divided by n into one set, and call that set $[0]_n$, all integers that leave a remainder of 1 into one set, and call that $[1]_n$, and so on, to get n new sets $[0]_n, [1]_n, \dots, [n-1]_n$. We define an addition on these sets by the rule $[r]_n+[s]_n = [r+s]_n$ (for $0 \leq r, s, < n$), and similarly, a multiplication by the rule $[r]_n[s]_n = [rs]_n$ (with the understanding that if $r + s$ or rs are $\geq n$, then we first replace them by their remainder on dividing by n). This new "number system" is known as *the integers mod n* and is denoted $\mathbb{Z}/n\mathbb{Z}$.

The addition and multiplication rules described in Definition 1.4.1 above are the analogs of the rules for the $n = 2$ case like "odd integer times odd integer gives you an odd integer," or "an even integer plus an even integer gives you an even integer." (Exercise 1.15 at the end of the chapter provides some of the conceptual underpinnings behind these number systems.)

You will work out the case $n = 3$ completely in Exercise 1.4.1 below. But here are some random examples to give you a feel for these number systems:

(1) In $\mathbb{Z}/4\mathbb{Z}$, $[2]_4+[2]_4 = [2+2]_4 = [4]_4$, but the understanding is that since $4 \geq 4$, we replace 4 by its remainder on dividing by 4, which is 0. So we find $[2]_4+[2]_4 = [0]_4$.

(2) In $\mathbb{Z}/5\mathbb{Z}$, $[3]_5 \cdot [4]_5 = [3 \cdot 4]_5 = [12]_5$, and since $12 \geq 5$, we replace 12 by its remainder on dividing by 5, which is 2. So we find $[3]_5 \cdot [4]_5 = [2]_5$.

(3) In $\mathbb{Z}/23\mathbb{Z}$, $[7]_{23} + [8]_{23} = [15]_{23}$, while $[7]_{23} \cdot [8]_{23} = [10]_{23}$. (Work this out!)

Exercise 1.4.1. As described earlier in this example, we will collect all integers that leave a remainder of 0 when divided by 3 and view it as a single number denoted $[0]_3$, all that leave a remainder of 1 and view it as a single number $[1]_3$, and likewise for the new number $[2]_3$. Using the rules of addition and multiplication we have introduced earlier in this example (see Exercise 1.15.2 above), we find, for instance, $[0]_3 + [1]_3 = [1]_3$, $[0]_3 \cdot [1]_3 = [0]_3$, $[2]_3 \cdot [2]_3 = [1]_3$, etc. Fill out the following addition and multiplication tables for this new number system:

"+"	$[0]_3$	$[1]_3$	$[2]_3$
$[0]_3$			
$[1]_3$			
$[2]_3$			

"•"	$[0]_3$	$[1]_3$	$[2]_3$
$[0]_3$			
$[1]_3$			
$[2]_3$			

This new number system, denoted $\mathbb{Z}/3\mathbb{Z}$, is known as "the integers mod 3," and the individual "numbers" in this system are referred to as "zero mod 3,"one mod 3," and "two mod 3." (Once again, if the context of $\mathbb{Z}/3\mathbb{Z}$ is clear, these numbers are loosely called "zero," "one," and "two," but the context needs to be very clear to justify this laziness!)

Remark 1.5. Example 1.4 leads to new notation and vocabulary. If an integer a leaves a remainder r when divided by n, we say that a is *congruent* to r mod n. We write this as $a \equiv r \bmod n$. More generally, if integers a and b both leave the same remainder on dividing by n, we say a is congruent to b mod n, and write this as $a \equiv b \bmod n$. Thus, $5 \equiv 2 \bmod 3$, and also $5 \equiv 8 \bmod 3$.

Remark 1.6. Of course, we have not really told you what a "number system" is, but have happily referred to $\mathbb{Z}/n\mathbb{Z}$ in Example 1.4 above (and earlier to $\mathbb{Z}/2\mathbb{Z}$ in Example 1.3) as number systems! This is an introductory chapter, where we are just giving you glimpses of mathematics that will come ahead, so we will leave the term undefined. Our goal here is simply to make you aware that there are "numbers" other than just the real numbers and the complex numbers, and to make you learn to compute with numbers in systems like $\mathbb{Z}/n\mathbb{Z}$ to gain familiarity with them. Later, in courses in abstract algebra, you will learn about rings and fields, and will have a clearer picture of what a number system is!

1.1 Further Exercises

Exercise 1.7. Practice Exercises:

> **Exercise 1.7.1.** What number serves as "zero" in $\mathbb{Z}/2\mathbb{Z}$? (Note: "zero" should be that number x that when added to any number y in the system gives you back that same number y.)
>
> **Exercise 1.7.2.** What number serves as "one" in $\mathbb{Z}/2\mathbb{Z}$? (Note: "one" should be that number x that when multiplied by any number y in the system gives you back that same number y.)
>
> **Exercise 1.7.3.** What number serves as the negative of $[1]_2$ in $\mathbb{Z}/2\mathbb{Z}$? (Note: the negative of a number x should be that number y that when added to x gives you zero!)
>
> **Exercise 1.7.4.** What number serves as the reciprocal of $[1]_2$ in $\mathbb{Z}/2\mathbb{Z}$? (Note: the reciprocal of a number x should be that number y such that xy equals one!)
>
> **Exercise 1.7.5.** Repeat Exercises 1.7.1 through 1.7.4 for the number system $\mathbb{Z}/3\mathbb{Z}$.
>
> **Exercise 1.7.6.** What number serves as the reciprocal of two in $\mathbb{Z}/3\mathbb{Z}$?
>
> **Exercise 1.7.7.** Find two "numbers" in $\mathbb{Z}/4\mathbb{Z}$ that are not zero, but multiply out to zero! (Your multiplication table should show that these numbers do not have a reciprocal in $\mathbb{Z}/4\mathbb{Z}$. That is, if you call either of these numbers you have found $[x]_4$, there should be no $[y]_4$ such that $[x]_4[y]_4 = [1]_4$. See Exercise 1.16 ahead.)
>
> **Exercise 1.7.8.** In the number systems you have used so far ($\mathbb{Z}$, $\mathbb{Q}$, etc.) there have been two and only two distinct numbers, namely 1 and -1, whose square is 1. Verify that in the number system $\mathbb{Z}/8\mathbb{Z}$, there are *four* distinct numbers whose square is 1, whereas in $\mathbb{Z}/2\mathbb{Z}$, there is just one number whose square is 1.

Exercise 1.8. Prove that the sum of the squares of three odd integers is odd.

Exercise 1.9. Prove that the sum of the squares of four odd integers is even.

Exercise 1.10. Let k be a positive integer. Prove that the sum of the squares of k odd integers is odd if k is odd, and even if k is even.

Exercise 1.11. Prove that the sum of the cubes of two odd integers is even.

Exercise 1.12. Prove that for any integer m, $m^2 - m$ is even.

Exercise 1.13. Use the multiplication table developed in Exercise 1.4.1 to show that 33999935 is not the square of any integer. (Hint: Consider the remainder when dividing by 3. What does the table in Exercise 1.4.1 tell you about the remainder left by integers that are squares on dividing by 3?)

Exercise 1.14. Show that 33999935 cannot be written as the sum of two square integers, that is, one cannot find integers x and y such that $x^2 + y^2 = 33999935$. (Hint: Consider the remainder when dividing by 4. Study the multiplication table for $\mathbb{Z}/4\mathbb{Z}$ and determine what possible values $[x]_4^2 + [y]_4^2$ could take as $[x]_4$ and $[y]_4$ vary.)

Exercise 1.15. In this exercise we will prove two results that provide some of the foundations for the number systems $\mathbb{Z}/n\mathbb{Z}$. Let n be given (n an integer ≥ 2). As utilized extensively in Example 1.4, we may write any integer x using integer division as $x = np + r$, where p is the quotient on dividing x by n and r is the remainder—thus, r is in the range $0 \leq r < n$. Let y be a second integer; we may similarly write $y = nq + s$, where again, q is the quotient on dividing y by n, and s is the remainder—so s is in the range $0 \leq s < n$.

Exercise 1.15.1. While it is clear from our experience that the remainder that the integer x leaves on dividing by n is uniquely determined by x and n (that is, two different algorithms for integer division, as long as they are correct, will produce the same result), this is worth establishing properly. (After all, if to the contrary what remainder you get depends on which algorithm you use, the number systems $\mathbb{Z}/n\mathbb{Z}$, which have all been based on remainders, will turn out to be very wobbly structures!) So, suppose that $x = np + r$, and also $x = np' + r'$, where $0 \leq r, r' < n$. (Thus, r and r' are two possible remainders one can get on dividing x by n, depending on which algorithm you use.) Show that $r = r'$. (Hint: Observe that the given equalities yield that $n(p - p') = r' - r$. Consider absolute values.)

Exercise 1.15.2. Using your knowledge of integer addition, multiplication, and division, show that the sum $x + y$ leaves the same remainder on dividing by n as that left by $r+s$, and that the product xy leaves the same remainder on dividing by n as that left by the product rs. (Hint: For instance, when considering the sum $x + y$, first divide $r + s$ by n and write $r + s = nk + l$, where $0 \leq l < n$. Then $x + y = np + nq + r + s = np + nq + nk + l = n(p + q + k) + l$.)

Remark 1.15.1. The result above provides the analog of results like "the sum of two odd integers is even" and "the product of an even integer and an odd integer is even" for remainders on dividing by a general n—these analogous results are the basis of the number systems $\mathbb{Z}/n\mathbb{Z}$. For instance, when $n = 3$, it shows that any integer that gives a remainder of 1 on dividing by 3, when added to any integer that gives a remainder of 2 on dividing by 3, yields an integer that leaves the same remainder as $1 + 2 = 3$ on dividing by 3, that is, it leaves a remainder of 0. It is this family of results that motivates the addition and multiplication operations in $\mathbb{Z}/n\mathbb{Z}$.

Exercise 1.16. Suppose n is a (positive) composite integer. (Recall that this means that $n = ab$ for some positive integers a and b, with neither a nor b equal to 1—we say in such a situation that "n has a nontrivial factorization," that is, a factorization other than the obvious or "trivial" factorization $n = n \cdot 1$.) Show that there exist two "numbers" $[x]_n$ and $[y]_n$ in $\mathbb{Z}/n\mathbb{Z}$ such that neither equals $[0]_n$, yet, $[x]_n[y]_n = [0]_n$. Now show that neither $[x]_n$ nor $[y]_n$ can have a reciprocal in $\mathbb{Z}/n\mathbb{Z}$. (See Exercise 1.7.7 for a special case of this phenomenon.)

Exercise 1.17. This exercise is in contrast to Exercise 1.16 above! Show that if p is prime integer, then there do not exist nonzero $[x]_p$ and $[y]_p$ in $\mathbb{Z}/p\mathbb{Z}$ such that $[x]_p[y]_p = [0]_p$. (Hint: what do you know from school about a prime dividing a product of two integers? We will prove this suggested result formally in Proposition 10.21 in Chapter 10, but this is an introductory chapter and you should feel free to use this result.)

Exercise 1.18. We know that in $\mathbb{Z}/2\mathbb{Z}$, since $[0]_2^2 = [0]_2$ and $[1]_2^2 = [1]_2$, both $[0]_2$ and $[1]_2$ are squares. In this exercise we will show that if p is a prime not equal to 2, then not all elements in $\mathbb{Z}/p\mathbb{Z}$ are squares; in fact, we will show that precisely $(p+1)/2$ elements in $\mathbb{Z}/p\mathbb{Z}$ are squares.

(1) Observe that $[0]_p$ is already a square since $[0]_p^2 = [0]_p$. (That is correct: you do not have to do anything in this part of the exercise except observe!)

(2) Let us now focus on the *nonzero* elements of $\mathbb{Z}/p\mathbb{Z}$. Consider the assignment that sends each nonzero element of $\mathbb{Z}/p\mathbb{Z}$ to its square (thus, $[a]_p$ goes to $[a]_p^2$). Use the result of Exercise 1.17 above to show that $[a]_p^2 = [b]_p^2$ if and only if either $[a]_p = [b]_p$ or else $[a]_p = -[b]_p$. (Hint: You can observe informally—you have to wait till a formal course in abstract algebra for a rigorous proof—that the relation $x^2 - y^2 = (x-y)(x+y)$ continues to be true in number systems like $\mathbb{Z}/p\mathbb{Z}$.)

(3) Show that we cannot simultaneously have $[a]_p = [b]_p$ *and* $[a]_p = -[b]_p$. (Hint: argue that if this were to happen, $[2]_p[a]_p$ must be zero. Use the result of Exercise 1.17 above to show that $[2]_p[a]_p = 0$ is impossible.)

(4) Conclude from Part 3 by a simple count applied to the assignment in Part 2 that the number of nonzero elements of $\mathbb{Z}/p\mathbb{Z}$ that are squares is $(p-1)/2$.

(5) Conclude from Parts 1 and 4 that there are $(p+1)/2$ elements in $\mathbb{Z}/p\mathbb{Z}$ that are squares.

Exercise 1.19. A consequence of Exercise 1.18 above is that for odd primes p, not all elements of $\mathbb{Z}/p\mathbb{Z}$ are squares, since $\mathbb{Z}/p\mathbb{Z}$ has p elements, $(p+1)/2$ of them are squares, and $p > (p+1)/2$. We will show in this exercise that, however, every element of $\mathbb{Z}/p\mathbb{Z}$ (for odd p) can be written as the *sum* of two squares. Thus, given arbitrary $[a]_p \in \mathbb{Z}/p\mathbb{Z}$, we will show that there exist $[x]_p$ and $[y]_p$ in $\mathbb{Z}/p\mathbb{Z}$ such that $[x]_p^2 + [y]_p^2 = [a]_p$.

(1) Let T be the set of elements of the form $[a]_p - [z]_p$ in $\mathbb{Z}/p\mathbb{Z}$ with the property that $[z]_p$ is a square. Show that T also has $(p+1)/2$ elements. (Hint: Let S be the set of squares in $\mathbb{Z}/p\mathbb{Z}$, and consider the assignment that takes $[a]_p - [z]_p$ in T to $[z]_p$. Show that this pairs each element of T with an element of S, no two elements of T are paired with the same element of S, and every element of S takes part in the pairing. This will show that T and S have the same number of elements. In the language of Definition 5.17, Part (5) of Chapter 5 ahead, our assignment creates a bijection between T and S.)

(2) As in the hint to part (1) above, let S be the set of squares in $\mathbb{Z}/p\mathbb{Z}$. Show that S and T have nonempty intersection. Conclude that there exist $[x]_p$ and $[y]_p$ in $\mathbb{Z}/p\mathbb{Z}$ such that $[x]_p^2 + [y]_p^2 = [a]_p$.

Remark 1.19.1. Of course, it is trivial to check that every element of $\mathbb{Z}/2\mathbb{Z}$ is also a sum of squares: $[0]_2 = [0]_2^2 + [0]_2^2$, and $[1]_2 = [1]_2^2 + [0]_2^2$.

Exercise 1.20. Let p be a prime, and let a be an integer. Define the object $\left(\frac{a}{p}\right)$ (known as the *Legendre symbol*) to stand for 1 if $p \nmid a$ and $[a]_p$ is a square in $\mathbb{Z}/p\mathbb{Z}$, -1 if $p \nmid a$ and $[a]_p$ is not a square, and 0 if $p|a$. Assume now that p is odd. Show that $\sum_{a=1}^{p-1}\left(\frac{a}{p}\right) = \sum_{a=0}^{p-1}\left(\frac{a}{p}\right) = 0$. (Hint: Part (4) of Exercise 1.18 above.)

Exercise 1.21. This exercise is for those who have some familiarity with real-valued functions defined on the real line: a typical calculus course and a willingness to expand your repertoire of functions to those that are not just defined by a single formula should do. Write $\mathcal{F}$ for the collection of all real-valued functions on the real line. Write $0_{\mathcal{F}}$ for the function that sends every $x \in \mathbb{R}$ to 0, and $1_{\mathcal{F}}$ for the function that sends every $x \in \mathbb{R}$ to 1. Recall that the sum of two functions f and g in $\mathcal{F}$, denoted "$f+g$," is the new function defined by $(f+g)(x) = f(x) + g(x)$, and the product of f and g, denoted "fg," is the new function $(fg)(x) = f(x)g(x)$. We may thus think of the set of real-valued functions on the real line as a "number system" in its own right!

(1) Construct infinitely many pairs of functions (f, g) from $\mathcal{F}$, $f \neq 0_{\mathcal{F}}$ and $g \neq 0_{\mathcal{F}}$, such that $fg = 0_{\mathcal{F}}$. (Hint: Go beyond functions defined by single formulas. For instance, consider functions like the following:

$$f(x) = \begin{cases} x & \text{if } x \geq 0, \\ 0 & \text{if } x < 0. \end{cases}$$

Play with these.)

(2) Construct infinitely many pairs of functions (f, g) from $\mathcal{F}$ such that $fg = 1_{\mathcal{F}}$. Try to find pairs for which neither f nor g are constant functions, that is, they are not of the form $f(x)$ (or $g(x)$) $= c$ for some fixed real number c.

(3) Suppose that functions f and g in $\mathcal{F}$ satisfy $fg = 1_{\mathcal{F}}$. Show that $f(x) \neq 0$ for any $x \in \mathbb{R}$. (Hint: if $f(x) = 0$ for some $x \in \mathbb{R}$, what can you say about $f(x)g(x)$?)

(4) Show that if f, g, and h in $\mathcal{F}$ satisfy $fg = 1_{\mathcal{F}}$ and $fh = 1_{\mathcal{F}}$, then g must equal h as functions. (Hint: for any $x \in \mathbb{R}$, we find $f(x)g(x) = 1$ and $f(x)h(x) = 1$ from the definition of the product of two functions and the definition of the function $1_{\mathcal{F}}$. Observe from these that $f(x)(g(x) - h(x)) = 0$, and invoke part (3) above.)

Exercise 1.22. In Example 1.4, and earlier in Example 1.3, we formed new number systems from the integers, using remainders on division by some integer $n \geq 2$. There is another number system beside the integers where we have a nice notion of division: the set of polynomials in a single variable with coefficients in $\mathbb{R}$. We can play a similar game as in Examples 1.4 and 1.3, and create new number systems from remainders from polynomial division.

Let us write $\mathbb{R}[x]$ for the set of polynomials in the variable x with real coefficients. (So, these are objects of the form $a_0 + a_1x + \cdots + a_nx^n$, where $n \geq 0$, and the a_i are real numbers.) We will refer to individual polynomials as f, g, q, r, etc. Recall that given polynomials f and g, with g of degree at least 1, we can do long division and divide f by g to produce a quotient q and a remainder r, with the remainder either being the zero polynomial or of lower degree than g. In this exercise, just to keep things simple, we will focus on the case where $g = x^2 + 1$, although suitably modified, the same game can be played for any choice of g of degree at least 1. The choice of $x^2 + 1$ will lead to a significant example, as we will see.

With this choice of g, long division shows that any polynomial f in $\mathbb{R}[x]$ can be written as $(x^2+1)q+(ax+b)$ for a quotient polynomial q and a remainder polynomial $ax + b$ of degree at most 1 (so a and b are suitable real numbers, and each can be 0).

The following are analogs of Exercises 1.15.1 and 1.15.2:

(1) Show that the remainder $ax + b$ is uniquely determined by f. In other words, suppose $f = (x^2 + 1)q + (ax + b)$ and also $f = (x^2 + 1)q' + (a'x + b)$. Show that $a = a'$ and $b = b'$. (Hint: Proceed as in the hint of Exercise 1.15.1, except work with degrees of polynomials instead of absolute values of integers. Your proof will also show that $q = q'$.)

(2) Suppose f and g are two polynomials in $\mathbb{R}[x]$, and suppose $f = (x^2+1)q+(ax+b)$ and $g = (x^2+1)s+(cx+d)$. By part (1) above, the remainders $ax+b$ and $cx + d$ are uniquely determined by f and g (respectively). Using your knowledge of polynomial addition, multiplication, and division, show that the sum $f + g$ leaves the remainder $(a + c)x + (b + d)$ on dividing by $x^2 + 1$, and that the product of fg leaves the same remainder on dividing by x^2+1 as that left by the product $(ax + b)(cx + d)$. (Hint: The proof for the sum

should be clear. For the product, write $(ax+b)(cx+d)$ as $(x^2+1)t+(ex+f)$ for a suitable polynomial t (t will in fact be a constant) and proceed as in Exercise 1.15.2.)

The collection of remainders on dividing polynomials by x^2+1 is clearly the set of polynomials of the form $ax+b$, as a and b vary in $\mathbb{R}$. As with $\mathbb{Z}/n\mathbb{Z}$ (Exercise 1.15), we will create a new number system with these remainders. First, to focus on the fact that they are remainders, we will (just as with $\mathbb{Z}/n\mathbb{Z}$) denote the remainders as $[ax+b]_{x^2+1}$, as a and b vary in $\mathbb{R}$, and then, to simplify our notation, we will simply denote these as $[ax+b]$. (Thus, in analogy with $\mathbb{Z}/n\mathbb{Z}$, $[ax+b]$ stands for the collection of *all* polynomials f that leave the remainder $ax+b$ on dividing by x^2+1.) These remainders $[ax+b]$ will be the "numbers" of our new number system. Note that unlike in $\mathbb{Z}/n\mathbb{Z}$, there are now infinitely many numbers, as a and b vary in $\mathbb{R}$. So instead of a table that shows how every pair of remainders add (or multiply), we will define directly how the general pair of remainders $[ax+b]$ and $[cx+d]$ add and multiply.

Definition 1.22.1. Addition is easy: $[ax+b]+[cx+d]$ should be (in analogy with $\mathbb{Z}/n\mathbb{Z}$) the remainder of $(ax+b)+(cx+d)$ on dividing by x^2+1, which by part (1) above is $(a+c)x+(b+c)$. Hence, we define $[ax+b]+[cx+d]$ to be $[(a+c)x+(b+c)]$. As for multiplication, $[ax+b]\cdot[cx+d]$ should be (again in analogy with $\mathbb{Z}/n\mathbb{Z}$) the remainder of $(ax+b)\cdot(cx+d)$ on dividing by x^2+1. Now, $(ax+b)\cdot(cx+d) = acx^2+(ad+bc)x+bd$. We can divide this by x^2+1 using long division, but it is easy to observe instead that we may write the right side as $ac(x^2+1)-ac+(ad+bc)x+bd = (x^2+1)ac+(ad+bc)x+(bd-ac)$. It is clear now that the remainder on dividing by x^2+1 is $(ad+bc)x+(bd-ac)$. We hence define $[ax+b]\cdot[cx+d]$ to be $[(ad+bc)x+(bd-ac)]$.
We denote this new number system as $\mathbb{R}[x]/\langle x^2+1\rangle$ (read as "R X mod x^2+1").

Before doing the following, think about this multiplication rule above. Where else have you seen this sort of multiplication?

(1) Verify that $[ax+b]+[0]=[ax+b]$ and $[ax+b]\cdot[1]=[ax+b]$.

(2) Verify that $[-ax-b]$ serves as the negative (or *additive inverse*) of $[ax+b]$.

(3) Assume that at least one of a or b is not zero. Show that $\left[\frac{-a}{a^2+b^2}x+\frac{b}{a^2+b^2}\right]$ serves as the reciprocal of $[ax+b]$.

(4) Note that x $(=1\cdot x+0)$ is a possible remainder, as is r $(=0\cdot x+r)$ for any real number r. Verify that $[x]^2=-[1]$.

(5) Assign the remainder $[ax+b]$ to the complex number $ai+b$ and the remainder $[cx+d]$ to the complex number $ci+d$. (Thus, $[x]$ $(=[1\cdot x+0])$ gets assigned to i.) What similarities can you observe about how $[ax+b]$ and $[cx+d]$ add and multiply in $\mathbb{R}[x]/\langle x^2+1\rangle$ and how the complex numbers $ai+b$ and $ci+d$ add and multiply?

Intuitively, part (5) above shows that the number system $\mathbb{R}[x]/\langle x^2+1\rangle$ is "the same" as the familiar number system $\mathbb{C}$, once $[x]$ has been identified with i. The point here is that we now have a *different* way of constructing a number system containing $\mathbb{R}$ in which there is a square root of -1. Instead of having to say "let us invent a new number i and decree that $i^2 = -1$," we see here that a square root of -1 appears naturally when we consider the arithmetic of the remainders of polynomials with real coefficients on dividing by x^2+1.

2

The Pigeonhole Principle

To understand mathematics beyond introductory computationally oriented material, you need to teach yourself to work with ideas, instead of just with formulas. Thus, just as you played around with different formulas ("$x^2 - y^2 = (x + y)(x - y)$," or "the derivative of x^n is nx^{n-1}"), from now on, you will be playing around with ideas, taking them apart to understand them, combining them to form more complex ideas, and so on. This is of course much harder, but you see deeper, and you learn more and more beautiful mathematics, all of which only makes it that much more fun!

Necessarily, there is a lot of discovery that happens when you study advanced mathematics. Solutions cannot be written down algorithmically, as in computational calculus for instance. Instead, you have to play with the material given to you and discover various results for yourself and then combine them with known ideas to generate solutions to problems. Thus, doing advanced mathematics entails a good deal of creativity, which of course makes it all the more enjoyable. But all of this takes some getting used to, and, as always, a necessary ingredient for learning to think this way, as for learning anything, is practice.

In this chapter we will consider a beautiful but extremely elementary and obvious principle known as the Pigeonhole Principle (which we will abbreviate as PHP). We'll see that the very simple idea behind this principle is powerful enough to allow us to solve some seemingly very difficult problems. Trying to discover solutions to these problems is a lot of fun, and it can be very satisfying when you have successfully solved one of these problems. They provide excellent practice for working with ideas—there are no formulas anywhere in sight for you to manipulate!

2.1 Pigeonhole Principle (PHP)

It is easiest to start with an example that illustrates the PHP:

Example 2.1. Prove that in a group of 13 people, there must be at least two people who were born in the same month of the year.

The idea behind the proof is the following: there are 12 possible months of the year, January through December. Think of each month as a room, and place each person in the room corresponding to the month of the year in which the person was born. Then it is clear that because there are 13 people but only 12 rooms, there must be at least one room in which more than just one person has been placed. That proves it.

As it turns out, that was just the PHP at work! Pigeonholes are open compartments on a desk or in a cupboard where letters are placed. They were apparently named after actual boxes used in medieval times where pigeons would nest:

Figure 2.1. Used under the terms of the GNU Free Documentation License, licensed under Creative Commons Attribution-Share Alike 3.0 Unported `https://creativecommons.org/licenses/by-sa/3.0/deed.en` License, Author BenFrantzDale commonswiki.

Principle 2.2. *Pigeonhole Principle*: Let n be a positive integer. If more than n letters are distributed among n pigeonholes, then at least one pigeonhole must contain two or more letters.

Thus in Example 2.1 above, the months represent the pigeonholes, and the persons represent the letters. There are more letters (persons) than pigeonholes (months), so at least one pigeonhole (month) should contain two or more letters (persons). The fun here lay in setting up the problem as a PHP problem, by thinking of a person's month of birth as a pigeonhole into which we are placing that person!

Here is another example:

Example 2.3. Prove that among any 6 integers, there must be at least one pair whose difference is divisible by 5.

This is an example where the choice of the pigeonholes requires a bit of cleverness —they are not one of the objects initially given to you in the statement of the problem. This is an instance of the kind of transition to be made when studying advanced mathematics: you will have to discover for yourself the right ideas to apply in a given situation. The ideas needed to solve a particular problem need not be very complicated, but they often require some deep thought for you to discover them. *But the effort is very rewarding,* since you will then understand the situation much more deeply than if it were a mere "plug-this-value-of-x-into-the-formula" type of problem!

Proof: Let $a_1, \dots, a_6$ be the six integers. Denote by $r_1, \dots, r_6$ the remainders when $a_1, \dots, a_6$ respectively are divided by 5. Note that each r_i is either 0, 1, 2, 3, or 4. Let us apply PHP, choosing the possible remainders $(0, 1, \dots, 4)$ as the pigeonholes and the actual remainders ($r_1, \ldots, r_6$) as the letters. Since the number of letters is greater than the number of pigeonholes, at least two of the r_i must be equal. Suppose for instance that r_2 and r_5 are equal. Then a_2 and a_5 leave the same remainder on dividing by 5, so when $a_2 - a_5$ is divided by 5, these two remainders cancel, and $a_2 - a_5$ will be divisible by 5. (Described more precisely, a_2 is of the form $5k+r_2$ for some integer k since it leaves a remainder of r_2 when divided by 5, and a_5 similarly is of the form $5l + r_5$ for some integer l. Hence, $a_2 - a_5 = 5(k-l) + (r_2 - r_5)$. Since $r_2 = r_5$, we find $a_2 - a_5 = 5(k-l)$ it is thus a multiple of 5!) Obviously, the same idea applies to any two r_i and r_j that are equal: the difference of the corresponding a_i and a_j will be divisible by 5, thus proving the result.

Now remember the advice in the introduction about always going back and thinking about your solution to any problem? Let us go back and ask what made Examples 2.1 and 2.3 work. Example 2.1 was essentially a direct application of PHP; all that we needed to do was to recognize that one of the sets of objects given in the problem (the months of the year) should be thought of as pigeonholes, and the other set of objects (the people) as letters. In Example 2.3 on the other hand, we had to derive for ourselves one new set of objects from the given objects (the actual remainders $r_1, \ldots, r_6$ on dividing by 5) and view them as the letters, and view a second set of derived objects (the five possible remainders $0, 1, \ldots, 4$) as the pigeonholes. As a mathematics student, you should now sit back and think of various situations that would be analogous to these two examples; in other words, *you should create your own exercises similar to these two examples.* This will help immensely in your understanding of the situation and will give you confidence in the subject.

Here are further examples to help you get started:

Example 2.4. These are analogous to Example 2.1:

(1) In a group of 32 (or more) people, two people will be born on the same day of the month. (Do you see that this is just a mildly different version of the situation considered in Example 2.1? Notice that if there were 33 or 78 or a million people instead of 32 people, the same ideas would apply. What other obvious manifestations of the same example can you think of?)

(2) If you pick 101 integers at random, two of them will have the same units digit and the same tens digit. (Here the pigeonholes are the possible last two digits of the integers; there are a hundred possibilities.)

(3) If you have six shirts, wear any given shirt from morning to night, and never go shirtless on any day, then there will be two days of the week when you wear the same shirt.

Example 2.5. These are analogous to Example 2.3:

(1) If you pick 7 integers at random, the difference of some two of them will be divisible by 6. (This is just a straightforward modification of Example 2.3. Notice that if you pick 8 integers or 35 integers or a million integers, you can still argue that the difference of some two of them will be divisible by 6.)

(2) What is the most general way of expressing the idea behind Example 2.3? Clearly, what the example is showing is that if you are considering divisibility by a given

positive integer k, then as long as you have *at least one more than* k integers, there will be more integers than possible remainders on dividing by k, so two of them will have the same remainder and their difference will therefore be divisible by k. Thus, what Example 2.3 is showing is that if k is any positive integer, and if you are given any $k+1$ (or more) integers, then the difference of some two of these integers will be divisible by k.

2.2 PHP Generalized Form

Now let us consider a mild generalization of the Pigeonhole Principle as stated in Principle 2.2 above. Let us consider the following example:

Example 2.6. Show that if there are 61 people in a room, there must be at least six people who were born in the same month of the year. The proof idea is the same as that enshrined in the PHP: if we view the twelve possible months of the year as the pigeonholes, and the 61 people as the letters, then of course, at least one pigeonhole must contain at least two or more letters, but now, much more is true: at least one pigeonhole must contain six or more letters because otherwise, if each pigeonhole contains at most five letters, then the total number of letters will be at most 12 (the number of pigeonholes) times 5 (the maximum number of letters in each), i.e., the total number of letters will be at most 60. But we know that we have distributed 61 letters in all, so indeed, at least one pigeonhole must contain six or more letters.

As before, you should think of how to extend this example to create new exercises. For instance, do you see that if you had 73 people, at least 7 must have been born in the same month of the year? If you had 85 people, at least 8 must have been born in the same month of the year? We'll give this generalization of the PHP a name:

Principle 2.7. *Generalized Pigeonhole Principle.* Let k and n be positive integers. If $kn+1$ (or more) letters are distributed among n pigeonholes, then at least one pigeonhole must contain $k+1$ or more letters.

Thus, in Example 2.6, k is 5 and n is 12. There are $5 \cdot 12 + 1 = 61$ people (whom we take as letters) and $n = 12$ months (which we take as pigeonholes), so we find that at least one pigeonhole (month) should contain at least $k + 1 = 6$ letters (people).

The following exercise should be easy now; it is a combination of Examples 2.3 and 2.6.

> **Exercise 2.8.** Show that from any set of 100 integers one can pick 15 integers such that the difference of any two of these is divisible by 7.

Now we move to a different type of example involving some elementary geometry:

Example 2.9. Show that if 5 points are placed at random on or inside an equilateral triangle of side 2 units, at least two of them must be no more than 1 unit apart.

Based on our experience with PHP problems so far, we expect that the 5 points must be the letters, that they should be placed in 4 (or fewer!) pigeonholes, and that the pigeonholes must somehow be designed so that if there are two letters (points) in the same pigeonhole, then these points cannot be more than 1 unit apart! But how to do this?

A little playing around with equilateral triangles should suggest something. The pigeonholes should clearly be various regions inside the equilateral triangle of side 2 units. Thus, we want to divide the equilateral triangle into four regions: that way, if we place 5 points inside the triangle, two of the points at least should land in the same region (PHP: $5 > 4$!!!). The only thing to do is to choose the regions in such a way that any two points in any region must be no more than 1 unit apart.

Now obviously, we can divide the triangle into four regions in infinitely many ways, but here is a general rule: when confronted with many choices, and you are not sure which one to pick, it is a good idea to start with the choice *that has the most symmetry.* Thus, try to choose the four regions so that they "look the same," or in other words, are congruent to each other. The figure below shows how to do it:

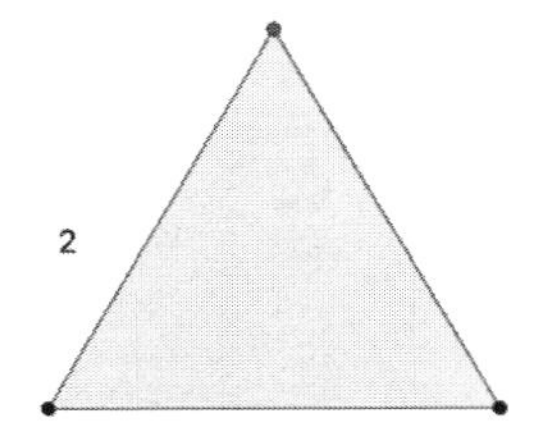

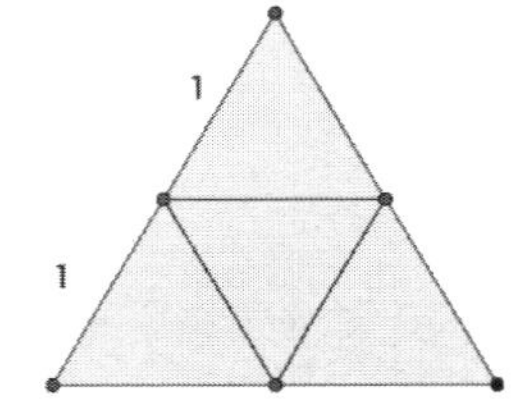

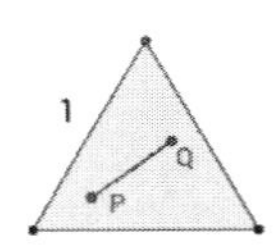

Each smaller triangle is now of side 1. As can be seen, any two points inside an equilateral triangle of side 1 can be at most 1 unit apart. (For instance, given the segment PQ shown in the rightmost triangle, extend the segment PQ so that one end is on one side of the triangle and the other is on another, as in the picture below:

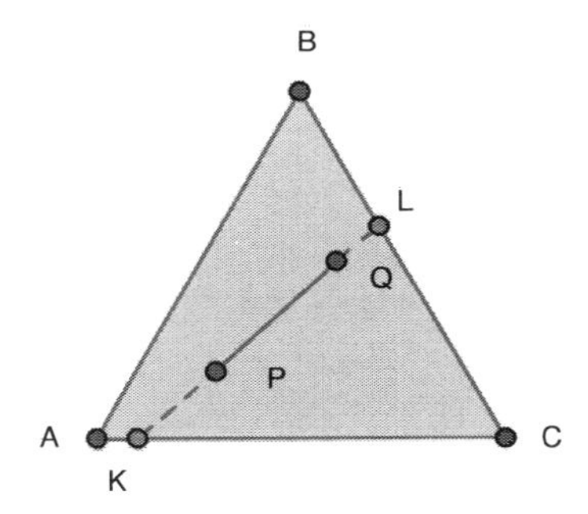

We have $|PQ| \leq |KL|$, where by $|PQ|$ and $|KL|$ we mean the lengths of the respective line segments. Next, $|KL| \leq |AL|$, and then $|AL| \leq |AB| = 1$.)

This does it! To summarize: by PHP, at least two points must lie on or inside one of the four smaller right triangles. These two points can then be no more than 1 unit apart!

2.3 Further Exercises

Exercise 2.10. Practice Exercises:

Exercise 2.10.1. Suppose that there are 5 pairs of socks in a drawer, each pair of a different color, but all jumbled up. Suppose you were to reach into the drawer and pick out socks at random, one at a time. How many must you pick to be certain that you will have a pair of matching socks?

Exercise 2.10.2. Show that if you take eleven integers at random, at least two of them will have the same digit in the units place.

Exercise 2.10.3. Show that if you take over a hundred integers at random, at least two of them will have the same digit in the units place and the same digit in the tens place.

Exercise 2.10.4. Ten students in a space science class had to choose from among Jupiter, Venus, Mars, Saturn, Uranus, Neptune, and also Ganymede, Io, and Europa on which to write research reports. Show that at least one of these objects in the solar system will have two or more reports written about it.

Exercise 2.10.5. Suppose that City X has a population of over ten million. Suppose someone told you that the maximum number of hairs on a human head is half a million. Prove that at least two people in City X must have the same number of hairs on their head!

Exercise 2.10.6. In the setup of Exercise 2.10.5 show that there must be at least twenty-one people in City X that have the same number of hairs on their head!

Exercise 2.10.7. In the setup of Exercise 2.10.4 suppose that twenty more students from neighboring schools joined this class. Show that at least one of these objects in the solar system will have four or more reports written about it.

Exercise 2.10.8. Suppose that 27 students must be assigned to five teachers. Show that at least one teacher will be responsible for six or more students.

Exercise 2.11. Let k be a positive integer. Show that there exist at least two distinct powers of 2 (i.e., two integers of the form 2^m and 2^n for distinct positive integers m and n) whose difference is divisible by k. Next, show that there are infinitely many distinct pairs of powers of 2 whose difference is divisible by k. Would the results in this problem still be true if "powers of 2" is replaced by "powers of 3?" How about if "powers of 2" is replaced by "elements of any infinite collection of integers?"

Exercise 2.12. Suppose you choose 22 integers at random from the set $\{1, 2, \ldots, 42\}$. Show that at least two of them must differ by 7. (Hint: Apply the generalized PHP to find a suitable subset of your selected integers that all leave the same remainder on dividing by 7. Now invoke the fact that these integers come from the set $\{1, 2, \ldots, 42\}$ to conclude that at least one pair must differ by *exactly* 7.)

Devise other triples (k, l, m) with the property that if we choose l integers from the set $\{1, 2, \ldots, m\}$, some pair must differ by k.

Exercise 2.13. Given any integer $n \geq 1$, show that there exists an integer, whose digits are all either 0 or 1, that is divisible by n. (Hint: you know that if you take any set of $n + 1$ integers, the difference of some two of them must be divisible by n. Now try to choose your set of $n + 1$ integers carefully!)

Exercise 2.14. Given 12 different 2-digit positive integers, show that one can choose two of them so that their difference is a two-digit integer with identical digits in the tens and units places. (Hint: determine what characterizes two digit integers with identical digits in the tens and units places.)

Exercise 2.15. Write down any 10 integers in a list. Label them a_1, a_2, $\ldots$, a_{10}. Prove that some sequence of consecutive terms a_i, a_{i+1}, a_{i+2}, $\ldots$, a_j in your list (with $1 \leq i \leq j \leq 10$) is sure to add to a multiple of 10. (Note: it is possible that $j = i$. Hint: let a_0 be an arbitrary integer, and consider the expanded list of integers a_0, a_1, $\ldots$, a_{10}. Form various sums of these integers such that if you think of the sums as letters to be stuffed into appropriate pigeonholes, you will be able to solve your problem.)

Exercise 2.16. Suppose you have a hexagonal table whose six sides are marked A, B, $\ldots$, F. Six people, also labeled A, B, $\ldots$, F, are sitting around this table, one person at each of the six sides. Initially, no one is sitting at the side whose label corresponds to their own label. Show that one can rotate the table so that at least two persons will now be sitting at the side corresponding to their label. (Hint: to each person, associate the number of sides, counted say in the counterclockwise direction, by which they are off from where they should be.)

Exercise 2.17. Suppose you have a circle of radius 1 with center at the origin. Suppose that 181 distinct points are picked on the circle, of integer degrees, where by degree of a point, we mean the degree measure of the angle made by the radius through that point with respect to the positive x-axis. Show that at least two points must be *antipodal,* i.e., at opposite ends of a diameter of the circle. Show that there is a selection of 180 points of integer degrees in which there will be no pair of antipodal points.

Exercise 2.18. Continuing with the situation in Exercise 2.17 above, suppose that 241 points of integer degrees are picked on the circle. Show that some three of them will form the vertices of an equilateral triangle. Now show that there is a selection of 240 points of integer degrees such that no three of them will form the vertices of an equilateral triangle. (Hint: show using elementary geometry that three points of integer degrees will form an equilateral triangle if and only if they are of degrees θ, $\theta + 120^\circ$ and $\theta + 240^\circ$ for some integer θ, $0 \leq \theta \leq 359$.)

More generally, show that for any divisor of 360, selecting $(360/k)\cdot(k-1)+1$ points will guarantee that some k of them will form a regular k-gon, but there is a selection of $(360/k)\cdot(k-1)$ points for which no k of them will form a regular k-gon.

Exercise 2.19. Show that if there are ten circular discs of radius 1 inside a circle of radius 3, at least two discs must overlap.

Exercise 2.20. Suppose you place 1000 points at random in a square region measuring one foot by one foot. Show that at least seven of them must lie within $\sqrt{2}$ inches of one another.

Exercise 2.21. Let a be an *irrational* number. It is clear that no (nonzero) integer multiple of a can equal an integer, for, if say $ma = n$ for integers m and n with $m \neq 0$, then we would find $a = n/m$, violating the fact that a is irrational. However, as we will show in this exercise, integer multiples of a can come *arbitrarily close* to being an integer. First a definition: Given any real number r, we know from our knowledge of the real numbers that we can find an integer n such that $n \leq r < n+1$ (although, as we will see in Corollary 12.20 in Chapter 12 ahead, this innocuous sounding result is actually a consequence of a very deep property of the real numbers known as the Least Upper Bound property). Let us write r as $n+\alpha$, where $0 \leq \alpha < 1$. We will call α the *fractional part* of r, and denote it as $\{r\}$. (For instance, $\{\pi\} = \{3.14159\ldots\} = 0.14159\ldots$, while $\{-\pi\} = \{-3.14159\ldots\} = \{-4 + 0.85840\ldots\} = 0.85840\ldots$.)

(1) Given our irrational a, show that if m and n are distinct integers, then $\{ma\} \neq \{na\}$.

(2) Given a positive integer N, show that there are *distinct* integers m and n, with $1 \leq m, n \leq 10^N + 1$, and an integer s with $0 \leq s \leq 10^N - 1$ such that $s/10^N < \{ma\} < (s+1)/10^N$ and $s/10^N < \{na\} < (s+1)/10^N$. (Hint: Divide the interval $[0,1]$ into 10^N intervals of equal width. Consider the various integer multiples a, $2a$, $\ldots$, $(10^N + 1)a$, and use part (1).)

(3) Conclude from part (2) that the fractional part of either $(m-n)a$ or $(n-m)a$ is less than $1/10^N$.

Thus, one of $(m-n)a$ or $(n-m)a$, each of which is an integer multiple of a, is of the form integer plus a fractional part that is less than $1/10^N$. Since N was arbitrary, this shows that integer multiples of a come arbitrarily close to being an integer.

(4) Conclude from part (3) that given any irrational a and a positive integer N, there exists a *rational* number q such that $|a-q| < 1/10^N$. (This shows that we may approximate irrational numbers by rational numbers to arbitrary closeness.)

Exercise 2.22. There are 10 people in a room. Some pairs of people may be acquainted, other pairs of people may not be. (Note that we assume that if Person A knows Person B, then Person B knows Person A, too. Moreover, we do not consider any person to be his or her own acquaintance.)

(1) Assume that everybody in the room knows at least one other person. Show that there must be at least two people in the room who have the same number of acquaintances.

(2) Now drop the assumption in part (1) above. Show that even without that assumption, there must be at least two people in the room who have the same number of acquaintances.

(3) Will the conclusions in parts (1) and (2) continue to be true if instead of 10, there were k people in the room, for arbitrary integer $k \geq 2$?

Exercise 2.23. Each point in the plane is randomly assigned one of two colors, red or green. Show that for any real number $l > 0$, there exist two points on the plane a distance l apart, and that they are either both assigned red or both assigned green. (Hint: consider equilateral triangles.)

Exercise 2.24. Suppose six points are chosen on a plane, no three of which are collinear. Line segments between each pair of points are drawn, some in red and some in green. Show that there exists a triangle formed by some three of these points, all of whose edges are of the same color. (Hint: Pick any one point. Consider the colors of the line segments that connect this point and the remaining points, and apply the Generalized PHP suitably to find three segments of the same color. Proceed.)

Exercise 2.25. This exercise builds on Exercise 2.24 above. Suppose, instead, seventeen points are chosen on a plane, no three of which are collinear, and suppose that line segments between each pair of points are drawn, some in red, some in green, *and,* some in blue. Show once more that that there exists a triangle formed by some three of these points, all of whose edges are of the same color. (Hint: Start as in Exercise 2.24 by picking any one point and considering the colors of the line segments that connect this point and the remaining points. You may find yourself invoking the result of Exercise 2.24 at some point!)

Exercise 2.26. Study how the solution of Exercise 2.25 reduces to Exercise 2.24 above. Define the integers F_k, $k = 2, 3, \ldots$, by the rule $F_2 = 6$, $F_k = k(F_{k-1} - 1) + 2$. (In particular, $F_3 = 3(6-1) + 2 = 17$.) Show that if F_k points are chosen on a plane, no three of which are collinear, and if line segments between each pair of points are drawn, each segment in one of k different colors, then there exists a triangle formed by some three of these points, all of whose edges are of the same color. (Hint: Use induction, which you may be familiar with. If you are not, not to worry, we will study induction in Chapter 7. Come back and do this exercise after you have worked through that chapter!)

3

Statements

As we have said many times, advanced mathematics deals with ideas, and before any progress can be made in the subject, it is vital that we be precise in what we say. Language thus becomes crucial: terms cannot be allowed to be ambiguous, and all situations considered must be described with such exactness that nobody is left puzzled about any meaning. The tools commonly used for this are statements, sets, and functions. Any student wishing to study mathematics beyond high school, and certainly beyond calculus, must be comfortable with the basic aspects of these three objects. We will try to gain sufficient understanding of these objects so that we can use them the way a working mathematician uses them (a deeper study needs us to venture into the field of logic, and we won't go there at this introductory juncture!).

In this chapter we will study statements.

3.1 Statements

We will take the following as our definition of the term:

Definition 3.1. A *statement* is a sentence which can be either true or false.

To help us understand this definition, here are examples of both sentences that are statements and those that are not. Study each one; the examples that are *not* statements can give you as much or sometimes even more insight into what makes something a statement as the examples that are statements!

Example 3.2.

(1) "Dude, can I borrow your car?" This is NOT a statement! A statement needs to be either true or false. A statement is thus an *assertion:* somebody is *declaring* that something is true (and, of course, it is quite possible that their assertion or declaration is false). "Dude, can I borrow your car?" is a question, and a question is not an assertion at all, it cannot be classified as being true or false, so it cannot be a statement.

(2) "Five is less than four." This is indeed a statement. There is an assertion here: somebody is claiming that five is less than four. As with any assertion, this assertion could be true, or it could be false. (As you know, the assertion is indeed false! But that's OK, what makes it a statement is that the sentence can be categorized as true or false.)

(3) "The sky is beautiful today!" This is NOT a statement. Before we can say whether this is true or not, we first need to understand what is being claimed. The problem is that the word "beautiful" does not have a precise meaning. What one person considers beautiful another could consider ugly (and the other way around as well). Thus, all words in a sentence must be unambiguous before we can consider whether it is a statement or not: if there is any ambiguity, it cannot be a statement. We use the term *well-defined* specifically in mathematics for being unambigous.

(4) "If Alicia is registered for Math 320 then she must have received a score of 70% or higher in her Math 150B final exam." This is indeed a statement. The logic behind the sentence is a bit more involved (we'll see later that this is an example of a compound statement), but if you analyze the sentence, you'll see that all terms are unambiguous, and the entire sentence is an assertion, that could be either true or false.

(5) "For any integer n, $n^2 > n$." This is indeed a statement. Certainly all terms here are unambiguous. If the "n" here bothers you and you feel that perhaps it is not defined, note that the sentence could have been reworded, without changing its meaning, as "For any integer you pick, you will find that the square of the integer you picked is greater than that integer you picked"—the "n" has now disappeared!

The disappearance of the "n" in the rewording should not be surprising: it was only acting as a *placeholder* or *shorthand* for the more unwieldy expressions "integer you pick" in the first part of the sentence and "the integer you picked" in the second part of the sentence. Actually, what we have here is *a family* of statements, one for each integer. Thus, "For any integer n, $n^2 > n$" is really just a compact way of saying "$0^2 > 0$, and $1^2 > 1$, and $(-1)^2 > -1$, and $2^2 > 2$, and $(-2)^2 > -2$, and ...." See Remark 3.20 on Page 35 ahead.

(6) "Two divides $x + 1$." This is NOT a statement, since x is undefined. (Recall from Example 3.2(3) above that all words in a statement must be unambigous.) Now contrast this with the sentence "For all integers x, two divides $x + 1$." This new sentence is indeed a statement: it is an unambiguous assertion (that just happens to be false!).

(7) "It will rain next Wednesday." This is indeed a statement: it is an assertion that can be either true or false. Of course, it cannot be determined until sometime in the future whether this statement is true or false, but that is not crucial. For us, it is sufficient that, if only based on a future event, the sentence can be classified as true or false.

It would be instructive now if you were to do the following exercise, to test your understanding of the definition of a statement:

Exercise 3.3. Determine if the following sentences represent statements.

(1) "I am SO happy today!"

(2) "Oh My God!"

(3) "There are exactly six prime integers."

(4) "Abdul is at least 6 feet tall."

(5) "$x > x - 1$."

(6) "The square of the hypotenuse of a right triangle equals the sum of the squares of the other two sides."

(7) "Mustafa and Ayesha arrived simultaneously."

(8) "Mustafa and Ayesha did not arrive simultaneously."

(9) "All my children are daughters."

(10) "I have at least one son."

(11) "For all integers x, $x + 2$ is even."

(12) "There exists an integer x such that $x + 2$ is odd."

(13) "I will see a movie tonight."

(14) "You will feel happy tonight."

Remark 3.4. You will routinely come across sentences in mathematics that read "Prove ⟨*something*⟩." For instance, you could come across the statement "If a and b are the two legs of a right triangle, and if c is the hypotenuse, prove that $a^2 + b^2 = c^2$." If we analyze this, what we find first is the statement "The two legs a and b of a right triangle and its hypotenuse c together satisfy the equation $a^2 + b^2 = c^2$." At this point, it is just a statement, it could be either true or false: it has the potential of being either. The phrasing "prove ... " means that you are required to show that indeed this is a true statement and not a false statement. In the general phrasing "Prove ⟨*something*⟩," ⟨*something*⟩ is a statement, that could either be true or false, and you are required to show that ⟨*something*⟩ is indeed true.

Remark 3.5. A *theorem* in mathematics is any statement that is true. Now, while every true statement is a theorem, the label "theorem" is actually used quite judiciously. For instance, "the sun rises in the east" is a true statement. However, we do not refer to it as a theorem because by now it is an everyday fact that the sun rises in the east. Typically, mathematicians use the word theorem to indicate a (true) statement for which deep study and specialized analysis is required before the truth of the statement can be established. Obviously, there is some personal judgement here in what can be labeled a theorem. Mathematicians also use terms like *proposition* and *lemma* for theorems. A proposition is a true statement that someone has deemed to be less significant in the development of a subject than a full-blown theorem. This choice to decide that something is less significant than something else is obviously very personal as well. A lemma is a true statement that precedes a theorem and whose primary role is to provide some preliminary results that will be useful in establishing the truth of the statement that follows.

When mathematicians say "let us prove this theorem" they mean that they will now establish the truth of this statement, so that it can then rightly be labeled as a theorem!

Remark 3.6. If S is a statement, we say that the *truth value* of S is "True" (or "T") if S is a true statement, and "False" (or "F") if S is a false statement. The truth values "True" and "False" are often described as *opposites* of each other.

3.2 Negation of a Statement

Notice that statements (7) and (8) of Exercise 3.3 above (they are indeed statements!) are closely related: they seem *negations* of each other. We need to flesh out exactly what this means, and we will do so in this section, and consider several examples of negations of statements.

Notice that if statement (7) of Exercise 3.3 is true, then statement (8) of that same exercise will be false, and if statement (7) is false, then statement (8) will be true! Table 3.1 shows the situation. (Such a table is known as a *truth table:* it depicts the various possibilities for the true/false status of the depicted statements.)

Table 3.1. Truth table for statements (7) and (8), Exercise 3.3.

Statement (7) (Arrived simultaneously)	Statement (8) (Did not arrive simultaneously)
True	False
False	True

This observation about statements (7) and (8) of Exercise 3.3 will be the basis of the following:

Definition 3.7. Two statements S and T are considered to be *logically opposite* if whenever S is true, then T is false, and whenever S is false, then T is true. When this happens, we also say S is a *logical opposite* of T, or T is a *logical opposite* of S.

Remark 3.8. Definition 3.7 above appears to give primacy to S: it starts with the truth values (see Remark 3.6 above) of S and requires corresponding truth values of T. However, there are hidden implications in the definition that makes the situation symmetric with respect to S and T. Suppose that S and T are logically opposite as per the definition above. What if T is false? Then S must be true, since if S were false, the definition above would say that T must have been true, which it is not! Similarly, what if T is true? Then S must be false, since if S were true, the definition above would say that T must have been false, which it is not! Thus, to say that S and T are logically opposite is to say that their truth values are in *opposition to each other.* Described thusly, we see that the condition of being logically opposite is indeed symmetric with respect to S and T. In particular, we could just as easily have defined S and T to be logically opposite if whenever T is true, then S is false, and whenever T is false, then S is true.

Based on our observations just before Definition 3.7 about statements (7) and (8) of Exercise 3.3, we would be tempted to define the negation of a statement to be a statement that is its logical opposite. As we will see ahead, this would not be incorrect, but there are a few subtleties we should consider. For instance, continue to take S and T to be statements (7) and (8) of Exercise 3.3. As observed, S and T are logical opposites of each other. Now consider instead the following statement: T': "It is not true that Mustafa and Ayesha arrived simultaneously." It is clear that T' too is false if S is true and T' too is true if S is false, so T' and S are also logical opposites of each other. Thus, there can be more than one statement that is the logical opposite of a given statement! So which one should we take to be the definition of a negation?

To help us, let us consider one other almost obvious fact: If T and T' are both logical opposites of S, then whenever T is true, T' is also true, and whenever T is false, then T' is also false. This can be seen by invoking the fact that both are logical opposites of S: If T is true, then S must be false as S and T are logical opposites. But then T' must be true as T' and S are also logical opposites! Thus, if T is true, T' must also be true. Likewise, if T is false, T' must also be false. This leads to the following definition:

Definition 3.9. Two statements S and T are said to be *logically equivalent* if whenever S is true, T is true, and whenever S is false, T is false. When this happens, we also say S is *logically equivalent* to T, or T is *logically equivalent* to S.

The truth table for logically equivalent statements is shown in Table 3.2.

Table 3.2. Truth table for logically equivalent statements.

Statement S	Statement logically equivalent to S
True	True
False	False

Remark 3.10. As with the definition of logically opposite (see Remark 3.8 above) the definition of logically equivalent seems to give primacy to S: it starts with the truth values of S and requires corresponding truth values of T. But it is easy to see, just as in Remark 3.8, that we could have just as easily defined S and T to be logically equivalent if whenever T is true, S is true, and whenever T is false, S is false. To say that S and T are logically equivalent is to say that their truth values are in lockstep with each other—they are *simultaneously* true or *simultaneously* false.

Now that we are in possession of the concepts of logically opposite and logically equivalent statements, we are in a position to craft our definition of the negation of a statement. But first, another fact that is almost obvious: If T' and T are logically equivalent to each other, and if T and S are logical opposites of each other, then T' and S are also logical opposites of each other. This falls out from our considerations in Remarks 3.8 and 3.10: The truth values of S and T are opposite, while the truth values of T and T' are the same, so the truth values of S and T' are also opposite.

Definition 3.11. The *negation* of a statement S is any statement that is a logical opposite of S. Equivalently, the negation of S is any statement logically equivalent to the new statement "S is not true". It is denoted $\neg S$.

The truth table for logically equivalent statements is shown in Table 3.3.

Table 3.3. Truth table for the negation of a statement.

Statement S	Negation of S
True	False
False	True

Definition 3.11 captures fully our intuitive sense of what a negation ought to be, based on our examination of statements (7) and (8) of Exercise 3.3. On the one hand, the negation of S ought to be something that is logically opposite S. On the other hand, there could be several statements that are logically opposite S, so we will choose to call every one of them a negation of S. Moreover, from our various discussions above Definition 3.11 we know that two statements T and T' are both logically opposite to S if and only if T and T' are logically equivalent to each other. One such statement that is logically opposite to S is the statement "S is not true." Thus, the negation of S could also be taken to be any statement that is logically equivalent to "S is not true."

Here is something for you to wrap your head around:

Exercise 3.12. Convince yourself that the negation of the negation of S is S itself!

Remark 3.13. Note that "S is not true" (or, what is the same thing, "S is false") is itself a negation of S. However, in many situations, it is useful to formulate the negation of a statement S not as "S is not true," but as a logically equivalent statement, since this equivalent statement may be immediately more meaningful (even though, at base, it is equivalent to "S is not true.")

It is critical to be able to construct negations of a statement correctly. We will look at some examples:

Example 3.14. Let us look at statement (9) in Exercise 3.3 above; call it S. One negation of S is the statement "S is false." But what does it really mean to say that the statement "All my children are daughters" is false? Let us think about this. It certainly could mean that all my children are sons, but it need not *necessarily* mean that: it could be that I have two sons and two daughters! But it need not necessarily mean that either: it could be that I have one son and one daughter, or maybe just one son! What must *necessarily* be true if "All my children are daughters" is false? It is the statement "I have *at least one* son," in other words, statement (10) of that exercise! (Notice that this statement carries immediate meaning to us, in a way that the statement "S is false" does not — although the two are of course logically equivalent).

Example 3.15. Now let us look at statement (11) in Exercise 3.3 above, call it S again. As always, one negation of S is the statement "S is false." But what does it really mean to say "It is not true that for all integers x, $x + 2$ is even?" It certainly could mean that for all integers x, $x + 2$ is odd, but it does not *necessarily* have to mean that: it could be that for most integers x, $x + 2$ is indeed even, but there are these, say, five holdout integers x for which $x + 2$ happens to be odd. But it need not necessarily mean that either, perhaps there are only four holdout integers x for which $x + 2$ is odd. What must *necessarily* be true if it is not true that for all integers x, $x + 2$ is even? It is that for *at least one* integer x, $x + 2$ is odd, that is, there exists an integer x such that $x + 2$ is odd. This is just statement (12) of that exercise.

Example 3.16. Consider the following statement S (convince yourself first that it is indeed a statement!): "During the past month, whenever it rained, I carried my umbrella." Which of the following do you think would be its negation?

(1) During the past month whenever it did not rain, I carried my umbrella.

(2) During the past month whenever it rained, I did not carry my umbrella.

(3) During the past month there were one or more occasions when it rained when I did not carry my umbrella.

The negation of S by Definition 3.11 is the statement "It is not true that during the past month whenever it rained, I carried my umbrella" (or any logically equivalent statement). What does it mean to say that it is not true that whenever it rained during the past month, I carried my umbrella? It certainly could be that whenever it rained, I did not carry my umbrella (part (2)), but this need not *necessarily* be so: it may have been that whenever it rained on a weekday I carried my umbrella, but whenever it rained on a weekend I did not carry my umbrella. But it need not necessarily mean that either: Perhaps whenever it rained on a weekday I carried my umbrella, and also whenever it rained on the first two weekends, but whenever it rained on the last two weekends, I did not carry my umbrella. What must *necessarily* be true if it is not true that whenever it rained during the past month I carried my umbrella? It is that there was at least one occasion last month when it rained but I did not carry my umbrella. Thus, the statement in Part 3 is the correct negation. (Another equivalent way to express the negation is to say "There were times when it rained during the past month when I did not carry my umbrella." Also, note that the statement in part (1) is obviously not the logical opposite of S. For instance, both S and the statement in part (1) could be true: I could be an obsessive compulsive person who never goes anywhere without an umbrella, no matter whether it rains or not.)

Example 3.17. Consider the following statement (convince yourself first that it is indeed a statement!): "For all integers x, $x^2 \geq x$." Which of the following do you think would be its negation?

(1) For no integer x is $x^2 \geq x$.

(2) There exists some integer x such that $x^2 \leq x$.

(3) There exists some integer x such that $x^2 < x$.

(4) For all integers x, $x^2 < x$.

(5) It is not true that for all integers x, $x^2 \geq x$.

Write S for the statement "for all integers x, $x^2 \geq x$." The negation of S is the statement "It is not true that for all integers x, $x^2 \geq x$" (part (5)) or any statement logically equivalent to this. As with Examples 3.14, 3.15, and 3.16, let us think about what "S is not true" means. It could certainly mean that for all integers x, x^2 is not greater than or equal to x, or what is the same thing, for all integers x, $x^2 < x$ (part (4)). But it need not *necessarily* be so: it could be that for almost all integers x, x^2 is indeed greater than or equal to x, but again, there are these five holdout integers for which $x^2 < x$. But this need not necessarily be true either, perhaps there are only four holdout integers for which $x^2 < x$. What must *necessarily* be true if it is not true that for all integers x, $x^2 \geq x$? It is that for at least one integer x, $x^2 < x$, that is, there exists an integer (or in equivalent phrasing: there exists *some* integer) x such that $x^2 < x$ (part (3)).

Thus, the statements in parts (5) and (3) are both negations of S. The statement in part (4) is not a negation as we saw in the discussion above. The statement in part (4) is logically equivalent to the statement in part (1), so this latter statement is also not a negation. The statement in part (2) is not a logical opposite of S, so it cannot be a negation of S. For instance, it could be true that for all integers $x^2 \geq x$, and that there are some of these for which $x^2 = x$. Thus, both S and the statement in part (2) could be true.

This is now a good point for you to test your understanding of the negation of a sentence by working on the following:

Exercise 3.18. For each of the statements below, pick the correct negation(s) among the choices shown.

(1) Xavier played his flute nonstop from 8AM to 8PM.

  (a) Xavier played his guitar nonstop from 8AM to 8PM.
  (b) Xavier did not play his flute nonstop from 8AM to 8PM.
  (c) Xavier played his flute only from 8PM to 8AM.
  (d) There was some time between 8AM to 8PM when Xavier did not play his flute.

(2) Every integer greater than 1 can be factored into a product of primes.

  (a) There is an integer greater than 1 that cannot be factored into a product of primes.
  (b) No integer greater than 1 can be factored into a product of primes.
  (c) The collection of integers greater than 1 that can be factored into a product of primes does not contain every integer greater than 1.
  (d) Every integer less than -1 can be factored into a product of primes.

(3) Scott goes to work by car every day.

  (a) Scott goes to play golf by car every day.
  (b) Scott goes to work by car every other day.
  (c) Scott does not go to work every day.
  (d) Scott goes to work every day but not always by car.
  (e) Either Scott does not go to work every day, or, if he does, he does not always do so by car.

3.3 Compound Statements

Given two statements S and T, we routinely combine them together in day-to-day usage to form new statements. We will study four different ways of making compound statements that form the underpinning of mathematics:

(1) *And* (technically known as a *conjunction*),

(2) *Or* (technically known as a *disjunction*),

(3) *Implies*, or *if then* (technically known as *conditional*), and

(4) *Implies and is implied by*, or *if and only if*—often abbreviated to *iff* (technically known as *biconditional*).

We will look at each of these compound statements below:

***S* and *T*.** Given statements S and T, the statement "S and T" is the new statement that is true when both S and T are true, and false if even one of S and T is false. To understand what this means, let us consider the following:

Example 3.19. Take S to be the statement "I will play basketball this evening" and T to be the statement "Patricia will play tennis this evening." (Confirm first for yourself that both S and T are indeed statements—see Example 3.2, part (7).) Then "S and T" is the statement "I will play basketball this evening and Patricia will play tennis this evening," which under standard English usage, gets shortened to "I will play basketball and Patricia will play tennis this evening." This new sentence is indeed a statement: it *declares* something to be true, and this declaration could either turn out to be true or it could turn out to be false! Moreover, it will be a true declaration *precisely* if both S and T turn out to be true, that is, if I play basketball this evening *and* Patricia plays tennis this evening. If I fail to play basketball this evening, or if Patricia fails to play tennis this evening, this new compound statement will be false.

To understand when compound sentences become true and when they become false, it is very helpful to construct the truth table for the compound statement *and*. See Table 3.4. The table is essentially self-explanatory: it looks at every possible combination of the true/false status of the component statements S and T and writes out the true/false status of the compound statement "S and T."

Table 3.4. Truth table for the *and* statement.

S	T	S and T
True	True	True
True	False	False
False	True	False
False	False	False

Remark 3.20. *Families of Statements:* We have already seen in Example 3.2, part (5) an instance of *a family of statements:* "For any integer n, $n^2 > n$." As described there, this is really just a compact way of saying "$0^2 > 0$, and $1^2 > 1$, and $(-1)^2 > -1$, and $2^2 > 2$, and $(-2)^2 > -2$, and ...." We will suggestively label by $S(n)$ the statement "For any integer n, $n^2 > n$." What is implicit in this labeling is that $S(n)$ is a family of statements, one for each integer n. Thus, $S(0)$ is the statement, "$0^2 > 0$," $S(1)$ is the statement "$1^2 > 1$," etc. These individual statements are then connected by the *and* construction we have just studied. Thus, to say $S(n)$ is true is to say that $S(0)$ is true, *and* $S(1)$ is true, *and* $S(-1)$ is true, *and* $S(2)$ is true, *and* $S(-2)$ is true, and so on.

Another example is in Exercise 3.3, part (11): "For all integers x, $x + 2$ is even." Here too, we have a family of statements which we may label $S(x)$, one for each integer x. Thus, $S(0)$ is the statement "$0 + 2(= 2)$ is even," $S(1)$ is the statement "3 is even," $S(-1)$ is the statement "1 is even," and so on. Once again, these individual statements are connected by the *and* construction: to say that $S(x)$ is true for all x is to say that 2 is even, *and* 3 is even, *and* -1 is even, and so on.

The technical term for a family of statements, one for each integer, is "a family of statements *indexed* by the integers." We could consider families of statements indexed by other kinds of numbers as well (and more generally, indexed by other kinds of sets). For instance, the statement "for every positive real number x, there exists a positive real number y whose square equals x" is a family of statements *indexed* by the positive real numbers (in other words, there is one statement for every positive real number x), and these are connected by the *and* construction. We may label this family of statements by $S(x)$, one for every positive real number x. So for instance, $S(2)$ is the statement that there exists a positive real number whose square is 2, etc.

Statements indexed by the positive integers are often amenable to proofs by induction, something we will look at in Chapter 7 ahead.

***S or T*.** Given statements S and T, the statement "S or T" is the new statement that is true when at least one of S and T is true and false if both S and T are false. To understand this better, let us consider the following:

Example 3.21. Take S to be the statement "I will eat corn tortillas tonight" and T to be the statement "I will eat rice tonight." Then "S or T" is the statement "I will eat corn tortillas tonight or I will eat rice tonight." This new sentence is indeed a statement: it *declares* something to be true, and this declaration could either turn out to be true or it could turn out to be false!

A little care needs to be taken, however, in figuring out when this compound statement is true. In day-to-day usage, we often expect the statement "I will eat corn tortillas tonight or I will eat rice tonight" to be true when either one or the other — *but not both* — is true. Thus, we expect the compound statement to be true if I eat corn tortillas but not rice, or if I eat rice but not corn tortillas. However, if I (the glutton) end up eating *both* corn tortillas and rice, we often expect the compound statement "I will eat corn tortillas tonight or I will eat rice tonight" to be false. This is, however, NOT how we interpret the compound statement *or* in mathematics. In mathematics, the statement "I will eat corn tortillas tonight or I will eat rice tonight" is considered to be true even if I end up eating both corn tortillas and rice. Thus, the truth table for the compound statement *or* is as follows:

Table 3.5. Truth table for the *or* statement.

S	T	S or T
True	True	True
True	False	True
False	True	True
False	False	False

In technical terms, the *or* form of making compound statements that we use in mathematics is known as the *inclusive or*. Its name is suggestive: we declare "S or T" to be true even if both are true. (By contrast, the day-to-day usage we alluded to above, where we often consider S or T to be false if both S and T are true, is known as the *exclusive or*. Mathematicians like to be inclusive.)

S implies T. Given statements S and T, the new statement "S implies T" is just a common way of writing the assertion "if S is true then T must be true." The meaning of this assertion is of course clear. "S implies T" will be true if it turns out that whenever S is true, T is indeed true as asserted. Let us consider the following:

Example 3.22. Let S be the statement "It will be raining tonight," and T the statement "I will be carrying my umbrella with me tonight" (these are indeed statements; see Example 3.2, part (7)). Then "S implies T" is the statement "If it will be raining tonight, then I will be carrying my umbrella with me tonight," or, as is the usual usage in English, "If it rains tonight then I will carry my umbrella." This statement will be true if it turns out that it is raining tonight, and I am carrying my umbrella with me. And of course, if it is raining tonight, but I am not carrying my umbrella with me, then the statement "If it is raining tonight then I will carry my umbrella" will be false.

But there are more cases to consider, and these are notable. Do we take "If it is raining tonight then I will carry my umbrella" as a true statement or a false statement if it is not raining tonight? Counterintuitive as this might sound, in the field of logic, the statement above is taken to be *true* if it is not raining tonight, *irrespective of whether I carry my umbrella or not!*

As already manifest in Example 3.22 above, the truth table of the compound statement *implies* is as follows:

Observe in the truth table that, just as in the discussion of Example 3.22 above, if S is false, "S implies T" is taken to be true independent of whether T is true or not!

Table 3.6. Truth table for the *implies* statement.

S	T	S implies T
True	True	True
True	False	False
False	True	True
False	False	True

(One advantage of taking S implies T to be true when both are false is that this ensures that the contrapositive of "S implies T" becomes logically equivalent to "S implies T." See Definition 3.25 ahead and also see Table 3.7.) This leads to an interesting question: what if S were a statement that is *always* false? Then this would mean that "S implies T" is always true! In such a situation, we say that "S implies T" is *vacuously* true. For instance, take S to be the statement "The sun rises in the west," and T to be "All dogs have eight legs." Then the compound statement "If the sun rises in the west then all dogs have eight legs" is considered a true statement!

Notice too from the truth table that "S implies T" is true if T is always true, independent of whether S is true or not! Thus, taking T to be the (true) statement "The earth rotates once in every day" and S to be the statement "My face is red," the statement "If my face is red then the earth rotates once in every day," is a true statement, whether my face is red or not!

But we do not have to worry about such counterintuitive situations: what is significant in mathematics about the compound statement *implies* is contained in the first two rows of the truth table above, and corresponds exactly to our day-to-day usage: If whenever S is true T turns out be true, the statement "S implies T" is true, and if there is even one instance where S is true but T turns out to be false, then "S implies T" is false.

The symbol $\implies$ is used in mathematics to stand for the word implies: thus, "S implies T" is written in mathematics as $S \implies T$. Almost all of mathematics involves the compound statement *implies!* Most theorems in mathematics read as follows: If ⟨*stuff*⟩ is true, then ⟨*other stuff*⟩ is true!

3.4 Statements Related to the Conditional

There are two statements related to the conditional statement "$S \implies T$" that are critical in mathematics: the *converse*, and the *contrapositive*.

Definition 3.23. The *converse* of the conditional statement "$S \implies T$" is the new statement "$T \implies S$."

Note that "$S \implies T$" and "$T \implies S$" are two logically independent statements: you cannot conclude that either is true (or false) because the other is true (or false). Here is an example:

Example 3.24. As in Example 3.22 above, let S be the statement "It will be raining tonight," and T the statement "I will be carrying my umbrella with me tonight." Then "$S \implies T$" is the statement (in common English usage) "If it rains tonight then I will carry my umbrella." On the other hand, "$T \implies S$" is the statement (again, in common

English usage) "If I carry my umbrella tonight then it is raining." Convince yourself that these statements are logically independent. (For instance, suppose "$S \implies T$" is a true statement. Then, if I see you with your umbrella tonight, this does not have to imply that it is raining: it could be that there is no rain at all, just that someone who had borrowed your umbrella earlier had returned it to you a moment ago! Thus, "$T \implies S$" need not be a true statement. Similarly, "$T \implies S$" need not be a false statement either. You could just be a sensible person who carries an umbrella if it rains, but will not carry one if it doesn't rain.)

Definition 3.25. The *contrapositive* of the conditional statement "$S \implies T$" is the new statement "$\neg T \implies \neg S$."

Note that the contrapositive of "$\neg T \implies \neg S$" is "$S \implies T$" again! (This follows from the fact that $\neg\neg S$ is S itself; see Exercise 3.12.) But what is used extensively in mathematics is the following: the statement "$S \implies T$" and its contrapositive "$\neg T \implies \neg S$" are *logically equivalent* statements! This should be intuitively clear, for instance from the example below, but we'll write out a truth table just ahead too:

Example 3.26. Again as in Example 3.22 above, let S be the statement "It will be raining tonight," and T the statement "I will be carrying my umbrella with me tonight." Then "$\neg T \implies \neg S$" is the statement (in common English usage) "If I do not carry my umbrella tonight then it is not raining tonight."

Suppose "$S \implies T$" is a true statement, and suppose I do not carry my umbrella tonight. Then it certainly cannot be raining, since had it been raining, the true statement "$S \implies T$" would have said that I will carry my umbrella with me tonight! Hence, if "$S \implies T$" is true, then "$\neg T \implies \neg S$" is also true.

Similarly, suppose "$S \implies T$" is false. This means that it is raining tonight but I do not carry my umbrella (see Table 3.6). The contrapositive "$\neg T \implies \neg S$" reads "if I do not carry my umbrella tonight then it is not raining." This is patently false, as the previous sentence says the that opposite happens! Thus the contrapositive is also false.

Table 3.7 is the truth table for "$S \implies T$" and its contrapositive "$\neg T \implies \neg S$." (Note that the statement "$S \implies T$" is automatically true if S is false; see the truth table following Example 3.22. Observe that defining "$S \implies T$" to be true if S is false and T is false ensures that "$\neg T \implies \neg S$" is equivalent to "$S \implies T$".)

Table 3.7. Truth tables for $S \implies T$ and $\neg T \implies \neg S$.

S	T	$S \implies T$	$\neg T$	$\neg S$	$\neg T \implies \neg S$
True	True	True	False	False	True
True	False	False	True	False	False
False	True	True	False	True	True
False	False	True	True	True	True

Remark 3.27. In mathematics, very often, when trying to prove that "$S \implies T$" is a true statement, we instead choose to prove that the contrapositive "$\neg T \implies \neg S$" is a true statement. Since the two statements are logically equivalent, proving that the contrapositive is a true statement is equivalent to proving that the original statement is true.

3.5 Remarks on the Implies Statement: Alternative Phrasing, Negations

Remark 3.28. When mathematics is written in English, the phrase "if S is true" is often replaced with "if S holds." Here is a typical usage—do not try to decipher the exact mathematical meaning, there's too much there probably that you haven't seen before, but focus just on the logical structure: "If $p \nmid n$ holds, then n is invertible modulo p." Here, S is the statement "$p \nmid n$," and T is the statement "n is invertible modulo p." (Again, don't worry about what S and T really mean at this point!) The assertion is that if S is true, then T is true. See also Remark 3.30 ahead.

Remark 3.29. *Negation of "$S \implies T$:"* Care must be taken while negating the implies statement. The negation of "$S \implies T$" is the statement "it is not true that if S is true then T is also true." What this negation means, if you take it apart carefully, is this: there is *at least one instance* where S is true but T is false. Note that this negation is *not* saying that in *all* instances where S is true T is false. This distinction appears when S can become true in several ways. Consider the following:

> **Example 3.29.1.** Take S to be the statement "the integer x is even," and T to be the statement "the integer x^2 is even". The statement "$S \implies T$" in common usage is written as "if x is an even integer, then x^2 is even." The negation of this statement is "there exists an even integer x such that x^2 is not even," and *not* "for all even integers x, x^2 is not even." Note that the statement "the integer x is even" can be made true in several ways (e.g., by taking $x = 2$ or $x = 24$ or $\dots$). The negation only says that for at least one such instance the integer x^2 is not even.

Remark 3.30. In the English language, mathematical constructs such as the if-then statement are often written in forms slightly different from the straightforward "S implies T" or "if S is true then T is true" forms. These forms arise from the way English is used: words such as *provided, given, whenever, only if, as long as* are often used instead to denote the implies construction, and care must be taken to decipher what the constituent statements S and T are. Consider the following:

> **Example 3.30.1.** "For any integer x, x^2 is even provided x is even." Here, S is the statement "The integer x is even" and T the statement "the integer x^2 is even. The statement is really saying, in the form of the implies statement that we have considered so far: "the integer x is even implies x^2 is even." Notice that in the version "for any integer x, x^2 is even provided x is even" the statement T has come first and S has come second. This is just an artifact of the English language; you must learn to look past these distractions to decipher the logical structure. "Provided ⟨*stuff*⟩..." means "if ⟨*stuff*⟩... is true."

Example 3.30.2. "Given that the integer x is even, x^2 must be even. Here, S and T are as in Example 3.30.1 above, and except for English usage, the statement is the same statement as in Example 3.30.1. It too is really saying "the integer x is even implies x^2 is even." In usage such as this, "given ⟨*stuff*⟩..." means "if ⟨*stuff*⟩... is true."
Note: The word "given" also shows up in mathematics in sentences such as the following: "given an integer x, its square is also an integer." Here, there is no logical implication going on, this is just another way of saying "the square of an integer is also an integer." The phrase "given an integer x" is merely defining x to be some integer. Thus, careful attention needs to be paid to how the word "given" is being used; sometimes it is used as a substitute for "if" and sometimes it is used just to define some object.

Example 3.30.3. "Whenever an integer x is even, x^2 is also even." Again, S and T are as in Example 3.30.1 above, and again, the statement is really saying "the integer x is even implies x^2 is even." "Whenever ⟨*stuff*⟩..." means "if ⟨*stuff*⟩... is true."

Example 3.30.4. "For any integer x, x is even only if x^2 is even." This construction can be a bit confusing, and to understand it, it is best to take it quite literally: it is saying that if x^2 is not even, then x is not even. This is just the contrapositive of "if x is even then x^2 is even," and hence is logically equivalent to it. So, with S and T as in Example 3.30.1 above, our statement here in this example is saying "$\neg T \implies \neg S$," which is the contrapositive of "$S \implies T$" and is thus logically equivalent to it. We can hence take our initial given statement to read "$S \implies T$," or "For any integer x, if x is even then x^2 is even." Thus, in general, "⟨*stuff*⟩... only if ⟨*other stuff*⟩" means "if ⟨*stuff*⟩... then ⟨*other stuff*⟩...."

Example 3.30.5. "As long as the real number x is a rational number, $2x$ will also be rational." Here, "as long as" really stands for "if." The statement is just saying "if the real number x is rational, then $2x$ is also rational." Thus, "as long as *stuff*..." means "if *stuff*... is true."

S implies and is implied by T. If S and T are statements, the new statement "S implies and is implied by T" is just a common way of writing the assertion "if S is true then T must be true and if T is true then S must be true." Thus, $S \iff T$ is the compound statement "$S \implies T$ and $T \implies S$." Symbolically, we write "$S \iff T$."

The meaning of "$S \iff T$" is clear, but note that there are hidden assertions in this statement. For instance, what if S is false? Then the statement implicitly asserts that T must also be false, since if T were true, the statement has already asserted that S must be true. Similarly, what if T is false? Again, the statement implicitly asserts that S must also be false, since if S were true, the statement has already asserted that T must be true. Thus, S and T must either *simultaneously* be true, or *simultaneously* false. We used the term *logically equivalent* earlier (Definition 3.9 and Remark 3.10) to describe this situation. In that language, the statement "$S \iff T$" asserts that S and T are logically equivalent.

Remark 3.31. A standard English usage for $S \iff T$ is "S is true if and only if T is true." Note that there are two parts to this. The portion "S is true if T is true" is the statement $T \implies S$. The portion "S is true only if T is true" is the statement $S \implies T$, as discussed in Example 3.30.4 above. (Mathematics invents the word *iff* to shorten "S is true if and only if T is true" further to "S is true *iff* T is true," although, this usage is not universal.)

The truth table for the compound statement *implies and is implied by* is easy to write down, since S and T are logically equivalent. See Table 3.8.

Table 3.8. Truth table for the *implies and is implied by* statement.

S	T	$S \iff T$
True	True	True
True	False	False
False	True	False
False	False	True

Example 3.32. Here is an example of the if and only if construction: "I will carry my umbrella tonight if and only if it is raining." (The statement is standard English usage for the fuller sentence "I will be carrying my umbrella with me tonight if and only if it will be raining tonight.") There are two statements here: "I will be carrying my umbrella with me tonight," and "It will be raining tonight." The assertion is that the two statements are logically equivalent and that they have the same truth values. If the assertion is true (mind you, it need not be true!), then if somebody sees you tonight without an umbrella, they can conclude without looking outside that it is not raining! Likewise, if they see you with an umbrella tonight, they can conclude that it is raining!

3.6 Further Exercises

Exercise 3.33. Practice Exercises:

Exercise 3.33.1. Classify the following as statements or non-statements:

(1) I had a hearty laugh at the joke!

(2) He had a fever of 102° but still completed his task.

(3) Why would anyone want to do this?

(4) If you go to the city you will find what you are seeking.

Exercise 3.33.2. Write down a negation for each of the following statements:

(1) I will work on my homework tonight.

(2) $17 > 18$.

(3) If I were six inches taller, I would be taller than Jose.

(4) Lucy scored a perfect ten on her biology test.

Exercise 3.33.3. In the following compound statements, determine the constituent statements S and T:

(1) S and T: I will go to the station tonight and I will call my mother.

(2) S or T: John will pick up Thomas from the station or get a soda from the corner store.

(3) $S \implies T$: If I come there tonight we will speak about this.

(4) $S \iff T$: We will speak about this if and only if I come there tonight.

Exercise 3.33.4. Write down the converse of the following statements:

(1) If I receive a dollar bill I will give you a dollar bill.

(2) If all sides of a triangle are colored green then it is equilateral.

(3) If Divya chooses mathematics she will also choose biology.

(4) If I have toast for breakfast then I will also have eggs.

Exercise 3.33.5. Write down the contrapositives of the statements in Exercise 3.33.4 above.

Exercise 3.34. Using the expressions "for all integers x," or "for some integer x," or "there exists an integer x such that," or "for at least one x," or "there does not exist an integer x such that," write out complete sentences that will make each of the following sentences *true statements.* (As part of reading carefully, you should have noted that there are *two* requirements: the sentences should become *statements,* and next, they should become *true* statements!) Here are examples to start you out: if you are given the sentence $x > x+2$ (which is not a statement because x is not defined here), you would make it a true statement using the expressions above by saying "there does not exist an integer x such that $x > x+2$." On the other hand, if you are given the sentence $x < x+2$, you would make it a true statement by saying "for all integers x, $x < x+2$." (In some cases, there could be more than one choice that will make the given sentence a true statement. Pick the one that yields the most mathematical information. For instance, both "there exists an integer x such that $x < x+2$" and "for all integers x, $x < x+2$" are correct, but the statement "for all integers x, $x < x+2$" yields much more information.)

(1) x is even.

(2) x^2 is nonnegative.

(3) $x^2 + 5x + 6 = 0$.

(4) $x^2 + x + 1 = 0$.

(5) There are no positive integers less than x.

(6) There are infinitely many positive integers less than x.

(7) x is the hypotenuse of an isoceles right triangle with integer sides.

(8) $x = \tan(\theta)$ for some real number θ that depends on x.

(9) $x = e^y$ for some real number y that depends on x.

(10) $\cos(2\pi x) = 0$.

Exercise 3.35. Each of the following is a compound statement of the form $S \implies T$ or $S \iff T$. Determine which of the two it is, and determine the two component statements S and T in each. (See the examples in Remark 3.30 on Page 40 for the usage of words such as "as long as," "whenever," etc. Note that S or T could themselves be compound statements!)

(1) As long as x is an integer, the square of x will also be an integer.

(2) The square of an integer x is a multiple of 3 only if x is a multiple of 3.

(3) The square of an integer x is divisible by 2 provided x is divisible by 2, but also, given the square of an integer x is divisible by 2, then x is divisible by 2.

(4) Only if n is an integer is the value of $\sin(n\pi)$ zero.

(5) Whenever $\cos(n\pi) = -1$, n must be an odd integer, moreover, $\cos(n\pi) = -1$ provided n is an odd integer.

(6) Provided the integer p is prime, then whenever p divides a product of two integers, it must divide one or the other of these two integers.

(7) If the order of the finite group H does not divide the order of the finite group G, then H is not a subgroup of G. (Note: you do not need to know what the mathematical terms "group," "order," and "subgroup" mean to analyze this statement.)

(8) A subset of $\mathbb{R}^n$ is compact if it is closed and bounded, and the converse is true as well.

(9) A topological space X is compact if and only if whenever $\mathcal{A}$ is a family of closed subsets of X with $\cap\mathcal{A} = \emptyset$, there is some finite $\mathcal{F} \subset \mathcal{A}$ such that $\cap\mathcal{F} = \emptyset$.

(10) The identity $\int_{|z|=1} f(z)\,dz = 0$ holds for the complex function $f(z)$ provided $f(z)$ is analytic in an open set containing the closed set $\{|z| \leq 1\}$.

Exercise 3.36. Write down the contrapositive for each of the statements in Exercise 3.35 above that are of the form $S \implies T$.

Exercise 3.37. Each of the following is a family of statements, indexed by the positive integers (so you can assume that "n" has already been defined to be an integer). For which values of n are they each true? *Note that parts* (6) *through* (10) *are extremely difficult and are designed only for you to do some further reading on your own!*

(1) $P_1(n)$: "$\cos^2(n\pi) + \sin^2(n\pi) = 0$."

(2) $P_2(n)$: "$\cos(n\pi) + \sin(n\pi) = 1$."

(3) $P_3(n)$: "$\cos(n\pi/2)\sin(n\pi/2) = 0$."

(4) $P_4(n)$: "$\cos(n)$ is an integer." (You may assume that π is an irrational number.)

(5) $P_5(n)$: "$2\cos^2(2\pi/n) = 1$."

(6) $P_6(n)$: "There exist integers x, y, and z, none of them zero, such that the equation $x^n + y^n = z^n$ holds."

(7) $P_7(n)$: "The regular n-gon can be constructed by ruler and compass."

(8) $P_8(n)$: "$n = a^2 + b^2$ for suitable integers a and b."

(9) $P_9(n)$: "The n-th Fermat number is prime."

(10) $P_{10}(n)$: "There exist infinitely many consecutive primes that differ by no more than n."

Exercise 3.38. Write down mathematically meaningful negations of the following statements. In other words, do not simply answer "the given statement is not true!" (For instance, for a negation of "there exists an integer greater than one whose square is divisible by exactly one prime," you should say "for all integers x greater than one, x^2 is divisible by two or more primes" instead of "it is not true that there exists an integer greater than one whose square is divisible by exactly one prime!")

Note: You don't need to know what the various mathematical objects here mean to analyze the logical structure of this statement!

(1) "e^x is rational for some nonzero integer x."

(2) "For each positive integer n, the sphere S^n has a unique differentiable structure."

(3) "There exists a prime $p \leq 19$ such that the ring $\mathbb{Q}[\omega_p]$ is not a principal ideal domain."

(4) "There exists some semistable elliptic curves defined over the rationals that are not modular."

(5) "For all positive integers n, the symmetric group S_n is solvable."

(6) "All Fermat numbers F_n, $n = 1, 2, \ldots$, are prime."

(7) "The upper half-plane admits only finitely many automorphisms."

(8) "For some positive integer n, $\zeta(-2n)$ is nonzero."

(9) "There exist some holomorphic functions on the open set Ω that are not infinitely differentiable at all points of Ω."

(10) "No woman mathematician has won the Fields Medal up to now."

Exercise 3.39. *Logic puzzles.* Ideas in mathematics, when written out with formal proofs, are expressed as a series of *deductions:* we are given certain statements that are known to be true (example: "n is an odd integer"), and we *deduce* from this using logic and known mathematical results the truth of a different statement (for example "therefore n^2 is an odd integer"). There is a vast collection of puzzles available on the internet where you are given some unknowns, and some partial clues, and *purely by logical deduction* and no extraneous information or tricks, you are expected to determine the unknowns. These puzzles are good for honing skills of deduction! Besides being fun, they will be very useful for you in your pursuit of mathematics. Here are two sample puzzles, but you are encouraged to scour the internet and find more such puzzles to solve!

(1) Three prisoners are shown three black hats and two white hats, and then blindfolded and made to stand in increasing order of height. From these five hats, one hat is then placed on each prisoner's head. The blindfolds are removed. Each prisoner is allowed to see the hats of the persons in front, but is not allowed to turn back. They are asked to determine the color of their own hats. The tallest prisoner says "I don't know." Then the prisoner in the middle says "I don't know." Finally the prisoner at the front says "I know." What color hat was this prisoner wearing?

(2) Four friends Jasmine, Kavya, Larissa, and Susan pick oranges from a neighbor's garden (with permission of course!). Each uses a basket of a different color: the colors are blue, green, red and white. When done, the baskets are found to contain 20, 22, 24, and 26 oranges, not necessarily in order. You need to determine which girl carried which basket and picked how many oranges, if you are told:

(a) Larissa used the white basket.

(b) Larissa picked four more oranges than the girl who used the red basket.

(c) The girl that picked 20 oranges used either the red basket or the white basket.

(d) Kaavya picked four more oranges than Jasmine.

(e) Jasmine used the blue baseket

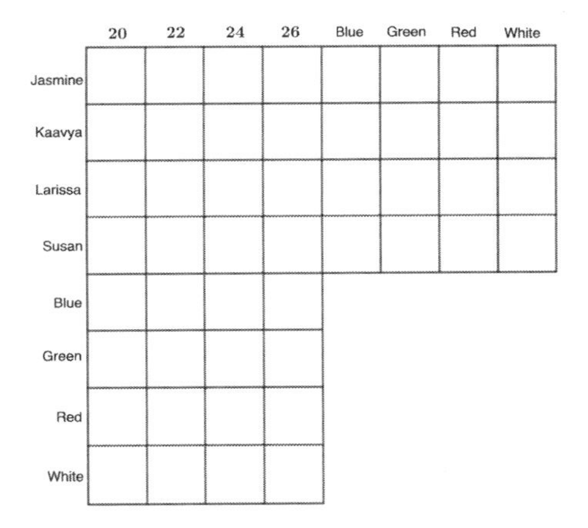

Figure 3.1. Grid for logic puzzle 3.39(2).

(Hint: In such puzzles it helps to draw grids and put a check mark in squares that you are sure of, cross out squares that are impossible, and also tentatively pencil in squares that are possibilities. Since there are three categories that need to be matched: name, number of oranges, and color of basket, one ideally needs a three-dimensional grid! But this can just as easily be done on the plane as in Figure 3.1 below. Statement (2)(a), for example, allows you to put a check mark on the square on the row labeled Larissa and on the column labeled white. The same statement shows that the remaining squares in that same column can be crossed out.)

4

Counting, Combinations

In this chapter we will review elementary counting arguments that you would have seen in high school, and we will see that using just these simple ideas, we can derive very pretty and powerful identities that are useful in all parts of mathematics. Many problems in diverse areas of mathematics involve the counting of a finite number of objects, and some practice with some simple counting techniques is in order. There are some basic principles, but the application of these principles to any specific problem can often involve a huge amount of ingenuity. Problems in counting are thus an excellent tool for developing mathematical creativity!

Let us start with an easy example:

Example 4.1. Suppose that Norma is working part-time and will take two courses this coming semester. She decides to complete her general education requirement, as part of which she needs to select one Chemistry course from a basket of three courses and one Africana Studies course from a basket of five. How many possible choices does she have for her study program this semester?

Notice that we can model Norma's selection as comprising two *events:* (1) Select a Chemistry course, and (2) Select an Africana Studies course, that are *independent* of each other. "Independence" here means that her selection of a Chemistry course will not affect her selection of an Africana Studies course, and likewise, her selection of an Africana Studies course will not affect her selection of a Chemistry course. (Although not stated explicitly, we can indeed assume the events are independent, based both on our own experience with college requirements and the fact that the two subjects are so unconnected!) The first event can be accomplished in three *distinct* ways. Similarly, the second event can be accomplished in five distinct ways. Elementary considerations show that she has $3 \cdot 5 = 15$ ways in which she can make up her study program, as follows: Let us label her Chemistry courses as C_1, C_2, and C_3, and her Africana Studies courses as A_1, A_2, A_3, A_4, and A_5. Then, if she picks C_1 for Chemistry, she can pick any of A_1 through A_5 for her Africana Studies, giving her the five possible combinations (C_1, A_1), (C_1, A_2), $\ldots$, (C_1, A_5). Similarly, if she picks C_2 for Chemistry she has five combinations: $(C_2, A_1), (C_2, A_2), \ldots, (C_2, A_5)$. If she picks C_3 for Chemistry again she has five combinations : $(C_3, A_1), (C_3, A_2), \ldots, (C_3, A_5)$. This accounts for the multiplication $3 \cdot 5$.

Now we consider a different example, in which the events are *not* independent, but which nevertheless shares a key feature with Example 4.1:

Example 4.2. Given seven cards in front of you, each containing a different letter of the English alphabet from A through G, how many two-letter words can you form by picking two cards out of the seven, assuming that we consider every possible combination of two letters as a word?

Here we have two events: E_1, the act of picking the first card, and E_2, the act of picking the second card *after the first card has been picked.* These two events are *not* independent, since if for instance the first card you picked was an A, the second card cannot be an A as it has already been picked! (And in fact, you now only have six cards to select from.) Nevertheless, here is a key feature of this example (italicized): The first card can be picked in 7 distinct ways (there are seven distinct ways because the cards all contain different letters), but *no matter what card you pick the first time around, you have a choice of six distinct cards to select from the next time around.* Of course, depending on which card you picked the first time, the choice of six cards you have the next time will be different: for instance, if you picked A the first time, you have a choice of B, C, D, E, F, and G the second time, but if you picked say F the first time, you have a choice of A, B, C, D, E, and G the second time. However, the number of choices you have the second time remains the same no matter what the outcome of E_1, namely 6.

Now, arguing exactly as in the previous example, if you picked A the first time, you have six distinct choices the second time, leading to the words (A, B), (A, C), (A, D), (A, E), (A, F), and (A, G). If you picked B the first time, you have six distinct choices the second time, leading to the words (B, A), (B, C), (B, D), (B, E), (B, F), and (B, G). And so on, giving you a total of $7 \cdot 6$ choices in all.

4.1 Fundamental Counting Principle

What is common to Examples 4.1 and 4.2 is that you have two events, and for each of several distinct ways of accomplishing the first event, you have a certain number of distinct ways of accomplishing the second event, and this second number *is the same no matter how you accomplish the first event.* We can extract the arguments in these two examples and encode it into a simple principle:

Definition 4.3. *Fundamental Counting Principle (FCP):* Suppose that Event E_1 can be accomplished in m_1 distinct ways, and for each way of accomplishing E_1, suppose that Event E_2 can be accomplished in m_2 distinct ways. Then the total number of ways in which E_1 and E_2 can be accomplished together is the product $m_1 \cdot m_2$.

Remark 4.4. The FCP can be extended to n events using exactly the arguments we used in Examples 4.1 and 4.2. Suppose one has Events $E_1, E_2, \ldots\ E_n$ $(n \geq 1)$. Suppose that E_1 can be accomplished in m_1 distinct ways, and for each way of accomplishing E_1, E_2 can be accomplished in m_2 distinct ways, and for each way of accomplishing E_1 and

E_2 together, suppose that E_3 can be accomplished in m_3 distinct ways, and so on, then the total number of ways in which E_1, E_2, ..., E_n can be accomplished together is the product $m_1 \cdot m_2 \cdot \cdots \cdot m_n$.

To illustrate the ideas that go into this, consider the same 7 cards in Example 4.2, but now consider the problem of forming *three*-letter words. Then if you pick A the first time, you can pick any of B, C, D, E, F or G the second time, and therefore, with A as the first choice, the six possible ways of picking the first two cards are (A, B), (A, C), (A, D), (A, E), (A, F), and (A, G) (as in Example 4.2). But now for the third card: *for each of these six possible ways* of picking the first and second card, you have five choices for picking the third card, as there are now five cards left. Of course, depending on what choices you made during the first two picks, you have different choices for the third pick, for instance, if you pick (A, B) in the first two picks, you have a choice of (A, B, C), (A, B, D), (A, B, E), (A, B, F), and (A, B, G), but if you pick say (A, F) in the first two picks, you have a choice of (A, F, B), (A, F, C), (A, F, D), (A, F, E), and (A, F, G). However, the number of choices the third time around is five for every choice of the first two picks. The total number of three-letter words you can form is therefore $7 \cdot 6 \cdot 5$.

Here are a few easy exercises that call for an application of the FCP:

Exercise 4.5. Apply the FCP to answer the following. In each, pay attention to what you think of as events, and pay attention to whether these events are independent of one another or whether there is a dependence.

(1) What is the number of possible three-letter words in the English language with distinct letters ("possible" here means that we consider every combination of three letters a word, just as in Example 4.2)?

(2) What is the number of choices of attire possible if I have 7 different colored ties, 4 different colored shirts, and 3 different colored pairs of pants? (You may assume that I have no sense of style and happily mix colors with no regard to whether they go together.)

(3) A committee comprising a president, a secretary, and a treasurer is to be formed from a pool of 10 candidates. How many different committees are possible?

One has to be careful in designating aspects of the problem as "events." Often, a hasty choice of labelling actions as events could lead to an overcount of the number of solutions to the problem! A factor to be kept in mind is that the FCP is only applicable to problems where you have so many distinct ways of performing Event 1, and for each such way, you have so many distinct ways of performing Event 2, and so on. Often, there may not be a proper choice of activities to label as events that will satisfy the requirements of the FCP. In such cases, one has to develop arguments that are of an *ad hoc* nature, native only to the given problem. Consider the following two modifications of Example 4.2:

Example 4.6. You are given seven cards. Two of them are labeled A, and the remaining five are labeled B, C, D, E, and F.

(1) How many possible two-letter words with distinct letters can you form by picking two cards out of the seven?

(2) How many possible two-letter words can you form by picking two cards out of the seven?

(Once again, just as in Example 4.2, a two-letter word is just any combination of two letters. In particular, the same two letters, one following the other, is a word.)

Let us consider part (1) first. Just as in Example 4.2, this problem too involves selecting cards from a pile, and it is tempting therefore to let events correspond to selecting cards. So, let us think of Event 1 as selecting the first card from the pile of seven, and Event 2 as selecting the second card from the remaining 6 cards. If A is selected first, a possible choice for the second card is another A, but this way of performing Event 1 and then Event 2 does not correspond to a solution of the problem, as the word AA is not allowed! Thus, we cannot count the number of solutions to the problem by counting the pairs of these events! Instead, what we can do is think of Event 1 as *forming the first letter of our word* and Event 2 as *forming the second letter of our word.* There is a subtle difference between selecting a letter from a pile and forming the letters of our word: choosing our event as forming the letters will incorporate the constraint given, namely that words should have distinct letters.

The number of *distinct* ways of forming the first letter of our word is now *six*, not seven. This is because, from the point of view of the final word we form, we cannot distinguish between the two cards that are both labeled A. Now, let us think carefully about the second letter. If the first letter was an A, the choices for the second are B, C, D, E, and F, as A cannot be repeated. Thus, with A as the first pick, we have 5 choices for the second. Now, if we select B as our first letter, we could *pick* either of two As, or any of C, D, E, and F, but as far as choices for the second letter is concerned, there are only five choices: A, C, D, E, and F. Thus there are five choices for the second letter if B is the first letter. The same is true if the first letter is any of C through F. We thus find that there are six distinct choices for the first letter, and for each of these, there are five distinct choices for the second letter, so the FCP applies, and together there are $6 \cdot 5 = 30$ picks.

In part (2), unlike in part (1), the word AA is allowed. In this case, if we view selecting the first card as Event 1 and selecting the second as Event 2, then we get *two* instances of the word AA: first by selecting the first A the first time around and the second A the second time, and then again, selecting the second A the first time and the first A the second time! So once again, selecting the cards is a poor choice of events.

There are two ways to analyze the situation in part (2). The easiest way to solve the second part is to note that the only possible word in which the letters are not distinct is AA. Thus, the answer to the second part must be one more than the answer to the first part, that is, $1 + 30 = 31$. Alternatively, we could just repeat our analysis of part (1): if we select A as the first letter of our word, we now have *six* choices for the second letter: all of A through F are valid, since we are allowed words in which letters are repeated. On the other hand, if we choose B for the first letter of our word, we can *pick* either of the two As, or any of C, D, E, and F, but as far as choices for the second letter is concerned, there are only five choices: A, C, D, E, and F. The same is true if the first letter is any of C through F. So, the total number of words is 6 (if you choose A the first time) plus $5 + 5 + 5 + 5 + 5$ (if you choose any of B through F); in other words 31. (Note that we have to add the number of cases together; we cannot invoke the FCP since the hypothesis of the FCP does not apply. Different ways of performing Event 1 lead to different numbers of ways of doing Event 2.)

4.2 Permutations and Combinations

Example 4.2 and its generalization in Remark 4.4 quickly lead to the notion of *permutations:* the rearrangements of a collection of distinct objects in a row. Suppose we have n distinct objects (for some integer $n \geq 1$, in fact, $n \geq 2$ to be interesting!), and for simplicity, suppose these are labeled as $t_1, t_2, \ldots, t_n$. The question is: in how many different ways can you arrange these n objects in a row? We can model the process of arranging these n objects as a collection of n events: E_1 is the event where you place one object in the first place in the row, E_2 is the event where, after placing one object in the first place, you now place another object in the second place in the row, and so on, until E_n, which is the event where, after placing $n-1$ objects in the first $n-1$ places, you place the remaining object in the last place. Exactly as in Example 4.2 and the generalization in Remark 4.4, E_1 can be accomplished in n ways, since any of the objects t_1 through t_n can be placed in the first spot, and then, after placing one object in the first place, E_2 can be accomplished in $n-1$ ways since there'll now be $n-1$ objects left. Proceeding, once an object has been placed into the first place and another into the second place, E_3 can be accomplished in $n-2$ ways, and so on, till we find E_{n-1} can be accomplished in 2 ways (at that point there will be only two objects left), and then E_n can be accomplished in just one way (at that point there will be only one object left). By the FCP, we find that the number of ways of arranging n objects in a row is given $n \cdot (n-1) \cdot (n-2) \cdots \cdot 2 \cdot 1$. You would have seen the special notation for this: $n!$ (read "n factorial"). Thus, $1! = 1$, $2! = 2 \cdot 1 = 2$, $3! = 3 \cdot 2 \cdot 1 = 6$, $4! = 24$, $5! = 120$, $6! = 720$, and so on. By convention, $0!$ is taken to be 1 (this value fits neatly in various formulas). The factorial of a negative integer is not defined.

We have thus proved the following:

Proposition 4.7. *The number of permutations of n distinct objects is $n!$.*

It is important to remember that the word permutations refers to the rearrangements of *distinct* objects in *a row*. There are other kinds of rearrangements that often need to be considered, and in such situations, arguments native to the situation need to be developed. Here is an example:

Example 4.8. Seven people are to be seated around a circular conference table that has exactly seven chairs. How many different seating arrangements are possible?

What is going to distinguish one seating arrangement from another is the order in which the people are seated. One is tempted at first to say that these 7 people can be rearranged in 7! different ways, so there are 7! different seating arrangements possible. But there is an overreach happening here: Our result in Proposition 4.7 above applies to permutations: rearrangements in a row. In this situation, because the table is circular, the seating arrangement doubles back on itself. So, for example, labeling the people as $t_1, \ldots, t_7$, the following two rearrangements of people in a row will determine the same seating arrangement, since there is nothing that will distinguish the start person and the end person: $t_1t_2t_3 \dots t_7$ and $t_7t_1t_2 \dots t_6$.

The good news though is that this example in the last line contains the idea needed to solve the problem! As this example shows, any two rearrangements of seven people in a row that differ from each other by only a "circular rearrangement" will yield the same seating arrangement around a circular table. Let us denote an arbitrary rearrangement of the persons $t_1, \dots, t_7$ in a row by $t_{i(1)}t_{i(2)} \dots t_{i(7)}$ (where $i(1), i(2), \dots, i(7)$ is a rearrangement of $1, 2, \dots, 7$). Then the various "circular rearrangements" of this rearrangement that will yield the same seating arrangement are $t_{i(7)}t_{i(1)}t_{i(2)} \dots t_{i(6)}$, $t_{i(6)}t_{i(7)}t_{i(1)} \dots t_{i(5)}, \dots, t_{i(2)}t_{i(3)} \dots t_{i(7)}t_{i(1)}$, along with the one we started with, namely, $t_{i(1)}t_{i(2)} \dots t_{i(7)}$, which we also count as a circular rearrangement. There are clearly 7 "circular rearrangements" of any given rearrangement. Thus, if we organize the various 7! permutations of the seven people into groups such that all the permutations in any one group are just circular rearrangements of a single permutation, we will have $7!/7 = 6!$ groups. Moreover, as you can easily see, if one permutation is not a circular permutation of another, then the two permutations, when strung into a seating arrangement around a circular table, will yield different seating arrangements. Thus, each group will yield a different seating arrangement, so there are at least 6! seating arrangements. To finish the counting, we just have to note that every arrangement of 7 people in a circle is obtained in exactly the same manner as we have followed: we take an arrangement of 7 people in a row and string them into a circle. There are thus exactly 6! possible seating arrangements.

(Later, with more sophisticated concepts under your belt, you'll realize that we have established a one-to-one correspondence between the set of equivalence classes of permutations of a 7 element set under the relation of being a circular rearrangement and the set of possible seating arrangements in a circle. But while that language certainly clarifies the process, you don't need that language to understand a relatively simple situation such as the one above.)

Now here is another example where we consider rearrangements of objects *that are not all distinct* in a row:

Example 4.9. How many seven-letter words can be formed from the letters A, A, B, C, D, E, F? (As in other such problems, we ignore the issue of whether some particular word we come up with actually exists or not!)

This is like Example 4.6(2), except that we are now asking to use all seven letters. We solve the problem as follows: temporarily think of the two As as distinct, for instance, by imagining that they are really two letters, A_1 and A_2. We thus have seven distinct letters, and these can be rearranged in 7! ways. But in any one rearrangement if say A_1 is in the i-th spot and A_2 is in the j-th spot (where $1 \le i, j \le 7, i \ne j$ of course), then the word formed from the rearrangement in which A_1 and A_2 are swapped, so A_1 is now in the j-th spot and A_2 is in the i-th spot, is the same word, since in reality, A_1 and A_2 are the same letter. Thus, if we arrange all 7! permutations of A_1, A_2, B, $\dots$, F into groups in which in each group, the positions of A_1 and A_2 are simply swapped, there will be $7!/2$ groups. Exactly as in Example 4.8, we can argue that the number of words will correspond to the number of groups, and therefore, there will be $7!/2$ words possible.

Remark 4.9.1. If instead, had the letters been A, A, A, B, C, D, E, F, so there are now eight letters but three of them are repeats of one another, we would first pretend that we had three distinct forms of A, namely A_1, A_2, and A_3, and get 8! rearrangements. But now, all possible internal rearrangements of A_1, A_2, and A_3 (in which the other letters B through F are not moved) all yield the same word. There are 3! (note: 3! not just 3) rearrangements of three letters. Thus, the total number of words would now be 8! /3!.

Exercise 4.10. Here are some exercises based on the ideas in Examples 4.8 and 4.9 above.

(1) In how many ways can you rearrange the letters of the word *AWESOME*?

(2) Repeat part (1) for the word *TOTALLY*.

(3) Repeat part (1) for the word *TOTALMENTE*.

(4) How many necklaces can you form with 7 identical size beads, three of which are red and the others are blue, white, yellow and green?

The idea in Example 4.9, where we account for duplication by first pretending that the duplicated objects are distinct and later viewing all internal rearrangements of the duplicated objects as the same, finds use in the formula for *combinations*. A combination (or *selection*) of three things from five things is just a choice of three things from five things in which the order of choosing is unimportant. Thus, if the five things are labeled A, B, C, D, and E, we view the choices ABC, ACB, BAC, BCA, CAB and CBA to all be the same. If we were now to write out all the combinations (or selections), we will find the following ten to be the full list: ABC, ABD, ABE, ACD, ACE, ADE, BCD, BCE, BDE, and CDE.

We denote the number of combinations of r things from n things $1 \leq r \leq n$ by the symbol $\binom{n}{r}$. (Notice that we did not specify that r and n are integers—but this is clear from the context since we are talking about a "number of things.") The symbol is also known as the *binomial symbol* because of its relation to the binomial theorem (4.16) that we will consider later in this chapter. It is convenient in various situations to have a definition for the binomial symbols $\binom{n}{0}$ (for $n \geq 1$) and also for $\binom{0}{0}$. We will *define* both of these to equal 1. This is not too artificial: we *can* say that the number of ways of selecting zero objects from n objects, or zero objects from zero objects, is indeed one—we perform our selection by simply doing nothing! But another reason for this definition is that it fits neatly into the second formula for $\binom{n}{r}$ in Proposition 4.11 below if we set $r = 0$: $\frac{n!}{0! \cdot (n-0)!} = \frac{n!}{1 \cdot n!} = 1$. (Note that this calculation works even if $n = 0$. This formula is an instance where our definition of 0! as 1—see the discussion just before Proposition 4.7—has come in useful!)

We have thus defined the symbol $\binom{n}{r}$ for the case $0 \leq r \leq n$. It is convenient as well to have a definition in the remaining cases: when $n < 0$ (any r), or when $r < 0$ (any n), or when $r > n$—we define $\binom{n}{r}$ to be zero in all these cases!

We have the following:

Proposition 4.11. *For $0 \le r \le n$, we have the formula*

$$\binom{n}{r} = \begin{cases} \dfrac{n \cdot (n-1) \cdot \dots \cdot (n-r+1)}{r!} & \text{if } r > 0, \\ 1 & \text{if } r = 0. \end{cases} \tag{4.1}$$

For both $r > 0$ and $r = 0$, this can also be written as

$$\binom{n}{r} = \frac{n!}{r! \cdot (n-r)!}. \tag{4.2}$$

Proof. To prove the first assertion of the theorem assume first that $r \ge 1$ (this forces $n \ge 1$ because of the condition $0 \le r \le n$). Our problem is to select r objects from n objects (so the order of choosing is not important), but instead, we first pretend that we are forming r-letter words from n distinct letters. We can pick the first letter in n ways, and for each choice of first letter, we can pick the second in $n-1$ ways, and so on, till we find that we can pick the r-th letter in $n-(r-1)$ ways (since at that point, $r-1$ letters would have been selected, leaving $n-(r-1)$ behind). Thus, by the FCP (see Remark 4.4 for the generalization to r events), the number of r letter words is the product $n \cdot (n-1) \cdot \dots \cdot (n-(r-1))$. (The last factor can of course be written as $n-r+1$.) Now, as in Example 4.9, we will observe that given any r-letter word, all rearrangements of that r-letter word represent the same combination (once again, to emphasize, a "combination" or "selection" treats different rearrangements as the same). There are $r!$ different rearrangements of an r letter-word, so if we organize all our r-letter words into groups where within each group all words are just rearrangements of each other, we will have $\frac{n \cdot (n-1) \cdot \dots \cdot (n-r+1)}{r!}$ groups. Each of these groups leads to a distinct selection of r objects from n objects, and further, every selection of r objects from n objects arises from one such group. We have thus shown that the number of combinations of r objects from n objects is precisely $\frac{n \cdot (n-1) \cdot \dots \cdot (n-r+1)}{r!}$.

When $r = 0$, the first assertion of the theorem falls out from the very definition of $\binom{n}{0}$, and there is nothing to prove!

For the second assertion of the proposition, just note that for $r > 0$,

$$\begin{aligned} \frac{n \cdot (n-1) \cdot \dots \cdot (n-r+1)}{r!} &= \frac{n \cdot (n-1) \cdot \dots \cdot (n-r+1) \cdot (n-r)!}{r! \cdot (n-r)!} \\ &= \frac{n!}{r! \cdot (n-r)!}. \end{aligned}$$

Also, as observed above the statement of the proposition, the expression $\frac{n!}{r! \cdot (n-r)!}$ reduces to 1 when $r = 0$, so the second assertion is certainly true if $r = 0$.

This proves the proposition. □

Remark 4.12. It is worth noting the values of $\binom{n}{r}$ for some key values of r: $\binom{n}{0}$ is just 1 by definition, $\binom{n}{n}$ comes out to be 1 by the formula (4.2) for instance, or just by definition—there is just one way to select n objects from n objects: select them all! $\binom{n}{1}$ comes out to be n by the formula (4.1) for instance, $\binom{n}{2}$ comes out to be $\frac{n(n-1)}{2}$, etc.

Example 4.13. A committee of 4 needs to be formed from 12 people, out of which 8 are men and 4 are women. How many committees can be formed if there needs to be at least one woman in any committee?

Note first that this is really a problem of combinations; the order of choosing of a committee is unimportant. A committee comprising A, B, and C is the same as a committee comprising B, A, and C. (By contrast, in Exercise 4.5, part (3), the order of choosing of the committee is important, because there each person in a committee plays a unique role: one is the president, one the secretary, and one the treasurer.) Had there been no restriction on the number of women, then the count is simple: we have 12 people, and from these we are selecting 4 people, so the number of ways of doing this is $\binom{12}{4}$. To take into account the restriction, we instead look at how many committees exist *without* any women. These would correspond to selecting all four members from the 8 men present, so there would be $\binom{8}{4}$ such members. Since there are $\binom{12}{4}$ committees in all, the difference $\binom{12}{4} - \binom{8}{4}$ represents the number of committees with at least one woman in it.

There is an alternative way to solve this problem. This involves more work, but all the same, is very instructive, since it gets you into the "nuts and bolts" of the selection process! We can form committees with at least one woman in four mutually exclusive ways: (1) form committees with exactly one woman, (2) with exactly two women, (3) with exactly three women, and (4) with all four being women. Since these are mutually exclusive, the total number of ways of forming committees with at least one woman will simply be the sum of the way of forming each of the four. We will just show you how the first can be computed: you should do the remaining cases yourself and confirm that you get the same number as in the previous paragraph.

To form a committee with exactly one woman member, we model our selection as comprising two independent events: the first involves making a selection of one woman from the pool of 4 women, and the second involves making a selection of 3 men (the balance on the committee) from the pool of 8 men. The first event can be accomplished in $\binom{4}{1}$ $(= 4)$ ways, while the second can be accomplished in $\binom{8}{3}$ ways. By the FCP, the number of committees with exactly one woman member is the product $\binom{4}{1} \cdot \binom{8}{3}$.

You can do the remaining cases similarly. (See also Exercise 4.33 ahead.)

Example 4.14. In this example, we will see a counting technique that is useful in a range of problems. Suppose you are asked to count the number of binary 8-bit words with three 1s, and in which no two of the 1s are next to each other. Translated into English from the world of computers, what you are being asked for is the number of distinct strings of length 8 you can form with three 1s, and therefore $8 - 3$ or five 0s, in which no two 1s appear next to each other. For instance, the string 10010010 is allowed, but the string 10011000 is not allowed, as it contains two 1s adjacent to each other. (We will assume here that it is perfectly alright for a string to start with one or more zeros. For instance 00010101 is a valid string.)

The trick is to place the five 0s in a row, with empty spaces ($\square$) around each one of them, as follows:

$$\square\ \boxed{0}\ \square\ \boxed{0}\ \square\ \boxed{0}\ \square\ \boxed{0}\ \square\ \boxed{0}\ \square \tag{4.3}$$

There are six empty spaces present. (Note that with five 0s, you will indeed get six empty spaces, since you need an empty space after every 0 and also one *before* the first 0.) Placing a 1 into any three of these empty spaces yields an eight-bit sequence with three 1s, no two of which are adjacent—this is guaranteed because each of the potential empty spaces into which you could place a 1 is separated from its neighbors by a 0. For instance, the following placement

$$\square\ \boxed{0}\ \boxed{1}\ \boxed{0}\ \square\ \boxed{0}\ \boxed{1}\ \boxed{0}\ \square\ \boxed{0}\ \boxed{1}$$

yields the 8-bit word 01001001. Moreover, different choices of the empty spaces to fill with 1s will yield different 8-bit words. Conversely, every 8-bit word with three 1s, no two adjacent, indeed arises from the configuration depicted in (4.3) above, with three empty spaces filled in. For instance, the word 10100010 comes by filling the first, the second, and the fifth empty box in (4.3) with 1s.

Therefore, to count the number of binary 8-bit words with three 1s, no two of which are adjacent, we can equivalently count the number of ways of filling in those six empty spaces in (4.3) with three 1s. But we know the answer to this one: we are simply making a selection of three spaces from six spaces (into which we'll fill a 1), and the answer is $\binom{6}{3}$.

(Using language that will come later, you'll realize that we have created a bijection between the set of binary 8-bit words with three 1s, no two of which are adjacent, and the set of arrangements of spaces in (4.3) in which three of the empty spaces are filled with 1. But while that language certainly clarifies, we do not need that language to understand what we have done here!)

4.3 Binomial Relations and Binomial Theorem

There are three classic relations concerning the binomial symbol $\binom{n}{r}$ that we will consider now:

Proposition 4.15. *We have the following:*

(1) *For* $0 \le r \le n$,

$$\binom{n}{r} = \binom{n}{n-r}.$$

(2) (*Pascal's Identity.*) *For* $0 < r < n$,

$$\binom{n}{r} = \binom{n-1}{r} + \binom{n-1}{r-1}.$$

(3) *For* $0 \le r \le n$,

$$\binom{n+1}{r+1} = \frac{n+1}{r+1}\binom{n}{r}.$$

Proof. Part (1) above can be seen from formula (4.2) in Proposition 4.11 above: By that formula,

$$\binom{n}{n-r} = \frac{n!}{(n-r)!\,(n-(n-r))!} = \frac{n!}{(n-r)!\,r!},$$

and this last expression is just $\binom{n}{r}$ by that same formula (4.2)!

Perhaps a more pleasing way to establish the relation $\binom{n}{r} = \binom{n}{n-r}$ is to note that for every selection of r objects taken from n objects, we leave behind a selection of $n-r$ objects from these same n objects. We can thus match the selections of r objects to the selections of $n-r$ objects in a one-to-one manner: for every selection of r objects we associate the selection of $n-r$ objects we leave behind, and for every selection of $n-r$ objects, we pretend that we selected these objects not by picking them out but by leaving them behind and instead throwing out the remaining r objects, and we then associate the thrown out selection of r objects to these $n-r$ objects. It follows that the number of selections of r objects from n objects must be the same as the number of selections of $n-r$ objects from n objects, because of this nice matching.

(In the language of sets and functions that you will study in Chapter 5, we have set up a bijection, or a one-to-one correspondence, between the set of selections of r objects from n objects and the set of selections of $n-r$ objects from n objects, so these two sets must have the same cardinality.)

As for the second relation, part (2), one method of proof is by symbol manipulation (we will provide a different proof too in just a bit). We start with the right side: Since $r < n$, we find $r \leq n-1$ and $r-1 < n-1$. Also, $0 \leq r-1$. Hence, we may apply formula (4.2):

$$\begin{aligned}\binom{n-1}{r} + \binom{n-1}{r-1} &= \frac{(n-1)!}{r!\,(n-1-r)!} + \frac{(n-1)!}{(r-1)!\,(n-1-(r-1))!} \\ = \frac{(n-1)!}{r!\,(n-1-r)!} + \frac{(n-1)!}{(r-1)!\,(n-r)!} &= \frac{(n-1)!\,(n-r)}{r!\,(n-1-r)!\,(n-r)} + \frac{(n-1)!\,(r)}{(r-1)!\,(r)(n-r)!} \\ = \frac{(n-1)!\,(n-r)}{r!\,(n-r)!} + \frac{(n-1)!\,(r)}{r!\,(n-r)!} &= \frac{(n-1)!\,(n-r+r)}{r!\,(n-r)!} = \frac{(n-1)!\,(n)}{r!\,(n-r)!} \\ &= \frac{n!}{r!\,(n-r)!} = \binom{n}{r}.\end{aligned}$$

But perhaps a more pleasing way to prove this relation is as follows: Pick one of the n objects and think of it as a distinguished object. Denote it by D. The selections of r objects from these n objects fall into two mutually exclusive classes: those that do not contain D, and those that contain D. Now selections of r objects from n objects in which D does not appear correspond to selections of r objects from the remaining $n-1$ objects (obtained by throwing out D). There are $\binom{n-1}{r}$ of these. On the other hand, selections of r objects in which one of the selected objects is D correspond to selections of $r-1$ objects from the remaining $n-1$ objects (again, obtained by throwing out D). There are $\binom{n-1}{r-1}$ of these. Putting these two mutually exclusive cases together, we find that the number of selections of r objects from n objects, namely $\binom{n}{r}$, equals the sum of $\binom{n-1}{r}$ and $\binom{n-1}{r-1}$.

Finally, the proof of the relation in (3) follows immediately from applying formula (4.2) to both sides, and recognizing that $(k+1)! = (k+1)\cdot\underbrace{k\cdot(k-1)\cdots 2\cdot 1} = (k+1)\cdot k!$. We find

$$\binom{n+1}{r+1} = \frac{(n+1)!}{(r+1)!\,(n+1-(r+1))!} = \frac{(n+1)\cdot n!}{(r+1)\cdot r!\,(n-r)!} = \frac{n+1}{r+1}\binom{n}{r}.$$

□

We close this chapter with a proof of the binomial theorem. We state it for any two variables x and y which commute with each other (i.e. $xy = yx$). In particular, substituting x and y to be any two complex numbers or real numbers or rational numbers or integers, we find the result to hold for any two complex numbers or real numbers or rational numbers or integers.

Theorem 4.16 (Binomial Theorem). *Let x and y be two variables that satisfy $xy = yx$, and let $n \geq 1$ be an integer. Then*

$$(x+y)^n = \binom{n}{0}x^n + \binom{n}{1}x^{n-1}y + \binom{n}{2}x^{n-2}y^2 + \cdots + \binom{n}{j}x^{n-j}y^j + \cdots \binom{n}{n-1}xy^{n-1} + \binom{n}{n}y^n. \quad (4.4)$$

Before we begin the proof, a few comments about the theorem. There are $n + 1$ summands on the right, with coefficients in a pattern: $\binom{n}{0}, \binom{n}{1}, \binom{n}{2}, \ldots, \binom{n}{n}$. But there is more pattern to these summands! Study the exponents of the variables. Remember that x^n is the same as $x^n y^0$ and y^n is the same as $x^0 y^n$. We can see that the exponent of x starts at n and goes down one by one to zero, while the exponent of y starts at 0 and goes up one by one to n. In particular, each term is of the form $x^{n-j}y^j$, and thus of total degree $n - j + j = n$; moreover, every term of the form $x^{n-j}y^j$, $j = 0, \ldots, n$, appears. Also, notice the symmetry of the coefficents: the coefficient of $x^{n-j}y^j$ and $x^j y^{n-j}$ are equal, because $\binom{n}{j}$ equals $\binom{n}{n} - j$ by Proposition 4.15. (Of course, the first and last coefficients are just 1, since $\binom{n}{0} = \binom{n}{n} = 1$.)

Next, let us recall how we multiply out expressions like $(a + b)(c + d)$, $(a + b)(c + d)(e + f)$, etc., where a, b, c, etc. are variables that “commute pairwise.” (What this means is that $ab = ba$, $ac = ca$, $bc = cb$, etc.) We have

$$(a + b)(c + d) = (a + b)c + (a + b)d = ac + bc + ad + bd.$$

We see that the product of $(a + b)$ and $(c + d)$ is the sum of all possible terms obtained by multiplying one summand from $(a + b)$ with one from $(c + d)$. We multiply a from the first factor with c from the second, then b from the first with c from the second, then multiply a from the first factor with d from the second, and then b from the first with d from the second.

Similarly, we have

$$\begin{aligned}(a + b)(c + d)(e + f) &= (a + b)(c + d)e + (a + b)(c + d)f \\ &= (ac + bc + ad + bd)e + (ac + bc + ad + bd)f \\ &= ace + bce + ade + bde + acf + bcf + adf + bdf.\end{aligned}$$

Once again, the product of $(a + b)$, $(c + d)$, and $(e + f)$ is the sum of all possible terms obtained by multiplying one summand from $(a+b)$ with one from $(c+d)$ and one from $(e + f)$.

The same idea applies when we multiply any number of expressions like $(a + b)$, $(c + d)$, $(e + f)$, $(g + h)$, $\cdots$, together. We will use this fact below in our proof.

Now for the proof of the binomial theorem:

Proof. When $n = 1$, the statement in the binomial theorem is clearly true: the left side reads $(x+y)$, while the right side also reads $(x+y)$. So we need to prove the theorem for $n \geq 2$. Thus, the expression $(x+y)^n$ on the left side represents a product of two or more $(x + y)$ terms. As discussed above, the product $(x + y)^n = \underbrace{(x + y) \cdot (x + y) \cdots (x + y)}_{n \text{ factors}}$ expands out as follows: we either take x or y from the first $(x + y)$ factor, either x or y from the second factor, and so on, till either x or y from the nth $(x + y)$ factor, and multiply them together. We do this for every possible choice of x or y from the first, every possible choice of x or y from the second, and so on till every possible choice of x or y from the n-th. We then add up all these products.

Now let us see what products we get. If we take y from some k of these $(x + y)$ factors, then from the remaining $n - k$ factors we would have taken an x, thus we would get an $x^{n-k}y^k$ term. In how many ways will such an $x^{n-k}y^k$ term appear? Well, this just corresponds to the number of ways we can choose k of $(x+y)$ factors from the n factors from which to take y. This number is just $\binom{n}{k}$. Adding all these $x^{n-k}y^k$ terms we thus get the summand $\binom{n}{k}x^{n-k}y^k$.

(For example, when $n = 3$, the term xy^2 arises from taking y from the first and second factors, as also from the first and third factors, and as also from the second and third factors. There are thus three xy^2 terms. This corresponds to the number of ways of selecting two factors (from which to pick y) from three, which is $\binom{3}{2} = 3$. Adding these three xy^2 terms, we get the summand $3xy^2$ in the expansion of $(x + y)^3$.)

This idea works for all k, from 0 to n. When $k = 0$, we select zero y's, and thus n x's, so we get the summand $\binom{n}{0}x^n y^0 = \binom{n}{0}x^n$. Of course, $\binom{n}{0}$ is just 1. When k is n, the situation is reversed: we select n y's and zero x's, so we get the summand $\binom{n}{n}x^0y^n = \binom{n}{n}y^n$. And of course, $\binom{n}{n}$ is just 1. For $k = 1$, for instance, we get the summand $\binom{n}{1}x^{n-1}y^1$, and so on.

Carrying out this expansion for all k from 0 to n, we get the expansion of $(x + y)^n$ in the form described in equation (4.4) above. □

Remark 4.17. It would be a good test of your understanding for you to figure out where and exactly how this proof will break down if we do not assume that $xy = yx$!

Remark 4.18. A different proof of the Binomial Theorem using induction can be found in Exercise 7.22 in Chapter 7 ahead.

Here is a quick exercise:

Exercise 4.19. What is the coefficient of $a^{15}b^4$ in the expansion of $(a^3 + b^4)^6$? (Hint: Write a^{15} as $(a^3)^5$.)

The binomial theorem allows us to derive lots of very pretty relations among the binomial symbols. These relations are every mathematician's delight! Typically, we get them by looking at the binomial expansion of $(x + y)^n$ for some specific values of x or y, sometimes by writing out the product in one or more ways. Let us start with the following obvious corollary to the binomial theorem obtained by putting $x = 1$:

Corollary 4.20. *For any integer $n \geq 1$,*

$$(1 + y)^n = \binom{n}{0} + \binom{n}{1}y + \binom{n}{2}y^2 + \binom{n}{3}y^3 + \cdots + \binom{n}{n-1}y^{n-1} + \binom{n}{n}y^n.$$

(Of course, the first and last coefficients on the left are just 1.) Here are some pretty identities we can derive from just Corollary 4.20 above.

Example 4.21. We have the following:

(1) For any integer $n \geq 1$, we have:

$$\binom{n}{0} + \binom{n}{1} + \binom{n}{2} + \cdots + \binom{n}{n-1} + \binom{n}{n} = 2^n.$$

This follows from simply putting $y = 1$ in Corollary 4.20!

(2) For any integer $n \geq 1$, we have:

$$\binom{n}{0} - \binom{n}{1} + \binom{n}{2} - \binom{n}{3} \pm \cdots + (-1)^{n-1}\binom{n}{n-1} + (-1)^n\binom{n}{n} = 0.$$

This follows from putting $y = -1$ in Corollary 4.20! The term $(1+y)^n$, after putting $y = -1$, becomes 0^n, which is 0 if $n \geq 1$.

(3) A very pretty result emerges from simply combining the identities in parts (1) and (2) above. We may rewrite the identity in part (2) as

$$\binom{n}{0} + \binom{n}{2} + \cdots = \binom{n}{1} + \binom{n}{3} + \cdots,$$

where on the left we consider all $\binom{n}{r}$ with r even and on the right we consider all $\binom{n}{r}$ with r odd. But by simply rearranging terms we can write the identity in part (1) as

$$\left[\binom{n}{0} + \binom{n}{2} + \cdots\right] + \left[\binom{n}{1} + \binom{n}{3} \pm \cdots\right] = 2^n.$$

It follows that each of the two quantitiies in the square brackets above is half of 2^n. Thus, we find

$$\binom{n}{0} + \binom{n}{2} + \cdots = \binom{n}{1} + \binom{n}{3} \pm \cdots = 2^{n-1}.$$

(4) (For those who know a little bit of calculus!) For any integer $n \geq 1$, we have:

$$\binom{n}{1} + 2 \cdot \binom{n}{2} + 3 \cdot \binom{n}{3} \cdots + (n-1) \cdot \binom{n}{n-1} + n \cdot \binom{n}{n} = n2^{n-1}.$$

We establish this by differentiating the equation in Corollary 4.20 with respect to y when $n \geq 1$ to obtain the following:

$$n \cdot (1+y)^{n-1} = \binom{n}{1} + 2 \cdot \binom{n}{2}y + 3 \cdot \binom{n}{3}y^2 + \cdots + (n-1) \cdot \binom{n}{n-1}y^{n-2} + n \cdot \binom{n}{n}y^{n-1}$$

If we put $y = 1$ now, we get the desired result.

(5) For integers k and n with $0 \leq k \leq n$, and $1 \leq n$,

$$\binom{2n}{k} = \sum_{i=0}^{k} \binom{n}{i}\binom{n}{k-i}$$

It is time for us to start using the summation notation: as you would know already, the right side is to be interpreted as "one by one, put in $i = 0$, $i = 1, \ldots, i = k$, in the expresssion $\binom{n}{i}\binom{n}{k-i}$ and sum them up." Thus, the right side really is

$$\binom{n}{0}\binom{n}{k} + \binom{n}{1}\binom{n}{k-1} + \cdots + \binom{n}{k}\binom{n}{0}.$$

We consider the binomial expansion $(1 + y)^{2n}$, and consider the coefficient of y^k. Directly from Corollary 4.20 (which we can apply here because the hypothesis $n \geq 1$ shows $2n \geq 2 > 1$), we find it is $\binom{2n}{k}$. On the other hand, view $(1 + y)^{2n}$ as $(1 + y)^n \cdot (1 + y)^n$. Expanding each factor using Corollary 4.20, we find

$$(1 + y)^{2n} = \left(\sum_{i=0}^{n} \binom{n}{i} y^i\right) \cdot \left(\sum_{j=0}^{n} \binom{n}{j} y^j\right).$$

Now, how do we get the term y^k out of the product on the right side above? Expanding the product, we will get the sum of all possible products of a summand from the first parenthesis and a summand from the second parenthesis. (See the discussion after the statement of the binomial theorem (Theorem 4.16) on how to multiply out two or more sums of terms.) So, the way we get a y^k term out of this product is by choosing the $\binom{n}{0}y^0$ term from the first and multiplying with the $\binom{n}{k}y^k$ term from the second, and by choosing the $\binom{n}{1}y^1$ term from the first and multiplying with the $\binom{n}{k-1}y^{k-1}$ term from the second, and so on, and then by choosing the $\binom{n}{k}y^k$ term from the first and multiplying with the $\binom{n}{0}y^0$ from the second. Doing this, and adding all the resultant y^k terms together (which just means adding all the resulting coefficients together), we find that we will get the term $\left(\sum_{i=0}^{k} \binom{n}{i}\binom{n}{k-i}\right) y^k$ on the right side. Since both coefficients of the y^k term must be the same on both sides, we find indeed that $\binom{2n}{k} = \sum_{i=0}^{k} \binom{n}{i}\binom{n}{k-i}$, as was desired.

Note that this problem can also be set in the context of selecting committees of men and women with restrictions on the number of women (or men) in the committee: see Example 4.13 earlier, as also Exercise 4.33 ahead, for a generalization of this problem.

Now is the time for you to attempt the following:

Exercise 4.22. Go discover new identities involving the binomial coefficients.

Now what is the point of this exercise above? Well, this is exactly how mathematicians do mathematics! They see some beautiful patterns coming out of the binomial theorem and they see some techniques for discovering them (as in Example 4.21), and they say to themselves: "Wow! That was neat! I wonder if I can find new things I can do with the binomial theorem myself. Maybe I can find even more identities!"

So how might you go about this? In the first two parts of Example 4.21, we substituted values in the binomial theorem. What would you get if you were to substitute other values? In the third part we differentiated, and in the fourth we took a binomial expansion and rewrote it as a product of two binomial expansions. What similar things can you do that might lead you to new identities? Can you combine two techniques and discover even more? Etc., etc. This is how to play with mathematics, and play is the basis for learning!

You will find more identities in the exercises ahead. But first, play with the binomial theorem and try to find identities for yourself!

4.4 Further Exercises

Exercise 4.23. Practice Exercises:

Exercise 4.23.1. How many five-letter words can you form in English, if you insist that all letters be distinct?

Exercise 4.23.2. How many five-letter words can you form in English, if you do not insist that all letters be distinct?

Exercise 4.23.3. Write down the answers for Exercises 4.23.1 and 4.23.2 but for six-letter words, then seven-letter words. Use the pattern you see to write down the number of k-letter words from an alphabet that contains n letters (so $n \geq k$), both when the letters are distinct and when they are not.

Exercise 4.23.4. How many two-letter words with distinct letters can you form by selecting from 11 cards with the letters $X, X, X, Y, Y, Z, A, B, C, D, E$ marked on them?

Exercise 4.23.5. How many two-letter words can you form by selecting from 11 cards with the letters $X, X, X, Y, Y, Z, A, B, C, D, E$ marked on them?

Exercise 4.23.6. How many committees of 4 people can be formed from 11 people?

Exercise 4.23.7. How many committees of 4 people, with one person designated as President, one person as Vice-President, one as Secretary, and one as Treasurer, be formed from 11 people?

Exercise 4.23.8. In how many ways can you rearrange the letters $X, X, X, Y, Y, Z, A, B, C, D, E$?

Exercise 4.23.9. In how many ways can you seat 12 people around a circular table?

Exercise 4.23.10. How many necklaces can you form with 25 beads of identical size, of which there are five beads in each of the five colors blue, red, orange, white, and black?

Exercise 4.23.11. What is the coefficient of x^8 in the expansion of $(1 + x^2)^{15}$?

Exercise 4.23.12. What is the coefficient of x^4y^9 in the expansion of $(x^2 + y^3)^5$?

Exercise 4.24. If $m = 2^p3^q$ where p and q are positive integers, how many permutations of $\{1, 2, \ldots, m\}$ start with a divisor of m? (Hint: First count the number of divisors–or factors–of m. Note that every divisor is of the form 2^a3^b, where $0 \leq a \leq p$ and $0 \leq b \leq q$.)

Exercise 4.25. Determine the number of positive integers with distinct digits. (Hint: First consider separately the number of integers with one digit, the number with two digits, etc.)

Exercise 4.26. Determine the number of integers between 1000 and 5555 with distinct digits. (Hint: One way to do this is the following. The first digit can be selected in 5 ways, but if it is any of 1, 2, 3, or 4, then there is no upper bound on what the remaining digits can be; of course they should be distinct. If the first digit is 5, then the second digit can only be one of 0, 1, 2, 3 or 4. Proceed.)

Exercise 4.27. Find the number of permutations of the letters $A, M, B, C, D, T, E, F, G, H$ that do not contain the letters $MATH$ in succession.

Exercise 4.28. Using the idea behind Example 4.14, show that the number of positive integer solutions to the equation $x_1 + x_2 + \cdots + x_n = M$ (where M is a positive integer) is $\binom{M-1}{n-1}$. (Hint: Draw M boxes in a row, and consider the $M - 1$ spaces *between* these M boxes.)

Exercise 4.29. Use the result of Exercise 4.28 above to show that the number of nonnegative integer solutions to equation $x_1 + x_2 + \cdots + x_n = M$ (for a given positive integer M) is $\binom{M+n-1}{n-1}$. (Hint: Set $y_i = x_i + 1$.)

Exercise 4.30. Suppose you had n bins, each of which contains an infinite supply of balls of the same color, and suppose that the colors of the balls in different bins are all different. Use the result of Exercise 4.29 to determine the number of selections of r balls from these n bins, where you are allowed to select multiple balls from the same bin.

If $x_1, \ldots, x_n$ are variables, a *monomial of degree* r in these n variables is an expression of the form $x_1^{i_1}x_2^{i_2} \cdots x_n^{i_n}$, where each $i_j \geq 0$ and $i_1 + i_2 + \cdots + i_n = r$. Apply the result you would have obtained in the last paragraph to determine the number of monomials of degree r in n variables.

Exercise 4.31. In how many ways can we seat 12 girls and 12 boys in a row if we insist that they must alternate: no two girls should be adjacent to each other, and no two boys should be adjacent to each other? How many different arrangements would we have if this same seating arrangement was to be done around a circular table?

Exercise 4.32. Establish the following identities. Here, n is a nonnegative integer, and $0 \leq r, a \leq r+a \leq n$:

$$\binom{n}{r+a}\binom{r+a}{r} = \binom{n}{r}\binom{n-r}{a}, \tag{4.5}$$

$$\binom{n}{r}\binom{n+r}{r} = \binom{n+r}{n-r}\binom{2r}{r}, \tag{4.6}$$

$$\binom{n}{r}\binom{2n}{n} = \binom{n+r}{r}\binom{2n}{n-r}. \tag{4.7}$$

Exercise 4.33. Let n and k be positive integers, and let r be an integer with $0 \leq r \leq n+k$. Using the ideas inherent in Example 4.13 or otherwise, show that

$$\binom{n+k}{r} = \binom{n}{r}\cdot\binom{k}{0} + \binom{n}{r-1}\cdot\binom{k}{1} + \cdots + \binom{n}{0}\cdot\binom{k}{r}.$$

Exercise 4.34. (For those who know a little bit of calculus! See Example 4.21(4) for inspiration.) Show that for any positive integer n

$$\sum_{i=0}^{n}(i+1)\cdot\binom{n}{i} = 2^{n-1}(n+2).$$

Exercise 4.35. (For those who know a little bit of calculus, just like in Exercise 4.34 above!) Show that for any integer $n \geq 2$

$$\sum_{i=2}^{n} i(i-1)\cdot\binom{n}{i} = 2^{n-2}n(n-1).$$

Note: When $n = 1$ or $n = 0$, you will be dealing with expressions like $\sum_{i=2}^{1}\langle\text{stuff}\rangle$ and $\sum_{i=2}^{0}\langle\text{stuff}\rangle$. Expressions like these, where the upper index of summation is less than the lower index, are considered *empty sums*, and their value is taken to be 0 by convention.

Exercise 4.36. Show that for any positive integer n

$$\sum_{i=1}^{n} i^2\cdot\binom{n}{i} = 2^{n-2}n(n+1).$$

Exercise 4.37. Show that for any positive integer n

$$\sum_{i=0}^{n}\frac{1}{(i+1)}\cdot\binom{n}{i} = \frac{2^{n+1}-1}{n+1}.$$

(Hint: If you know a bit of calculus, integrate both sides of the equality in Corollary 4.20 between 0 and 1. Otherwise, notice that the right side can be obtained by putting $x = 1$ in the expression $\frac{(1+x)^{n+1}-1}{n+1}$. Try to relate this to a binomial expansion. You may find Proposition 4.15(3) useful.)

Exercise 4.38. The following exercise is similar to Exercise 4.37. Show that for any positive integer n

$$\sum_{i=0}^{n}\frac{(-1)^i}{(i+1)}\cdot\binom{n}{i} = \frac{1}{n+1}.$$

(Hint: See the hint to Exercise 4.37. Try other values of x.)

Exercise 4.39. Show that for any positive integer n,

$$\sum_{i=0}^{n} \binom{2n}{i} = 2^{2n-1} + \frac{1}{2}\binom{2n}{n}.$$

(Hint: Write S for the sum on the left. Consider $2S$. Look to Example 4.21(1) for further inspiration!)

Exercise 4.40. The following is similar to Exercise 4.39 above. Show that for any positive integer n,

$$\sum_{i=0}^{n} \binom{4n}{2i} = 2^{4n-2} + \frac{1}{2}\binom{4n}{2n}.$$

(Hint: See the hint for Exercise 4.39 above. This time, part (3) of Example 4.21 might also help.)

Exercise 4.41. This exercise is for those who have studied calculus and are familiar with basic integration techniques. For any real number t with $t > 0$, the *Gamma Function* $\Gamma(t)$ is defined by the rule

$$\Gamma(t) = \int_0^\infty x^{t-1}e^{-x}\,dx.$$

It can be shown that $\Gamma(t)$ as defined is finite for all $t > 0$ (in other words, the integral converges). Show the following:

(1) $\Gamma(1) = 1$.

(2) $\Gamma(t+1) = t\Gamma(t)$. (Hint: Use integration by parts.)

(3) $\Gamma(n) = (n-1)!$ for all integers $n \geq 1$.

The significance of the result is the following: It can be shown that $\Gamma(t)$ is a *continuous* function of t for all $t > 0$. (Even if you have not taken calculus you will have an intuitive idea of a continuous function: for example, it is a function whose graph can be traced out without lifting one's pen off the paper. We will study continuity of functions in Chapter 11.) Thus, if we say define $\tilde{\Gamma}(t) = \Gamma(t+1)$ for $t > -1$, we find that $\tilde{\Gamma}(t)$ is a continuous function of t for all $t > -1$ with the property that $\tilde{\Gamma}(n) = n!$ whenever n is a nonnegative integer! In other words, the function $f(n) = n!$ defined on the nonnegative integers is merely a discrete snapshot of a continuous function defined on all real numbers greater than -1.

The Gamma function plays a significant role in number theory, and often appears paired with the Riemann Zeta function, which itself is at the center of the most famous unsolved mathematics problem, namely, the proof of the Riemann Hypothesis. You are encouraged to read about this!

5

Sets and Functions

In Chapter 3 we studied the first key ingredient of mathematical vocabulary, namely statements. Here we will study the second and third key ingredients: sets, and the functions between them. Together, these three form the bedrock of mathematics. In fact, according to one viewpoint, mathematics can be described most broadly as the discovery of theorems (true statements) about sets and the functions between them!

As in Chapter 3, our goal here is to understand sets and functions in the intuitive way a mathematician understands them. Any deeper study (particularly of sets) will involve venturing into the field of mathematical logic. At this introductory level, we will not go there.

Accordingly, we will use what may be termed a *naive,* as opposed to *axiomatic,* approach to sets. But the term naive does not mean that our approach is somehow deficient for the purpose of doing mathematics. It is not! It is just that in certain technical situations that we will not confront at this beginning level, our intuitive approach can lead to internal contradictions. At that point we will need to adjust our approach with a firm set of axioms. Once we firm up our axioms and revisit what we do here, we will find that everything we do is correct after all! It is harmless therefore to proceed along more intuitive lines at this stage.

5.1 Sets

Definition 5.1. A *set* is a collection of objects. The members of the collection are called *elements* of the set. When an object x is an element of a set S, we write $x \in S$.

As with statements (see Example 3.2(3) of Chapter 3), objects must be unambiguous and well-defined.

Example 5.2. Let us consider some examples:

(1) We have considered the set of integers several times already, traditionally denoted $\mathbb{Z}$. We write this as $\mathbb{Z} = \{0, \pm 1, \pm 2, \pm 3, \dots\}$. Note the curly braces at the beginning

and the end. The dots indicate that the set is built up by continuing the obvious pattern.

Likewise, we have already considered $\mathbb{Q}$, the set of rational numbers, $\mathbb{R}$, the set of real numbers, and $\mathbb{C}$, the set of complex numbers.

(2) Let S denote the collection of all students in a room. Suppose the students are Amir, Bella, and Chris. We'd write this as $S = \{\text{Amir}, \text{Bella}, \text{Chris}\}$. The elements of S are Amir, Bella, and Chris. (Thus, symbolically, Amir $\in S$, Bella $\in S$, and Chris $\in S$.

(3) Consider the set $\mathbb{N}$ of all positive integers. One way to describe this set is $\mathbb{N} = \{x \mid x \in \mathbb{Z},\ x > 0\}$. The vertical bar "|" stands for "such that," or "which has the property that." We would read this description of $\mathbb{N}$ as "The set $\mathbb{N}$ is the collection of objects x such that x is an integer and x is positive." This method of describing a set is known as the *set builder notation,* for the obvious reason that it constructs a set by giving a body of rules that govern which objects are to be included in the set.

Similarly, using the set builder notation, we describe the set of rational numbers as $\mathbb{Q} = \{a/b \mid a, b \in \mathbb{Z},\ b > 0,\ \gcd(a, b) = 1\}$. Recall the definition of the *greatest common divisor*, $\gcd(a, b)$, from your high school: it is the largest integer that divides both a and b. (We will consider the notion of the greatest common divisor in greater depth later in Chapter 10.) The condition $\gcd(a, b) = 1$ in the set builder notation for $\mathbb{Q}$ above ensures that, for simplicity, we only consider reduced fractions. That is, we only consider fractions in which all common factors in the numerator and denominator have been cancelled already. Similarly, the condition $b > 0$ not only ensures that we don't do a "divide by zero," but also, that we restrict the negative signs to the numerator. This ensures, for instance, that we don't consider both $-1/-2$ and $1/2$. Once again, this is for simplicity.

(4) Sets can themselves be members of other sets! Thus, consider $S = \{\{1, 2\}, \{3, 4\}\}$. This set has two elements. The first is itself the set $\{1, 2\}$ consisting of the elements 1 and 2, and the second is the set $\{3, 4\}$ consisting of the elements 3 and 4. The elements of S are thus $\{1, 2\}$ and $\{3, 4\}$. Symbolically, we write $\{1, 2\} \in S$ and $\{3, 4\} \in S$.

Note that 1, 2, etc. are *not* elements of S, even though they are elements of the elements of S! In a similar vein, the set S is *not* the same as the set $T = \{1, 2, 3, 4\}$ (whose elements indeed are 1, 2, etc.).

(5) The *empty set* is the set with no elements in it! It is a very convenient gadget to have around for purposes of logic and to make formulas come out pretty. The empty set is denoted $\emptyset$.

Note that the set $S = \{\emptyset\}$ is not the same as the empty set! The set S just described is the set whose sole element is the empty set! Thus, $\emptyset \in S$, but S is not the same as $\emptyset$!

(6) The collection of tall people in this room: this is not a set, since the term "tall" is not well-defined. On the other hand, the collection of people in this room who are over 12 feet tall is indeed a set, since the objects are now well-defined. (This set is quite likely to be empty, but it is still a set!)

(7) Very often in mathematics we face situations like the following: we are simultaneously given sets of the form $A_n = \{1, 2, \ldots, n\}$ (n a positive integer). Or, we are simultaneously given sets of the form $B_d = \{x \in \mathbb{R} \mid -d \leq x \leq d\}$ (d a positive real number). Or else, we are simultaneously given sets of the form

$$C_s = \{\text{s's social security (SS) number, s's driver's license (DL) number}\}$$

as s ranges through, say, the set S of all the students in the classroom. What is common to these examples is a set of parameters—the positive integers in the case of the sets A_n, the positive real numbers in the case of the sets B_d, and the set S of students in the classroom in the case of the sets C_s, and then, the existence of one set for each value of the parameter from that set. Thus, in the case of the A_n, we have sets $A_1 = \{1\}$, $A_2 = \{1, 2\}$, $A_3 = \{1, 2, 3\}$, etc. In the case of the B_d, we have sets $B_1 = \{x \in \mathbb{R} \mid -1 \leq x \leq 1\}$, $B_{\sqrt{2}} = \{x \in \mathbb{R} \mid -\sqrt{2} \leq x \leq \sqrt{2}\}$, $B_\pi = \{x \in \mathbb{R} \mid -\pi \leq x \leq \pi\}$, etc. In the case of the C_s, we have sets like

$$\begin{aligned} C_{\text{Norma}} &= \{\text{Norma's SS number, Norma's DL number}\}, \\ C_{\text{Raju}} &= \{\text{Raju's SS number, Raju's DL number}\}, \\ C_{\text{Lisa}} &= \{\text{Lisa's SS number, Lisa's DL number}\}, \text{ etc.} \end{aligned}$$

Sets such as the A_n, the B_d, and the C_s above are known as *indexed sets*. These are families of sets where you have one set for each value of a parameter or *index*: the parameter n for the family A_n, d for B_d, and s for C_s. The set that the parameter varies through is known as the *index set*: the set $\mathbb{N}$ for A_n, $\mathbb{R}_{>0}$ for B_d, and the set of students in the classroom for C_s.

Remark 5.3. Note that the order in which the elements of a set are listed is considered immaterial. Thus, using example (2) for context, {Bella, Amir, Chris} denotes the same set as {Amir, Bella, Chris}. Further, repeating elements in the listing of a set is not considered to change the set; thus, {Amir, Amir, Bella, Chris, Chris, Bella, Amir} denotes the same set as {Amir, Bella, Chris}.

5.2 Equality of Sets, Subsets, Supersets

Definition 5.4. Let A and B be two sets, and let x denote an element of either A or B.

(1) We say A is a *subset* of B (written $A \subseteq B$) if $x \in A \implies x \in B$. In other words, every element of A is also an element of B. If A is a subset of B, we also say B is a *superset* of A (written $B \supseteq A$).

(2) If $A \subseteq B$ and there exists an element $x \in B$ such that $x \notin A$, we say A is a *proper subset* of B (written $A \subsetneq B$). If A is a proper subset of B, we also say that B is a *proper superset* of A (written $B \supsetneq A$).

Note that if A is a proper subset of B then A first of all is a subset of B. In the case of real numbers, we distinguish between the inequalities $a \leq b$ and $a < b$ by calling the second a *strict* inequality. Analogously, when A is a proper subset of B, we describe the containment of A in B as a *strict* containment.

Definition 5.5. Let A and B be two sets, and let x denote an element of either A or B. We say A is *equal* to B (written $A = B$) if $x \in A \iff x \in B$. In other words, every element of A is an element of B and every element of B is an element of A.

Note that the condition in the definition is symmetric: if $A = B$, then necessarily $B = A$ as well.

Example 5.6. Let us consider some examples:

(1) Let $A = \{1, 2\}$ and $B = \{1, 2, 3\}$. Then both elements 1 and 2 of A are contained in B. Thus A is a subset of B, which we write $A \subseteq B$. Note that A is a *proper* subset of B, since the element 3 that is in B is not an element of A. Thus we can also write $A \subsetneq B$; this notation describes the relation of being a subset more finely.

Further, we may say $B \supset A$ and also $B \supsetneq A$.

(2) Now let $A = \{1, 2, 3\}$ and $B = \{x \in \mathbb{Z} \,|x > 0 \text{ and } x < 4\}$. Working through the set builder notation, it becomes clear that the elements of B are 1, 2, and 3. Thus every element of A is an element of B, and every element of B is an element of A. Hence $A = B$.

Now note that it is also true that $A \subseteq B$, since every element of A is indeed an element of B. However, A is not a *proper* subset of B, because every element of B is also an element of A. Thus, to say that A is a subset of B but is not a proper subset of B is really to say that A and B are equal!

By the same token, it is also true that $B \subseteq A$ but B is not a proper subset of A.

(3) Suppose we are given two sets A and B that are equal. Perhaps as in part (2) above they are described differently. Then, just as in part (2), the fact that $A = B$ says that every element of A is an element of B, so $A \subseteq B$ is true. Similarly, the fact that $A = B$ says that every element of B is an element of A, so $B \subseteq A$ is also true. Thus, if $A = B$, then automatically, $A \subseteq B$ and $B \subseteq A$.

But let us turn this around now: suppose we are given two sets A and B such that $A \subseteq B$ and $B \subseteq A$. What can we conclude? To say that $A \subseteq B$ is to say that every element of A is an element of B, and to say that $B \subseteq A$ is to say that every element of B is an element of A. But this is what it means for A to equal B! Thus, the two simultaneous conditions $A \subseteq B$ and $B \subseteq A$ jointly say that $A = B$!

We have thus shown that the condition $A = B$ is equivalent to the two simultaneous conditions $A \subseteq B$ and $B \subseteq A$! Very often, the way we prove that two sets (described differently, such as in part (2) above) are equal is to show that one is a subset of the other and the other is a subset of the one.

(4) Here is an example that will illustrate this technique of showing that two sets are equal by checking that each is a subset of the other. Suppose we need to show that A, the set of all rational numbers whose decimal expansion has at most two digits to the *right* of the decimal place, and $B = \{x \in \mathbb{Q} |\ 100x \in \mathbb{Z}\}$ are equal. (Here, we will assume that any decimal expansion that ends in an infinite string of 9s has been written as a finite decimal, e.g., $0.49999\ldots$ has been rewriten as 0.5, $3.239999\ldots$ has been rewritten as 3.24.)

We first show that $A \subseteq B$. Suppose x has k digits to the left of the decimal place (where k could possibly be zero). Since x has at most two digits to the right of the decimal place, x can be written in decimal notation (for $k \geq 1$) as $\pm d_{k-1} \dots d_1 d_0.d_{-1}d_{-2}$, where the d_i, $i = -2, -1, 0, \dots, k-1$, are elements of the set $\{0, 1, \dots, 9\}$, and $d_{k-1} \neq 0$. (Thus, the d_i are the digits in the 10^i-th place.) Also, for $k = 0$, x can be written as $\pm 0.d_{-1}d_{-2}$. Note that d_{-2}, or for that matter both d_{-1} and d_{-2}, could be zero! (For instance, we would write the element 52.3 as 52.30, and the element 46 as 46.00.) Then it is clear that $100x$ is the integer whose decimal representation is $\pm d_{k-1} \dots d_1 d_0 d_{-1} d_{-2}$ (with the understanding that if $k = 0$ then $100x$ is the integer with decimal representation $\pm d_{-1}d_{-2}$). Thus, as $100x$ is an integer, $x \in B$.

Now we'll show $B \subseteq A$. Take $x \in B$. By definition, $100x = n$ for some integer n. If the number of digits in n is 1, and if $n = d_0$ (where $d_0 \in \{0, 1, \dots, 9\}$), then $x = d_0/100 = 0.0d_0$, so $x \in A$ by definition. If the number of digits in n is 2, and if $n = d_1 d_0$, then $x = 0.d_1 d_0$, so $x \in A$. Finally, if the number of digits in n is some $k \geq 3$ and if $n = d_{k-1} \dots d_2 d_1 d_0$, then $x = d_{k-1} \dots d_2.d_1 d_0$, and once again $x \in A$. So, as $x \in A$ in all cases, we find $B \subseteq A$.

Since we've shown that $A \subseteq B$ and $B \subseteq A$, we indeed find that $A = B$.

It would be helpful to do the following exercise at this stage:

Exercise 5.7. For each of the following, determine which of the following is true: $S \subseteq T$, $S \subsetneq T$, $T \subseteq S$, $T \subsetneq S$, $S \supseteq T$, $S \supsetneq T$, $T \supseteq S$, $T \supsetneq S$, $S = T$. (Justify your answers, of course!)

(1) Let S be the set of all right triangles in the plane and T be the set of all triangles in the plane.

(2) Let S be the set of all right triangles in the plane and T be the set of all triangles in the plane whose three sides a, b, and c (after being labeled in a suitable order) satisfy $c^2 = a^2 + b^2$.

(3) Let S be the set of all triangles in the plane and T be the set of all triangles in the plane whose three sides a, b, and c (after being labeled in a suitable order) satisfy $c^2 \leq a^2 + b^2$. (Hint: At least one angle must be $\leq 90°$. Use the cosine formula.)

5.3 New Sets From Old

Given one or more sets, there are several very useful sets that we can form from these. We will consider them here. In what follows, A and B will be arbitrary sets.

Definition 5.8. The *union* of A and B (written $A \cup B$) is the new set whose members are all those elements that belong to A or to B.

Remember that "or" in mathematics is the inclusive or—see Example 3.21 in Chapter 3. Thus, we include members that belong to both A and B in the set $A \cup B$.

Definition 5.9. The *intersection* of A and B (written $A \cap B$) is the new set whose members are all those elements that belong to both A and B.

Definition 5.10. The *difference* of A and B (written $A \setminus B$) is the new set whose members are all those elements that belong to A but do not belong to B.

Definition 5.11. The *power set* of A (written $\mathcal{P}(A)$) is the new set whose members are the various subsets of A.

Definition 5.12. The *direct product* of A and B (written $A \times B$) is the new set whose members are all pairs (a, b), where the first element a belongs to A and the second element b belongs to B.

To emphasize the fact that the first element in the pair (a, b) above should come from A and the second element from B, the pair is often referred to as an *ordered pair.* Thus, it represents more than just the process of grabbing something from A and sticking it with something from B—it represents grabbing something from A, *placing it in the first (left) slot,* then grabbing something from B, and *placing it in the second (right) slot!*

Note too that if $a \neq a'$ or $b \neq b'$ (remember we use the inclusive or in mathematics!), then (a, b) and (a', b') are considered to be different ordered pairs. Equivalently, $(a, b) = (a', b')$ only when $a = a'$ and $b = b'$.

The constructions above took in two sets A and B and gave you a third set. Here is a different construction that takes in just one set and gives another set:

Definition 5.13. Let U be a set which, in a particular context, is fixed as a background "reference" set. For a set $S \subseteq U$, the *complement* of S, variously denoted S^c or $\overline{S}$, or $\tilde{S}$, is the set $U \setminus S$.

This definition is very dependent on the context. The set U above is known, in that context, as the *universal* set. In a different context, a different set V may be fixed as the background universal set, and in that case, complements will be determined with reference to V. See Example 5.14(6) below.

These constructions, especially the union and intersection, are very familiar to us; here are examples:

Example 5.14. Take $A = \{1, 2, 3\}$ and $B = \{2, 3, 4, 5\}$, $U = \{1, 2, 3, 4, 5\}$, $V = \{1, 2, 3, 4\}$. Then

(1) $A \cup B = \{1, 2, 3, 4, 5\}$.

(2) $A \cap B = \{2, 3\}$.

(3) $A \setminus B = \{1\}$ and $B \setminus A = \{4, 5\}$.

(4) The power set of A,

$$\mathcal{P}(A) = \{\emptyset, \{1\}, \{2\}, \{3\}, \{1, 2\}, \{2, 3\}, \{1, 3\}, \{1, 2, 3\}\}.$$

Note that by convention, *the empty set is considered as a subset of every set,* so it is automatically an element of $\mathcal{P}(A)$. You can similarly write out the elements of $\mathcal{P}(B)$. If you do so, you will find that it has 16 elements.

(5) The direct product of A and B, denoted $A \times B$, is given by

$$A \times B = \{(1,2),(1,3),(1,4),(1,5),(2,2),\\ (2,3),(2,4),(2,5),(3,2),(3,3),(3,4),(3,5)\}.$$

By contrast,

$$B \times A = \{(2,1),(2,2),(2,3),(3,1),(3,2),(3,3),(4,1),(4,2),\\ (4,3),(5,1),(5,2),(5,3)\}.$$

Note that $A \times B$ and $B \times A$ are different sets.

(6) Taking U as the background universal set, the complement of A, denoted say A^c, is the set $U \setminus A = \{4,5\}$. But taking V as the universal set, $A^c = \{4\}$. This exemplifies how the definition of A^c very much depends on the choice of the universal set. Different contexts in mathematics naturally present candidates for universal sets. For instance, when studying mathematics in elementary school, a natural universal set would be $\mathbb{N}$, or perhaps $\mathbb{Z}_{\geq 0}$. While studying calculus, for example, a natural candidate would be $\mathbb{R}$.

Now is a good time for you to do the following to cement your understanding:

Exercise 5.15. In each of the following, give examples of sets A, B, and C such that:

(1) $A \cap C = B \cap C$, but $A \neq B$.

(2) $A \cup C = B \cup C$, but $A \neq B$.

In the examples on page 85 onwards, we will see many more instances of sets studied frequently in mathematics.

5.4 Functions Between Sets

You have seen functions several times already: your precalculus or calculus courses were mostly about functions defined on the real numbers via formulas like $f(x) = e^{-x}$ or $f(t) = v_0 t + \frac{gt^2}{2}$. Here is a more general definition:

Definition 5.16. Given two sets A and B, *a function "f" from A to B* is a subset of $A \times B$ in which each element $a \in A$ appears *once and only once* in the left slot of an ordered pair as (a,b) for some $b \in B$. It is denoted $f : A \to B$. We say *f sends A to B*, or *f maps A to B* (and also refer to f as a *map*). If (a,b) is in the subset f, then we write $f(a) = b$, and we say *f sends a to b*, or *f maps a to b*, or *b is the image of a under f*.

Thus, a function represents a connection between A and B in which *each* element of A is paired with *exactly one* element of B. This is intended to mimic, for instance, physical measurements. As an example, consider the measurement of heights of students in a certain class on a certain day. This would be a function from $A =$ {students in that class} to $B = \mathbb{R}$. Note that (1) We would like to measure the height of *every* student, so every $a \in A$ should show up once in the left slot of an ordered pair, and (2) No student can physically have two different heights, so to each student we can pair only one real number (his or her height) and not more than one. This is what we mean by saying that each $a \in A$ appears once and only once in the left slot.

We may describe a function either by writing it explicitly as a subset of $A \times B$, or by giving a rule which describes how an element $a \in A$ pairs with some element of B. Note in particular that *any* subset of $A \times B$ that has the property described in Definition 5.16 above is a function; it does not need to be given by a rule which describes the pairing.

Definition 5.17. Here are a few related definitions. In all of these, $f : A \to B$ is a function.

(1) The *domain of* f is the set A, while the *range of* f is the set $R = \{b \in B \mid f(a) = b \text{ for some } a \in A\}$. The range of f is thus a subset of B. The range is sometimes denoted $f(A)$. The set B is itself known as the *codomain of* f. (So the range of f is a subset of the codomain of f.) The range of f is also referred to as the *image of* f.

(2) For any $b \in B$, the *inverse image (or preimage)* of the *set* $\{b\}$, denoted $f^{-1}(\{b\})$, is the set $\{a \in A \mid f(a) = b\}$. (Often, $f^{-1}(\{b\})$ is loosely referred to as the inverse image or preimage of the *element* b.) More generally, given any subset $S \subseteq B$, the *inverse image (or preimage) of* S, denoted $f^{-1}(S)$, is the set $\{a \in A \mid f(a) = b \text{ for some } b \in S\}$.

(3) The function f is said to be *one-to-one* (or *injective*) if whenever $a_1 \neq a_2$, then $f(a_1) \neq f(a_2)$. Thus, a one-to-one (or injective) function takes distinct elements of A to distinct elements of B.

(4) The function f is said to be *onto* (or *surjective*) if given any $b \in B$, there exists $a \in A$ such that $f(a) = b$. (Note from the definitions of "onto" and "range" that the function f is onto *if and only if* the range of f equals all of B.)

(5) If the function f is both one-to-one and onto then it is said to be *bijective*. In such a case, f is often described as being, or providing, a *bijection* between A and B (simply abbreviated to "f is a bijection"). If $f : A \to B$ is a bijection, we also say that A and B are in *one-to-one correspondence* via f. Informally, we also say that f provides a *matching* or *pairing* between A and B.

(6) If $g : A \to B$ is another function, then f and g are considered to be equal (written $f = g$) if $f(a) = g(a)$ for all $a \in A$.

Now that is a handful of definitions, all at one time! But given that mathematics is all about sets and functions in at least one view, we will have plenty of opportunity to digest it all. Let us start with some very simple (and undoubtedly artificial) examples, designed to quickly illustrate these definitions.

Example 5.18. (See Figure 5.1.) Take $A = \{1, 2, 3\}$, $B = \{2, 3, 4, 5\}$. Take $f = \{(1, 3), (2, 4), (3, 5)\}$. Also, take $g = \{(1, 2), (2, 2), (1, 3), (3, 4)\}$. Next, take $h = \{(1, 3), (2, 3), (3, 4)\}$. Finally, take $c = \{(1, 3), (2, 3), (3, 3)\}$.

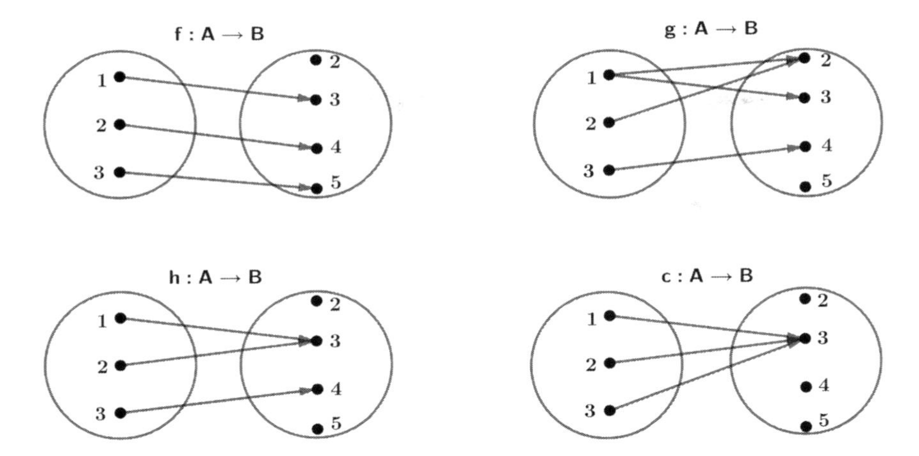

Figure 5.1. Examples of functions (and a non-function) from A to B.

(1) The subset f of $A \times B$ is indeed a function from A to B. Each of the elements of A, namely 1, 2, and 3, show up as the left slot of an ordered pair once and only once.

(2) By contrast, g is not a function from A to B. The element 1 of A shows up in both $(1, 2)$ and $(1, 3)$. (In the analogy given after Definition 5.16 above, a student in a class cannot have a height of both 2 and 3!)

(3) Back to f: The domain of f is of course A and the codomain of f is B. The range of f is the set of all "right-hand slots" in the ordered pairs representing f; thus, the range of f is the subset $\{3, 4, 5\}$ of B. Note that the range is a proper subset of the codomain. The inverse image of $\{3\}$, $f^{-1}(\{3\})$, equals $\{1\}$. (It is wrong to say $f^{-1}(\{3\}) = 1$: the inverse image is defined to be a set!) The inverse image of $\{2\}$, $f^{-1}(\{2\})$, is the empty set $\emptyset$, as no element of A is paired with 2. The inverse image of the subset $\{3, 4\}$ of B, $f^{-1}(\{3, 4\})$, is similarly $\{1, 2\}$. Of course, $f^{-1}(\{3, 4, 5\}) = A$, and as well, $f^{-1}(B) = A$. The function f is one-to-one, as 1, 2, and 3 all map to distinct elements of B. f is *not* onto, since the element 2 in B has no preimage in A.

(4) The subset h of $A \times B$ also represents a function from A to B, since each of 1, 2, and 3 show up once and only once as the left slot of the ordered pairs shown in the definition of h. Note something about h: the element 3 of B shows up as a *right* slot twice, once as $(1, 3)$ and again as $(2, 3)$. There is *nothing wrong with this!* The definition of function is not violated. (In the analogy given in Definition 5.16 above, two different students in a class could have the same height 3.) Of course, the very fact that 3 is the image of two distinct elements of A implies that h cannot be one-to-one. Clearly h is not onto either; both 2 and 5 have no preimages in A.

(5) The function c (it is indeed a function!) is a more extreme example of the phenomenon in part (4) above: *all* elements of A are paired to the same element $3 \in B$. (Again, in the analogy given in Definition 5.16 above, all students in a class could have the same height 3.) c is clearly neither one-to-one nor onto. Such a function

$f : A \to B$, where $f(a) = b$ for all $a \in A$, where b is some fixed element in B, is called a *constant* function.

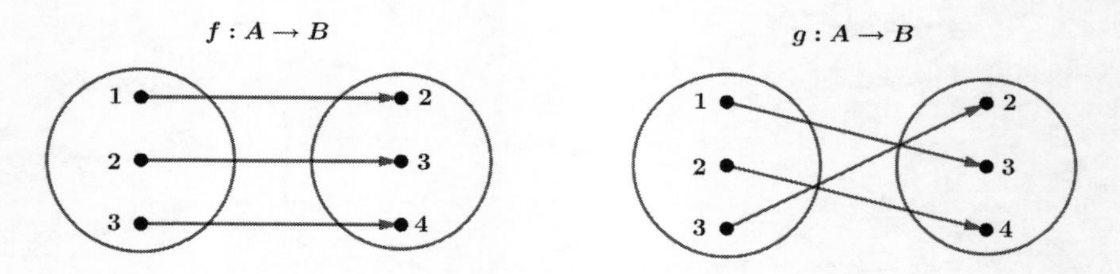

Figure 5.2. Examples of two one-to-one and onto functions from A to B.

Here is another example:

Example 5.19. Take $A = \{1,2,3\}$, $B = \{2,3,4\}$. Take $f = \{(1,2),(2,3),(3,4)\}$ and $g = \{(1,3),(2,4),(3,2)\}$. Then not only are both f and g functions from A to B, they are both one-to-one and onto (that is, they are both bijective). Thus, A and B are in one-to-one correspondence via f and also via g. See Figure 5.2.

Example 5.20. While previous examples were concocted artificially to quickly illustrate various concepts we have considered so far, here are some examples of functions that you would have already come across in your precalculus or calculus courses. Most of these are just described by some rule of the form $f(x) = \langle \textit{something} \rangle$, with no explicit indication of which set these functions are defined on, and to which set they take the domain to. In precalculus and calculus, the (unwritten) assumption when you see something like $f(x) = \langle \textit{something} \rangle$ is that these are functions whose domain is the largest subset of $\mathbb{R}$ on which the expression $\langle \textit{something} \rangle$ is defined, and whose codomain is $\mathbb{R}$. Thus, $f(x) = \langle \textit{something} \rangle$ is a function from a suitable *subset* of $\mathbb{R}$ to $\mathbb{R}$.

(1) $f(x) = x$. The understanding is that the domain is the largest subset of $\mathbb{R}$ on which the expression x is defined. Clearly x is defined for all real numbers (as opposed to something like $\frac{1}{x}$; see example (2) below). This function takes $\mathbb{R}$ to $\mathbb{R}$ (so the domain and codomain are both $\mathbb{R}$), and follows the simple rule: any real number goes to itself. Since any real number x can be written as $f(x)$, the range of f is $\mathbb{R}$. (In the notation introduced in Definition 5.24 ahead, this is the function $\mathbb{1}_{\mathbb{R}}$.)

(2) $f(x) = \frac{1}{x}$. Once again, the understanding is that the domain is the largest subset of $\mathbb{R}$ on which the expression $\frac{1}{x}$ is defined. We cannot find a reciprocal of 0, but can find the reciprocal of any other real number. So, the domain of f is the set $\mathbb{R} \setminus \{0\}$. Thus, f can be thought of as a function from $\mathbb{R}\setminus\{0\}$ to $\mathbb{R}$. The codomain is of course $\mathbb{R}$. Since 0 is not the reciprocal of any real number, 0 cannot be in the image of f. Any nonzero real number y can be written as a reciprocal: $y = \frac{1}{\frac{1}{y}}$. Hence the range of f is $\mathbb{R} \setminus \{0\}$.

(3) $f(x) = \sqrt{x}$. The domain of this function is the set $\mathbb{R}_{\geq 0}$ (why?). Note that by convention, the notation $\sqrt{x}$, where x is a positive real number, stands for the *positive* square root of x. Note too that any $y \geq 0$ is the square root of y^2. The range of f is therefore $\mathbb{R}_{\geq 0}$.

Remark 5.21. If $f : A \to B$ is a function, and $R \subseteq B$ is its range, then it would be just as valid to describe f as a function from A to R. Thus, in Example 5.18 above, it is valid to describe f as $f : A \to \{3, 4, 5\}$ and c as $c : A \to \{3\}$, for instance. If a function $f : A \to B$ with range $R \subseteq B$ is redescribed as $f : A \to R$, then as just redescribed, it is an onto (surjective) function, even though in its original description, it may not have been onto. Notice that if f is injective, then $f : A \to R$ becomes a bijection!

Remark 5.22. If A and B are sets with $A \subseteq B$, then the map $i : A \to B$ that sends a (viewed as an element of the set A) to a (viewed as an element of B using the fact that A is a subset of B) is an injective map. This map is known as the *inclusion map*.

Remark 5.23. If $f : A \to B$ is a function and if X is a subset of A, we write $f|_X$ to denote the function from X to B obtained simply by *restricting* the definition of f to elements of X. For instance, in Example 5.18 above, take $X = \{1, 2\}$. Then $f|_X$ is the function from X to B that sends 1 to 3 and 2 to 4, while $h|_X$ is the function from X to B that sends both 1 and 2 to 3.

5.5 Composition of Functions, Inverses

When a function $f : A \to B$ is both one-to-one and onto, there is a companion function g going in the *other direction*, from B to A, that we will now consider. But first, here is a special function:

Definition 5.24. If A is a set, the *identity function on* A is the function $f : A \mapsto A$ defined by $f(a) = a$ for all $a \in A$. It is denoted $\mathbb{1}_A$.

Thus, we may think of $\mathbb{1}_A$ as the *do nothing* function: it takes each $a \in A$ to itself.

Next, we need the notion of *composition of functions:*

Definition 5.25. If $f : A \to B$ and $g : B \to C$ are functions, the composition of f with g, written $g \circ f$, is the new function $A \to C$ given by $g \circ f\ (a) = g(f(a))$ for $a \in A$.

Note that we write the second function g before f in the expression "$g \circ f$". The order in which this is written is consistent with the notation $g(f(a))$: you act on a by f first, then by g.

Here's an example to illustrate the definition:

Example 5.26. (See Figure 5.3.) Take $A = \{1, 2, 3\}$ and $B = \{2, 3, 4\}$ as before, and now take $C = \{3, 4, 5\}$. Take $g : A \to B$ to be the function $\{(1, 3), (2, 4), (3, 2)\}$ as before, and take $h : B \to C$ to be the function $\{(2, 3), (3, 4), (4, 5)\}$. Then $h \circ g : A \to C$ is defined by $h \circ g\ (1) = h\,(g(1)) = h(3) = 4$, and similarly, $h \circ g\ (2) = h\,(g(2)) = h(4) = 5$, and $h \circ g\ (3) = h\,(g(3)) = h(2) = 3$.

Example 5.27. Here is another example of composition of functions, this time of functions that you would have seen in precalculus or calculus courses: Take $f(x) = x + 1$ and $g(x) = x^2$. (As in Example 5.20, the domain of these functions is the largest subset

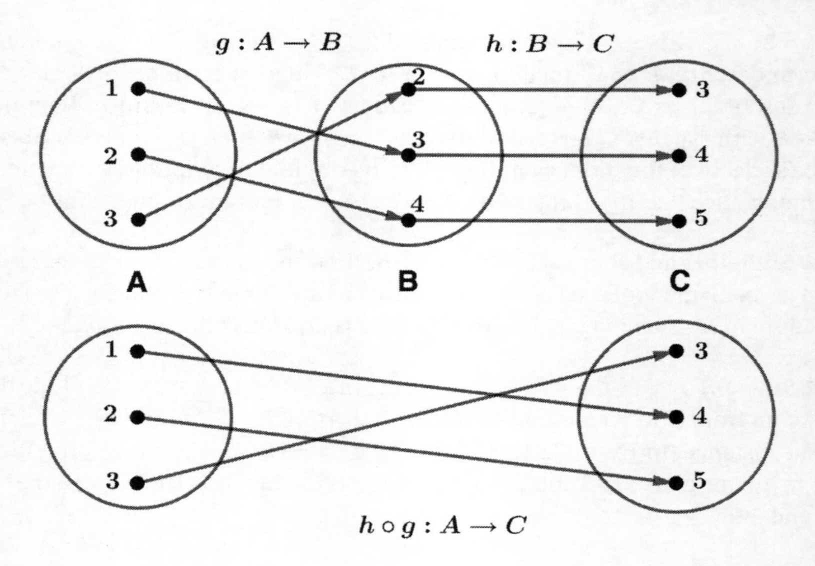

Figure 5.3. Example of composition of two functions.

of $\mathbb{R}$ on which these expressions are defined. Both $x + 1$ and x^2 are defined on all of $\mathbb{R}$, so these are both functions with domain $\mathbb{R}$ and codomain $\mathbb{R}$.) We find $f \circ g\ (x) = f(g(x)) = f(x^2) = x^2 + 1$. On the other hand, $g \circ f\ (x) = g(f(x)) = g(x+1) = (x+1)^2$. Clearly $f \circ g$ and $g \circ f$ are different functions!

Here is a quick result (that you will be asked to build on in Exercise 5.29 below):

Proposition 5.28. *Let $f : A \to B$ and $g : B \to C$ both be injective. Show that $g \circ f : A \to C$ is also injective.*

Proof. Write h for $g \circ f$. We wish to prove that if $a, a' \in A$ and $a \neq a'$, then $h(a) \neq h(a')$. We will prove the contrapositive: $h(a) = h(a') \implies a = a'$. So suppose that $h(a) = h(a')$ for $a, a' \in A$. We wish to show that a must equal a'. Now $h(a)$ is an element of c, and by the definition h, must equal $g(f(a))$. Similarly, $h(a') = g(f(a'))$. Thus, $h(a) = h(a')$ yields that $g(f(a)) = g(f(a'))$. But g is given to be injective, hence, this last equality allows us to conclude that $f(a) = f(a')$. Now we invoke the fact that f is injective to conclude that $a = a'$. Hence, h must be injective. □

Before proceeding further, test your understanding with the following extension of Proposition 5.28:

> **Exercise 5.29.** Let $f : A \to B$ and $g : B \to C$ both be surjective functions. Show that $g \circ f : A \to C$ is also a surjective function. Use this result along with Proposition 5.28 to conclude that if f and g are both bijective; then $g \circ f$ is also bijective.

Here is another test of your understanding. We will need this result later in Chapter 13.

Exercise 5.30. Suppose $f: A \to B$, $g: B \to C$, $h: C \to D$ are functions. Then $g \circ f$ is a function from A to C, and we can compose this function with h to obtain a function $h \circ (g \circ f)$ from A to C and then to D, and thus, a function from A to D. On the other hand, $h \circ g$ is a function from B to D, and we can compose f with this function to obtain a function $(h \circ g) \circ f$ from A to B and then to D, and thus, a function from A to D. Show that both of these composite functions are the same, that is, $h \circ (g \circ f) = (h \circ g) \circ f$. To describe this equality, we say that function composition is *associative.* (Hint: Simply follow an arbitrary element $a \in A$: let $f(a) = b \in B$, let $g(b) = c \in C$, and let $h(c) = d \in D$. Now compute $(h \circ (g \circ f))(a)$ and $((h \circ g) \circ f)(a)$.)

Now consider the function $f = \{(1,2),(2,3),(3,4)\}$ of Example 5.19 above (with $A = \{1,2,3\}$ and $B = \{2,3,4\}$). This is a bijective (one-to-one and onto) function. Let us see how to "undo" the effect of this function. Since f sends 1 to 2, to undo f, we need to send 2 back to 1. Similarly, f sends 2 to 3, we need to send 3 back to 2, and likewise, we need to send 4 back to 3. We have effectively created a new function from B back to A, suggestively denoted as f^{-1}, such that the composition $f^{-1} \circ f$, obtained by first sending A to B via the function f and then sending B back to A via the function f^{-1}, ends up doing nothing to A! (The composition sends 1 to 2 back to 1, and thus 1 to 1. It similarly sends 2 to 3 back to 2, so 2 to 2, then 3 to 4 back to 3, so 3 to 3.) Thus, $f^{-1} \circ f$ ends up acting as $\mathbb{1}_A$.

But it works in the other direction too! If we compose in the other direction, by first sending B to A via f^{-1}, and then A back to B via f, we get back the do-nothing function *on B!* (For instance, f^{-1} takes $2 \in B$ to $1 \in A$, and then f sends 1 back to 2, so $f \circ f^{-1}$ sends 2 to 2.) Putting this together with what we saw in the last paragraph, we find $f^{-1} \circ f = \mathbb{1}_A$ and $f \circ f^{-1} = \mathbb{1}_B$. (Recall Definition 5.17(6): two functions h and k from A to B are considered equal if their effect on every member of their common domain A is the same, i.e., if $h(a) = k(a)$ for all $a \in A$.)

The functions f and f^{-1} can thus be considered *inverses* of each other: Each undoes the effect of the other! They are inverses under composition. They are not to be thought of as multiplicative inverses (such as 5 and "5^{-1}" = 1/5)—there is no multiplication of functions going on!

Another note: It is common to use the notation "f^{-1}" as we did above for the function that undoes the effect of f, even though it is also common to use the same notation for the inverse image under f. This is an unfortunate cultural practice. See Remark 5.35 ahead.

We turn this situation around to define invertibility of a function abstractly, and then show that invertible functions are precisely like the kind that we have considered above.

Definition 5.31. A function $f: A \to B$ is said to be *invertible* if there exists a function $g: B \to A$ such that $g \circ f = \mathbb{1}_A$ and $f \circ g = \mathbb{1}_B$.

The definition above does not give you a specific construction for the function g that is brought in, it only says that g must have a certain property in relation to f. This leaves two questions open: the *existence* of such a function g, and the *uniqueness* of such a function g. You will find many such instances in mathematics where certain objects are defined via some property that they should possess, but where the question of the existence of such an object and its uniqueness, if it exists, is left undetermined. In such cases, existence and uniqueness of the objects need to be established independently. We will tackle both issues below.

Proposition 5.32. *Let $f : A \to B$ be a function. Then f is invertible if and only if f is bijective.*

Proof. There are two parts to the statement of this proposition. It says that if f is invertible, then f must be bijective. It also says that if f is bijective, then f must be invertible. Accordingly, our proof has to be in two parts. We first need to assume that f is invertible and then show that f must be bijective. Going the other way, we need to assume that f is bijective and then show that f must be invertible.

(This is a good place to review the compound statement "implies and is implied by" in Section 3.3 in Chapter 3.)

Accordingly, let us assume first that f is invertible. We need to show that f is both one-to-one and onto. Assume to the contrary that f is not one-to-one. We will arrive at a contradiction, which will force us to then conclude that our assumption must have been flawed, i.e., f is indeed one-to-one. (Such a proof is known as *proof by contradiction.* See Chapter 6 ahead.) Since f is assumed not one-to-one, there exist a_1 and a_2 in A, with $a_1 \neq a_2$, such that $f(a_1) = f(a_2)$. Denote by "b" this common value of $f(a_1)$ and $f(a_2)$. Consider the function $g : B \to A$ with the property that $g \circ f = \mathbb{1}_A$ and $f \circ g = \mathbb{1}_B$—such a function g exists by supposition of invertibility. We find $g(b) = g(f(a_1)) = (g \circ f)(a_1) = \mathbb{1}_A(a_1) = a_1$. On the other hand, b is also $f(a_2)$, and applying the same reasoning we find $g(b) = g(f(a_2)) = (g \circ f)(a_2) = \mathbb{1}_A(a_2) = a_2$. Putting it together, we find $g(b) = a_1 = a_2$, which contradicts the fact that a_1 and a_2 were distinct. Hence our original assumption must have been flawed, and indeed, f is one-to-one.

We also need to prove (still under the hypothesis that f is invertible) that f is onto. Consider any $b \in B$. We want to find an $a \in A$ such that $f(a) = b$. For this, let us invoke the fact that $f \circ g(b) = \mathbb{1}_B(b) = b$, where g is the same function from B to A we considered in the last paragraph. (Recall, g exists because of the assumption that f is invertible.) Thus, $f(g(b)) = b$. Now $g(b)$ is some element of A, so taking $a = g(b)$, we find $f(a) = b$. Thus, b is the image of some element of A under f, and since b was chosen arbitrarily, we find that indeed f is onto.

Now we go the other way in our if and only if proof. We will take as true the statement that f is both one-to-one and onto, and show that f is invertible. This can be done by a concrete construction, exactly as we constructed "f^{-1}" in the discussion just above Definition 5.31: we simply reverse the arrows from A to B! We proceed as follows: we first note that every $b \in B$ can be written as $f(a)$ for some $a \in A$ because f is onto. If say $f(a) = b$ and $f(a') = b$ for $a, a' \in A$, the injectivity of f would show that $a = a'$. Hence, this a is unique for this b. Now define $g(b) = a$. Note that g satisfies the definition of a function: it is defined for every $b \in B$, and given any $b \in B$, there is exactly one a that we can pick as a candidate for $g(b)$. You are invited at this point to check for yourself that indeed the two relations are met: $g \circ f = \mathbb{1}_A$ and $f \circ g = \mathbb{1}_B$. This is simply a matter of keeping track of how g is defined. □

Remark 5.33. One may wonder why it is necessary to stipulate in Definition 5.31 (the definition of the invertibility of a function $f : A \to B$) that the function $g : B \to A$ must satisfy both $g \circ f = \mathbb{1}_A$ and $f \circ g = \mathbb{1}_B$. After all, if g satisfied just the first relation $g \circ f = \mathbb{1}_A$, it would undo the effect of f–would this not be sufficient? Perhaps it would be sufficient in some limited situations, but in general, both aesthetics and practicality require symmetry of the relations. Let us consider the following example:

Example 5.33.1. Consider the functions $f : \mathbb{N} \to \mathbb{N}$ (that sends 1 to 2, 2 to 3, $\ldots$, n to $n+1$ $(n \geq 1)$, $\ldots$) and $g : \mathbb{N} \to \mathbb{N}$ (that sends 2 to 1, 3 to 2, $\ldots$, n to $n-1$ $(n \geq 2)$, $\ldots$, and 1 to 1). Then $g \circ f = \mathbb{1}_{\mathbb{N}}$ but $f \circ g \neq \mathbb{1}_{\mathbb{N}}$, as $f(g(1)) = f(1) = 2 \neq 1$.

In this example, the asymmetry is very stark. The two sets A and B of Definition 5.31 are equal here; both are $\mathbb{N}$. In such a situation, we do not wish to pick out f in the (f, g) pair as something distinguished, we wish to view it in the same light as g. So, finding that $g \circ f = \mathbb{1}_{\mathbb{N}}$ but $f \circ g \neq \mathbb{1}_{\mathbb{N}}$ is particularly disconcerting. This is one reason to insist in the definition that both $g \circ f = \mathbb{1}_A$ and $f \circ g = \mathbb{1}_B$.

But another reason is that this symmetry ensures that invertibility of a function is equivalent to its being both injective and surjective (Proposition 5.32), which is a very tidy result. If you study the proof of Proposition 5.32, you will see that we used both conditions $g \circ f = \mathbb{1}_A$ and $f \circ g = \mathbb{1}_B$ in Definition 5.31 to prove that a function satisfying Definition 5.31 must be injective and surjective. (And correspondingly, the function f of Example 5.33.1 for which $f \circ g \neq \mathbb{1}_{\mathbb{N}}$ is not surjective.)

We now justify our use of the expression *the* inverse, instead of *an* inverse).

Proposition 5.34. *Let $f : A \to B$ be a function. The inverse of f, if it exists, is unique.*

Proof. We wish to show that if $g : B \to A$ and $h : B \to A$ are two functions satisfying $g \circ f = \mathbb{1}_A$ and $f \circ g = \mathbb{1}_B$, as also $h \circ f = \mathbb{1}_A$ and $f \circ h = \mathbb{1}_B$, then $g = h$. Thus, we wish to show that $g(b) = h(b)$ for all $b \in B$. Take any $b \in B$. For this b, since $f(g(b)) = f \circ g(b) = \mathbb{1}_B(b) = b$, and similarly, since $f(h(b)) = f \circ h(b) = \mathbb{1}_B(b) = b$, we find that f sends both $g(b)$ and $h(b)$ (both members of A) to b. But f is invertible, and therefore by Proposition 5.32, f is injective. Since both $g(b)$ and $h(b)$ are sent to the same element by f, the injectivity of f says $g(b) = h(b)$. But b was chosen arbitrarily in B, so we find $g(b) = h(b)$ for all $b \in B$. That is, $g = h$. □

Remark 5.35. When a function $f : A \to B$ is invertible, we often write "f^{-1}" for the unique function $g : B \to A$ that satisfies $g \circ f = \mathbb{1}_A$ and $f \circ g = \mathbb{1}_B$. This has become the practice, even though, unfortunately, we also use the notation "f^{-1}" for the inverse image (Definition 5.17, see also Example 5.18(3)). Now, mathematicians are otherwise wonderful people, but they do exhibit annoying quirks every now and then, and this is one of those times. How should you know which f^{-1} is being referred to? Well, you have to judge that from the context. Note, first of all, that a given function f need not be invertible (recall from Proposition 5.32 that it needs to be bijective to be invertible), so if your function is not invertible, you can be assured that any "f^{-1}" you run across refers to the inverse image. But if your function is invertible, you have to look more closely at the context and figure out for yourself what was meant.

Remember that the inverse image kind of f^{-1} takes a *subset* of B and gives you back a *subset* of A. The inverse function kind of f^{-1} (when it exists), takes an *individual element* of B and gives you back an *individual element* of A.

Now is a good time for you to test your understanding by working on the following:

Exercise 5.36. Let $f : A \to B$ and $g : B \to C$ both be invertible. Show that

(1) f^{-1} is also a bijection.

(2) The inverse of f^{-1} is f.

(3) The composition $g \circ f$ is also invertible.

(4) The inverse of $g \circ f$ is $f^{-1} \circ g^{-1}$.

Example 5.37. We will consider some examples of invertible functions that you may have seen before:

(1) $f(x) = e^x$. The explicit way to write this so as to make it clear what is the domain of f, what is the codomain, and what is the rule for f, is as follows:

$$\begin{aligned} f : \mathbb{R} &\to \mathbb{R}, \\ x &\mapsto e^x. \end{aligned}$$

We know this function well; it takes each real number x to the number e^x (so indeed, f is a function), and its range is the set of positive real numbers. It is an injective function, but is not onto when considered as a function $f : \mathbb{R} \to \mathbb{R}$. However, when considered as a function from $\mathbb{R}$ to its range $\mathbb{R}_{>0}$, it is indeed an onto function. Thus, the function e^x provides a bijection between $\mathbb{R}$ and $\mathbb{R}_{>0}$. (In Example 13.43 in Chapter 13 we will see that e^x is more than a bijection, it preserves something called group structure.) By Proposition 5.32, e^x, considered as a function from $\mathbb{R}$ to $\mathbb{R}_{>0}$, is invertible. It's inverse, also naturally occurring, is the function $g(x) = \ln(x)$. (This function, known as the *natural logarithm* function, is defined only for positive real numbers, and has range all of $\mathbb{R}$, as you would already know.) See Figure 5.4.

(2) $f(x) = \frac{1}{x}$. See Example 5.20(2).
Formally, we would write f as

$$\begin{aligned} f : \mathbb{R} \setminus \{0\} &\to \mathbb{R} \text{ (or just } f : \mathbb{R} \setminus \{0\} \to \mathbb{R} \setminus \{0\}), \\ x &\mapsto \frac{1}{x}. \end{aligned}$$

This function is injective, with range $\mathbb{R} \setminus \{0\}$. Therefore, when considered as a function from $\mathbb{R} \setminus \{0\}$ to $\mathbb{R} \setminus \{0\}$, it is surjective as well. Thus, $\frac{1}{x}$ provides a bijection between $\mathbb{R} \setminus \{0\}$ to itself. Its inverse is itself: Composing $\frac{1}{x}$ with itself gives you back x!

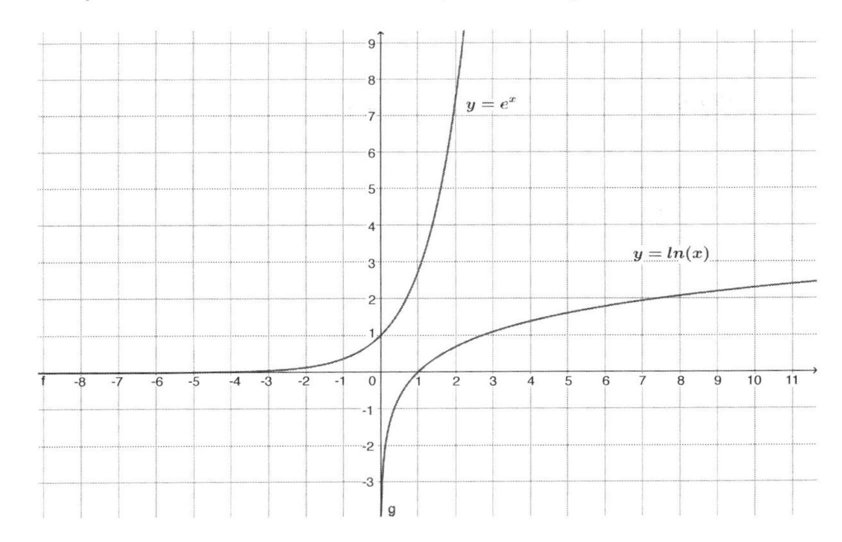

Figure 5.4. Graphs of $f(x) = e^x$ and $g(x) = \ln(x)$.

(3) $f(x) = \sqrt{1-x}$. See Example 5.20(3).

As in example (3), we must ensure that $1 - x \geq 0$ for the square root to be meaningful. Thus, we must have $x \leq 1$ for this function to be defined, so the domain is $\{x \mid x \leq 1\}$. This function is one-to-one, since if $\sqrt{1-a} = \sqrt{1-b}$, we find on squaring and canceling that $a = b$. We have remarked in Example 5.20(3) that "$\sqrt{a}$" for positive a refers to the positive square root. So the image of f is contained in $\mathbb{R}_{\geq 0}$. Given any $y \geq 0$, we can solve $\sqrt{1-x} = y$ to find $x = 1 - y^2$, so the range of this function is the entire set $\mathbb{R}_{\geq 0}$. Therefore, considered as a function from $\{x \mid x \leq 1\}$ to $\mathbb{R}_{\geq 0}$, this function is also onto. Thus, $\sqrt{1-x}$ provides a bijection between $\{x \mid x \leq 1\}$ and $\mathbb{R}_{\geq 0}$. See Figure 5.5. Its inverse can be calculated by the method we have already used: let $y = \sqrt{1-x}$, and square and transpose to find $x = 1-y^2$. In this formula we view x as a function of y; of course we already know that $x \leq 1$ and $y \geq 0$. Or, renaming x as y and y as x, we rewrite this as $y = 1 - x^2$, with $x \geq 0$ and $y \leq 1$. Thus, the inverse is the onto function $g : \mathbb{R}_{\geq 0} \to \{y \mid y \leq 1\}$ given by $g(x) = 1 - x^2$.

5.6 Examples of Some Sets Commonly Occurring in Mathematics

In this section we will consider examples of various sets that occur frequently in mathematics. These serve as the bedrock of much of mathematics! For instance, some of these sets occur throughout this text. You will see others in future courses you may take.

Example 5.38. We are familiar with the integers, written $\mathbb{Z} = \{0, \pm 1, \pm 2, \dots\}$. We have also introduced in Chapter 1 the natural numbers, written $\mathbb{N} = \{x \in \mathbb{Z} \mid x > 0\}$, as well as the set $\mathbb{Z}_{\geq 0} = \{x \in \mathbb{Z} \mid x \geq 0\}$. (It must be remarked, just in case you see this different

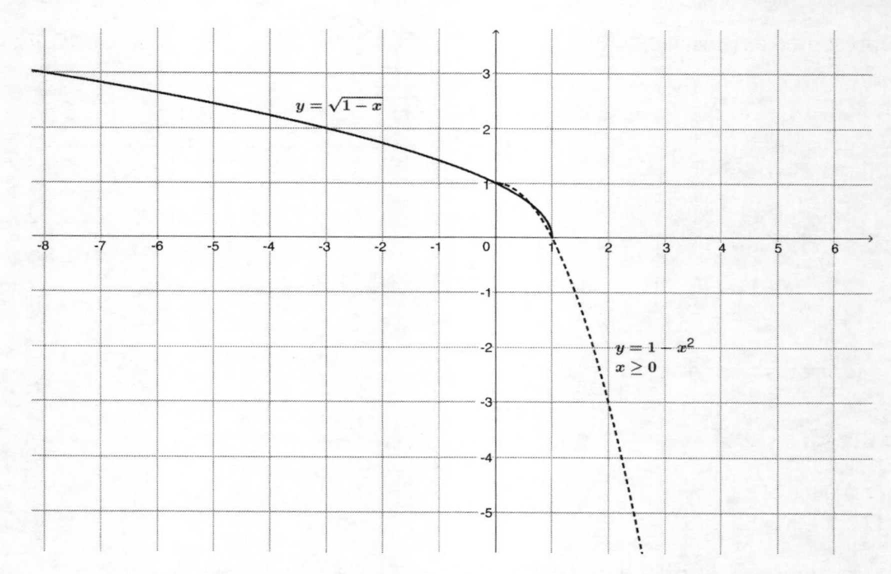

Figure 5.5. Graphs of $f(x) = \sqrt{1-x}$, defined only for $x \leq 1$, and its inverse $g(x) = 1-x^2$, shown dotted, defined for $x \geq 0$.

notation being used somewhere, that mathematicians occasionally refer to the larger set $\mathbb{Z}_{\geq 0}$ as the natural numbers. We will not adopt that convention in this book.)

For any $n \in \mathbb{Z}$, we will refer to the set of multiples of n as $n\mathbb{Z}$. Thus, $n\mathbb{Z} = \{x \in \mathbb{Z} \mid x = kn, \text{for some } k \in \mathbb{Z}\}$.

As you would know from high school, there are distinguished positive integers called *primes*. These are characterized as those positive integers ($\geq$ 2) whose only divisors are 1 and themselves. The key feature of the set of integers is the existence of unique prime factorization: Every integer $\geq$ 2 can be factored into a product of primes, and the primes that show up in any factorization are unique, except possibly for rearrangement. You may have studied all this in high school, but we will revisit these matters in Chapter 10.

Example 5.39. Similarly, we are familiar with the set of rational numbers (or reduced fractions with positive denominators). This set is typically denoted $\mathbb{Q}$ in mathematics. It is the set of all fractions $\frac{a}{b}$ (with $a, b, \in \mathbb{Z}, b > 0$), with the understanding that we consider $\frac{a}{b} = \frac{c}{d}$ if $ad = bc$. (Thus, in the language of equivalence relations that you will study in Chapter 9 ahead — see Example 9.34 in that chapter if you are interested — the set of rational numbers is the set of equivalence classes of fractions under the relation that $\frac{a}{b}$ is related to $\frac{c}{d}$ if $ad = bc$. Each equivalence class can be represented uniquely as a *reduced* fraction $\frac{a}{b}$, i.e., $\gcd(a, b) = 1$, with $b > 0$.) As with the integers, we will denote the set $\{x \in \mathbb{Q} \mid x > 0\}$ by $\mathbb{Q}_{>0}$, the set $\{x \in \mathbb{Q} \mid x \geq 0\}$ by $\mathbb{Q}_{\geq 0}$, and the set $\{x \in \mathbb{Q} \mid x < 0\}$ by $\mathbb{Q}_{<0}$.

Example 5.40. The set of real numbers $\mathbb{R}$ is of course well known from high school. You would certainly have also seen this set if you took any calculus courses. Real numbers are the numbers that correspond to (positive, negative, or zero) lengths, and can be represented as (possibly infinite, repeating or non-repeating) decimals. Although in this book we will treat the set of real numbers as a well-known entity that does not need further definition, it should be noted that real numbers can actually be constructed from the rational numbers $\mathbb{Q}$ by a process known as *completion*.

Just as with $\mathbb{Z}$ and $\mathbb{Q}$, we will denote the set $\{x \in \mathbb{R} \mid x > 0\}$ by $\mathbb{R}_{>0}$, the set $\{x \in \mathbb{R} \mid x \geq 0\}$ by $\mathbb{R}_{\geq 0}$, and the set $\{x \in \mathbb{R} \mid x < 0\}$ by $\mathbb{R}_{<0}$.

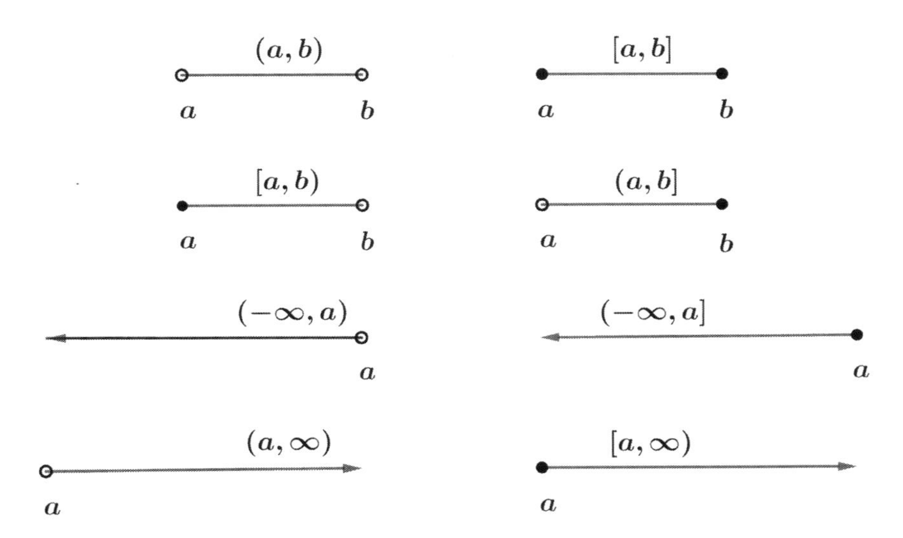

Figure 5.6. Intervals on the real line. An open circle at a signifies that a is *not* part of the interval; a closed circle signifies that a is part of the interval.

If $a < b$ are real numbers, we will denote the set $\{x \in \mathbb{R} \mid a < x < b\}$ by (a, b), and call it the *open interval* determined by a and b. See Figure 5.6. Similarly, we will denote the set $\{x \in \mathbb{R} \mid a \leq x \leq b\}$ by $[a, b]$, and call it the *closed interval* determined by a and b. The sets $\{x \in \mathbb{R} \mid a \leq x < b\}$ and $\{x \in \mathbb{R} \mid a < x \leq b\}$ are denoted $[a, b)$ and $(a, b]$ respectively, and are referred to as *half open* intervals determined by a and b.

If $a \in \mathbb{R}$, the set $\{x \in \mathbb{R} \mid x < a\}$ is denoted $(-\infty, a)$, the set $\{x \in \mathbb{R} \mid x \leq a\}$ is denoted $(-\infty, a]$, the set $\{x \in \mathbb{R} \mid a < x\}$ is denoted (a, ∞), and the set $\{x \in \mathbb{R} \mid a \leq x\}$ is denoted $[a, \infty)$. Consistent with this notation, $\mathbb{R}$ is also denoted $(-\infty, \infty)$. The intervals $(-\infty, a)$, (a, ∞) and $(-\infty, \infty)$ are all considered open intervals.

Example 5.41. The set of complex numbers is denoted $\mathbb{C}$ in mathematics. It is possible that you have seen this set in high school. It is a new system of numbers that contains the real numbers within it, in which -1 has a square root. Of course, this brings into question an oft-repeated principle that only positive numbers and zero can have square roots. On closer examination, it turns out that this statement is incomplete (and therefore not correct as stated). The correct form of the principle is that only positive numbers and zero can have square roots *within the real numbers.* To find square roots of negative real numbers, we have to search outside the set of real numbers. We have to go to some number system that is larger than the real numbers, and this is where complex numbers come in!

We will use i, as is standard, to denote a square root of -1. It turns out that we can represent any complex number as a real number plus i times another real number, that is, $\mathbb{C} = \{a + ib, \mid a, b \in \mathbb{R}\}$.

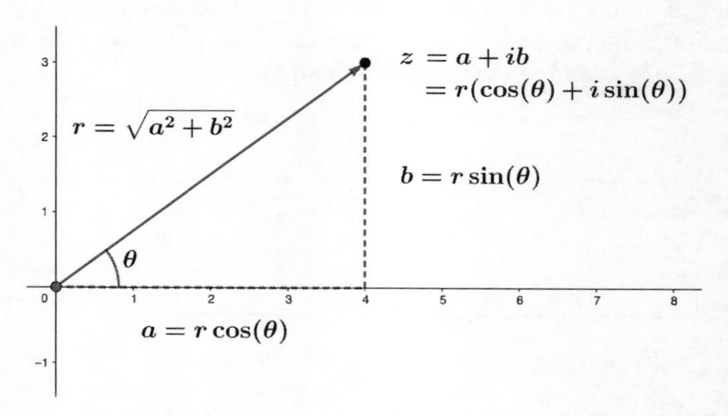

Figure 5.7. The complex number $z = a + ib$.

The set of complex numbers forms a "field." While we will not go into the precise meaning at this point of the word field, in practical terms it just means that you can add and subtract complex numbers, and you can divide a complex number by another (nonzero) complex number, just as you can with real numbers. The formulas are a bit different of course. We have the following:

$$\begin{aligned}(a + ib) \pm (c + id) &= (a \pm c) + i(b \pm d),\\ (a + ib) \cdot (c + id) &= (ac - bd) + i(ad + bc),\\ \frac{a + ib}{c + id} &= \frac{ac + bd}{c^2 + d^2} + i\frac{bc - ad}{c^2 + d^2}.\end{aligned} \tag{5.1}$$

The number "0"$= 0 + i0$ acts as the usual "zero," so $z + 0 = 0 + z$ for all complex numbers z. Similarly, the number "1"$= 1 + i0$ acts as the usual "one," so $z \cdot 1 = 1 \cdot z$ for all complex numbers z. The formulas will show that $z_1 + z_2 = z_2 + z_1$ and $z_1 z_2 = z_2 z_1$ for all pairs of complex numbers z_1, z_2. The subset of numbers of the form $a + i0$ is just the usual set of real numbers. The *modulus* of a complex number $a + ib$, denoted $|a + ib|$, is defined to be $\sqrt{a^2 + b^2}$. When the complex number $a + ib$ is represented as the point (a, b) in the plane, $|a + ib|$ is the distance of (a, b) from the origin. The same number can also be represented as $r(\cos(\theta) + i\sin(\theta))$, where r is the modulus $|a + ib|$ we just considered, and θ is the counterclockwise angle made by the ray joining the origin and (a, b) with the positive x-axis. See Figure 5.7.

The set of complex numbers has the property of being *algebraically closed.* This means that any equation of the form

$$a_n x^n + a_{n-1} x^{n-1} + \cdots + a_1 x + a_0 = 0,$$

where n is some positive integer, and the a_i are complex numbers, with $a_n \neq 0$, has a solution in $\mathbb{C}$. In contrast, $\mathbb{R}$ is not algebraically closed—the equation $x^2 + 1 = 0$ has no solution inside $\mathbb{R}$. (In fact, the set $\mathbb{C}$ was constructed precisely so that this particular equation could have a solution!) You will see the proof of this remarkable fact, for instance, in future courses on complex analysis.

Note that there is no "reasonable" notion of positive and negative complex numbers (see Exercise 5.41.1 below). As a result we can no longer talk of one complex number being "greater than" or "less than" another complex number.

> **Exercise 5.41.1.** The word "reasonable" above means a notion of positive and negative complex numbers—that is, some distribution of complex numbers into three mutually exclusive subsets: one labeled positive complex numbers, another labeled negative complex numbers, and the third being $\{0\}$—which is consistent with how positive and negative real numbers behave when we add or multiply them. For instance, for any nonzero complex number, we would like, in analogy with the real numbers, that $z^2 > 0$. Show that there cannot be any notion of positive and negative complex numbers that satisfies the property that $z^2 > 0$ for all nonzero $z \in \mathbb{C}$. (Hint: Play with some specific complex numbers for your choice of z.)

Example 5.42. Consider $\mathbb{R} \times \mathbb{R}$, the set of all ordered pairs of real numbers (a, b). Choosing a set of coordinate axes on the plane, we can identify $\mathbb{R} \times \mathbb{R}$ with the plane by identifying the ordered pair (a, b) with the point, also denoted (a, b), whose x-coordinate is a and y-coordinate is b. Similarly, we can identify $\mathbb{R} \times \mathbb{R} \times \mathbb{R}$, the set of all ordered triples of real numbers (a, b, c), with the three-dimensional space that we see around us. A triple (a, b, c) would correspond to the point whose x-coordinate is a, y-coordinate is b, and z-coordinate is c.

It is standard to denote $\mathbb{R}\times\mathbb{R}$ by $\mathbb{R}^2$ and $\mathbb{R}\times\mathbb{R}\times\mathbb{R}$ by $\mathbb{R}^3$. It is then straightforward to consider the sets $\mathbb{R}^n$ (for $n = 4, 5, \ldots$), whose elements are the "n-tuples" of real numbers $(a_1, a_2, \ldots, a_n)$. Once again, after selecting a set of n coordinate axes in n-dimensional space, we may view $\mathbb{R}^n$ as an algebraic model of *n-dimensional space*. Of course, these spaces are harder to visualize for $n \geq 4$!

We can play the same game with $\mathbb{C}^n$ $(n = 1, 2, \ldots)$. $\mathbb{C}^2$ for instance is the set of all ordered pairs (z_1, z_2), with z_1, z_2 in $\mathbb{C}$.

Example 5.43. Before we begin, a quick remark:

> *Remark* 5.43.1. If you have just learned about complex numbers and are still feeling a bit shaky about them, it is harmless for a first reading to consider pairs of *real* numbers in this example, and view them as *real-valued* 2-dimensional column vectors, with the same definitions of vector addition and multiplication by a *real number* λ. Similarly, in Example 5.44 below, for a first reading you can consider matrices whose elements are *real numbers* with the same definitions of matrix addition and multiplication. These matrices would constitute the set denoted $M_2(\mathbb{R})$.

You may have seen column vectors already, for instance in physics or elementary linear algebra. Consider the set of all pairs of complex numbers, namely $\mathbb{C}^2$ that we considered in Example 5.42 above. We will now write the typical elements (z_1, z_2) of $\mathbb{C}^2$ in *column* form as

$$\bar{v} = \begin{pmatrix} z_1 \\ z_2 \end{pmatrix}.$$

Ordered pairs of complex numbers, written one on top of the other as above, are known as (complex-valued 2-dimensional column) *vectors*, and z_1 and z_2 above are known as the *components* of the vector $\overline{v}$. Two vectors, $\overline{v}$ with components z_1 and z_2 and $\overline{w}$ with components u_1 and u_2, can be added by just adding the components:

$$\overline{v}+\overline{w}=\begin{pmatrix} z_1 \\ z_2 \end{pmatrix}+\begin{pmatrix} u_1 \\ u_2 \end{pmatrix}\stackrel{\text{def}}{=}\begin{pmatrix} z_1+u_1 \\ z_2+u_2 \end{pmatrix}.$$

Similarly, given a complex number λ (referred to in this context as a *scalar*), we can multiply a given vector $\overline{v}$ with λ by simply multiplying each component by λ: if the components of $\overline{v}$ are z_1 and z_2 then

$$\lambda\cdot\overline{v}=\lambda\cdot\begin{pmatrix} z_1 \\ z_2 \end{pmatrix}\stackrel{\text{def}}{=}\begin{pmatrix} \lambda z_1 \\ \lambda z_2 \end{pmatrix}.$$

The set of complex-valued 2-dimensional vectors is an enormously useful set, and in fact, the subset of vectors whose components are *real* is particularly useful. (For instance, positions of particles in a plane and velocities of particles moving in a plane can all be described by 2-dimensional vectors with real coordinates.)

To see the full advantage of viewing an ordered pair of complex numbers (z_1, z_2) as the 2-dimensional vector $\begin{pmatrix} z_1 \\ z_2 \end{pmatrix}$, we need to see how these vectors relate to another mathematical object, the 2×2 matrix! This will be our next example below.

It should come as no surprise that we can extend our notion of vectors to the n-dimensional situation: given an n-tuple $(z_1, z_2, \ldots, z_n)$ of complex numbers, we can stack them in a column to form an *n-dimensional vector.*

Example 5.44. (See Remark 5.43.1 above.) The set of 2×2 matrices with entries in $\mathbb{C}$, denoted $M_2(\mathbb{C})$:

$$\begin{pmatrix} a & b \\ c & d \end{pmatrix},$$

where a, b, c, and d are complex numbers.

We say that a is in the $(1,1)$ slot (or place) or that a is the $(1,1)$ entry. This refers to the fact that a is in the first row and first column. Similarly, b is in the $(1,2)$ slot, c is in the $(2,1)$ slot, and d is in the $(2,2)$ slot.

You may have seen these objects before. They are especially fun and easy to work with. They are also enormously useful and form the basis of several techniques used in physics, computer science and engineering.

Two such matrices are added *component-wise,* that is,

$$\begin{pmatrix} a & b \\ c & d \end{pmatrix}+\begin{pmatrix} e & f \\ g & h \end{pmatrix}\stackrel{\text{def}}{=}\begin{pmatrix} a+e & b+f \\ c+g & d+h \end{pmatrix}.$$

(Here, the symbol $\stackrel{\text{def}}{=}$ stands for "is defined to be equal to.")

$$\begin{pmatrix} a & b \\ c & d \end{pmatrix}\cdot\begin{pmatrix} e & f \\ g & h \end{pmatrix}\stackrel{\text{def}}{=}\begin{pmatrix} ae+bg & af+bh \\ ce+dg & cf+dh \end{pmatrix}.$$

The interesting thing about matrix multiplication is that it is not commutative! That is, the order in which you multiply two matrices matters. For instance, we find

$$\begin{pmatrix} 1 & 0 \\ 0 & 2 \end{pmatrix}\cdot\begin{pmatrix} a & b \\ c & d \end{pmatrix}=\begin{pmatrix} a & b \\ 2c & 2d \end{pmatrix},$$

while in the reverse order,

$$\begin{pmatrix} a & b \\ c & d \end{pmatrix} \cdot \begin{pmatrix} 1 & 0 \\ 0 & 2 \end{pmatrix} = \begin{pmatrix} a & 2b \\ c & 2d \end{pmatrix}.$$

So, for example, any choice of complex numbers a, b, c, and d with $b \neq 0$ will render the products in the two different orders unequal, since the $(1, 2)$ slots will be unequal: b cannot equal $2b$ if $b \neq 0$. (See Exercise 5.44.1 below too.)

Exercise 5.60 ahead explains the genesis of this somewhat unusual multiplication rule!

Exercise 5.44.1. Determine explicitly the subset of $M_2(\mathbb{C})$ whose elements all commute with the matrix $\begin{pmatrix} 1 & 0 \\ 0 & 2 \end{pmatrix}$ above. (You should be able to describe all elements of the desired set by a simple rule.)

The sets $M_n(\mathbb{C})$, $n = 3, 4, \ldots$, are defined analogously, with analogous addition and multiplication.

The advantage of viewing the set of ordered pairs of complex numbers, $\mathbb{C}^2$, as complex-valued 2-dimensional column vectors as in Example 5.43 above is that a typical 2×2 matrix from the set $M_2(\mathbb{C})$ in Example 5.44 above can be made to "act" on a typical 2-dimensional vector to give another 2-dimensional vector, via the following "multiplication" rule:

$$\begin{pmatrix} a & b \\ c & d \end{pmatrix} \cdot \begin{pmatrix} z_1 \\ z_2 \end{pmatrix} \stackrel{\text{def}}{=} \begin{pmatrix} az_1 + bz_2 \\ cz_1 + dz_2 \end{pmatrix}.$$

This action (that is, this process of obtaining a new vector by multiplying it with the given matrix) arises very naturally in mathematics and in various mathematical sciences, and has the very nice property of being *linear*. (See Exercise 5.58 ahead). This is the subject of *linear algebra*.

Exercise 5.59 ahead explains the rationale behind this formula above for the action of a matrix on a vector.

Example 5.45. We consider a particularly significant application of this notion of a matrix multiplying a vector that we saw at the end of Example 5.44 above. Suppose we wish to solve the simultaneous equations

$$ax + by = p,$$
$$cx + dy = q$$

for the variables x and y. (Here, a, b, c, d, p, and q are some given complex numbers.) We may write this using our discussion above as the product of a matrix and a vector equalling another vector, as follows:

$$\begin{pmatrix} a & b \\ c & d \end{pmatrix} \cdot \begin{pmatrix} x \\ y \end{pmatrix} = \begin{pmatrix} p \\ q \end{pmatrix}.$$

We can solve this single equation for the *vector* of variables $\begin{pmatrix} x \\ y \end{pmatrix}$ by working purely with the matrix on the left of the equation above. For instance, one can check if a number called the determinant of the matrix (see Exercise 5.62 ahead for the definition) is nonzero. If it is nonzero, we can multiply both sides of the equation on the left by something known as the inverse of the matrix and proceed to solve. (See Exercise 5.63 ahead for how this is done.) Alternatively, one can solve the equation by performing an operation known as *Gaussian elimination* on the matrix. (Gaussian elimination is a technique that you will study in future courses in linear algebra.) Now, in a 2×2 situation, the use of Gaussian elimination may seem unduly complicated, but the point is that the idea behind it works in exactly the same manner even if we had a system of, say, 10,000 linear equations in 10,000 variables! (The need for solving equations in so many variables occurs routinely in various applications.) The advantage of these matrix computations is that they are algorithmic, and can be programmed easily into a computer. The efficiency of Gaussian elimination (measured by the number of computer operations needed as a function of the number of equations) is such that solving such large systems of equations is quite easily done with today's computer speeds.

Remark 5.45.1. One can also view the typical ordered pair (z_1, z_2) in $\mathbb{C}^2$ as a *row* vector, with the typical 2×2 matrix acting on it by the following rule:

$$(z_1, z_2) \cdot \begin{pmatrix} a & b \\ c & d \end{pmatrix} \stackrel{\text{def}}{=} (az_1 + cz_2, bz_1 + dz_2)$$

Both points of view are useful.

Remark 5.45.2. As in Example 5.42, we may identify the column vector $\begin{pmatrix} z_1 \\ z_2 \end{pmatrix}$ with the point (z_1, z_2) *in complex 2-dimensional space* whose x-coordinate is z_1 and y-coordinate is z_2. Typically, one visualizes the vector $\begin{pmatrix} z_1 \\ z_2 \end{pmatrix}$ as an arrow with the tail at the origin and the tip at the point (z_1, z_2). However, this is all happening in complex 2-dimensional space, which is harder to visualize. This is because our brains seem to be wired only to visualize space coordinatized by *real* numbers, and that too only up to 3 dimensions, while complex 2-dimensional space space is 4-dimensional when it is coordinatized by real numbers (since each of the two complex coordinates needs two real numbers to specify it).

5.7 Further Exercises

Exercise 5.46. Practice Exercises:

Exercise 5.46.1. Let $U = \{1, 2, \dots, 10\}$, $X = \{n \mid 1 \le n \le 10$ and n is even$\}$, $Y = \{n \mid 1 \le n \le 10$ and n is odd$\}$, and $Z = \{n \mid 1 \le n \le 10$ and n is prime$\}$. Describe the following sets: $X \cup Y$, $X \cap Y$, $X \cup Z$, $X \cap Z$, $Y \cup Z$, $Y \cap Z$, X^c, Y^c, Z^c, $Y \times (Z \setminus X)$.

Exercise 5.46.2. Let $A = \{\text{Red}, \text{Blue}, \text{Green}\}$ and $B = X, Y, Z$. Determine if the following are functions:

(1) The subset $\{(\text{Green}, X), (\text{Red}, X), (\text{Blue}, X)\}$ of $A \times B$.

(2) The subset $\{(\text{Green}, X), (\text{Red}, X), (\text{Green}, Y), (\text{Blue}, Y)\}$ of $A \times B$.

(3) The subset $\{(\text{Green}, Z), (\text{Red}, X), (\text{Blue}, Y)\}$ of $A \times B$.

(4) The subset $\{(\text{Green}, X), (\text{Red}, X), (\text{Green}, X)\}$ of $A \times B$. (See Remark 5.3.)

Exercise 5.46.3. Let $X = \{0, \pm 1, \pm 2, \pm 3\}$, $Y = \{0, 1, 2, 3\}$, $U = \{1, 2, \dots, 10\}$. Let $f : X \to U$ be defined by $f(n) = n^2$. Answer the following:

(1) What is the domain of f?

(2) What is the codomain of f?

(3) What is the range of f?

(4) What is $f^{-1}(\{4\})$?

(5) What is $f|_Y^{-1}(\{4\})$? (See Remark 5.23 for the notation $f|_Y$.)

(6) Is f injective?

(7) Write Z for the range of f. Is $f|_Y : Y \to Z$ bijective? (See Remark 5.21.)

Exercise 5.46.4. Consider $f(x) = 2x$ and $g(x) = x^2$. Determine the domain and range of f and g, and then determine a formula for $f \circ g$ as well as $g \circ f$.

Exercise 5.46.5. For the function $f(x) = 2x$, determine the domain and range, verify that it is invertible as a function from the domain to the range, and determine a formula for its inverse.

Exercise 5.46.6. For the function $f(x) = \frac{1}{1-x}$, determine the domain and range, verify that it is invertible as a function from the domain to the range, and determine a formula for its inverse.

Exercise 5.46.7. Let z be the complex number $\frac{1}{\sqrt{2}} + i\frac{1}{\sqrt{2}}$.

(1) Write z as $r(\cos(\theta) + i\sin(\theta))$ for suitable r and θ.

(2) Write $\frac{1}{z}$ both as $a + ib$ for suitable a and b and as $s(\cos(\alpha) + i\sin(\alpha))$ for suitable s and α.

Exercise 5.46.8. For the matrices

$$X = \begin{pmatrix} 1 & 2 \\ 3 & 4 \end{pmatrix} \text{ and } Y = \begin{pmatrix} 5 & 6 \\ 7 & 8 \end{pmatrix},$$

calculate $X + Y, XY$, and YX.

Exercise 5.46.9. For the matrices

$$T_a = \begin{pmatrix} 1 & a \\ 0 & 1 \end{pmatrix} \text{ and } T_b = \begin{pmatrix} 1 & b \\ 0 & 1 \end{pmatrix},$$

where a and b are arbitrary complex (or real, if you prefer) numbers, verify that $T_a T_b = T_b T_a$ and use your computations to find c such that the matrix T_c satisfies

$$T_a T_c = T_c T_a = \begin{pmatrix} 1 & 0 \\ 0 & 1 \end{pmatrix}.$$

Exercise 5.46.10. Let $\overline{v}$ denote the vector

$$\begin{pmatrix} x \\ y \end{pmatrix},$$

where x and y are arbitrary complex (or real, if you prefer) numbers. For the matrix T_a introduced in Exercise 5.46.9 above, compute $T_a \cdot \overline{v}$.

Exercise 5.47. Given $n \in \mathbb{Z}$, recall the notation $n\mathbb{Z}$ from Example 5.38 for the set of all multiples of n, that is, $n\mathbb{Z} = \{nk \mid k \in \mathbb{Z}\}$. Write each of the following intersections as another set of the form $n\mathbb{Z}$, for suitable n, with proof of course:

(1) $4\mathbb{Z} \cap 5\mathbb{Z}$,

(2) $4\mathbb{Z} \cap 6\mathbb{Z}$.

Now let a and b be arbitrary integers. Based on the two examples above, and based on other concrete examples that you should make up yourself and study, for what n do you think $a\mathbb{Z} \cap b\mathbb{Z}$ will equal $n\mathbb{Z}$? Try to prove your conjecture.

Exercise 5.48. Let A, B and C be sets. Prove the following "distributive laws:" the first distributes an intersection across a union, and the second distributes a union across an intersection.

(1) $A \cap (B \cup C) = (A \cap B) \cup (A \cap C)$.

(2) $A \cup (B \cap C) = (A \cup B) \cap (A \cup C)$.

Exercise 5.49. Let A, B and C be sets. Here is another set of "distributive laws:" the first distributes a direct product across a union, the second distributes a direct product across an intersection, and the third distributes a direct product across a difference of sets. Prove the following;

(1) $A \times (B \cup C) = (A \times B) \cup (A \times C)$.

(2) $A \times (B \cap C) = (A \times B) \cap (A \times C)$.

(3) $A \times (B \setminus C) = (A \times B) \setminus (A \times C)$.

Exercise 5.50. Let A, B, and C be sets. Suppose that $A \cap C = B \cap C$ and $A \cup C = B \cup C$. Prove that $A = B$. (Hint, assume $A \neq B$. Then either $A \setminus B$ or $B \setminus A$ is nonempty. Suppose $A \setminus B$ is nonempty. Pick $x \in A \setminus B$. Apply the hypotheses to x to arrive at a contradiction.)

Exercise 5.51. Let I be an index set (see Example 5.2(7)) and let A_i, $i \in I$, be subsets of a fixed set U. Let A_i^c denote $U \setminus A_i$ as in Definition 5.13. Prove the following:

(1) $(\cup_{i \in I} A_i)^c = \cap_{i \in I} A_i^c$.

(2) $(\cap_{i \in I} A_i)^c = \cup_{i \in I} A_i^c$.

Exercise 5.52. Given two sets A and B, the *symmetric difference* of A and B is defined to the set $(A \setminus B) \cup (B \setminus A)$, and is often denoted $A \Delta B$.

(1) Write down explicitly the set $A \Delta B$ for the sets A and B of Example 5.14.

(2) For any two sets A and B, prove that $A \Delta B = (A \cup B) \setminus (A \cap B)$.

Exercise 5.53. Recall from Definition 5.11 that if S is a set, the power set $\mathcal{P}(S)$ of S is the set of all subsets of S. Assume that S is finite, of cardinality n. Show that $|\mathcal{P}(S)| = 2^n$. (Hint: See Example 4.21(1) in Chapter 4.)

Exercise 5.54. The following are functions from (subsets of) $\mathbb{R}$ to $\mathbb{R}$, where the domain is not explicitly described. In each of these: (1) Determine the largest subset of $\mathbb{R}$ on which the function can be defined, (2) Determine if the function is injective, (3) Determine the range of the function, (4) Determine the inverse function, if it exists.

(1)

$$f(x) = \frac{ax + b}{cx + d}, \qquad c \neq 0, ad - bc \neq 0.$$

(Hint: One can check for the injectivity directly, another way is to fold the test for injectivity into the test for the range by solving $f(x) = r$, where r is an arbitrary real number, and studying when this has a solution, and if it has a solution, how many solutions it has. Where is the condition $ad - bc \neq 0$ used?)

(2)
$$f(x) = \frac{e^x}{1+e^x}.$$
(Hint: The discussions in Example 1, Part 1 may be helpful.)

Exercise 5.55. We call a set $S \subseteq \mathbb{R}$ *open* if for each $b \in S$, there exists some $\epsilon > 0$ such that the open interval $(b-\epsilon, b+\epsilon)$ is contained in S. Further, the empty subset of $\mathbb{R}$ is *declared* to be open. We will explore a few properties of open sets in this exercise.

(1) Show that any open interval (see Example 5.40) is open, and in particular, that $\mathbb{R}$ is open.

(2) Show that any union of open intervals is open.

(3) Show that the intersection of a *finite* number of open sets is open.

(4) Give an example to show that the intersection of an *infinite* number of open sets need not be open. (Hint: Consider a suitable family of open sets $U_1 \supseteq U_2 \supseteq U_3 \supseteq \cdots$.)

Exercise 5.56. We call a set $S \subseteq \mathbb{R}$ *closed* if the complement $\mathbb{R} \setminus S$ is open. We will explore properties of closed sets analogous to those of open sets in Exercise 5.55 above. (You may assume the results of Exercise 5.55 but make sure you solve that exercise before you attempt this one!)

(1) Show that $\mathbb{R}$ and the empty set are closed sets.

(2) Show that any closed interval is closed.

(3) Show that a subset of $\mathbb{R}$ consisting of just one element is closed.

(4) Show that the intersection of closed sets is closed. (Hint: Combine Exercises 5.51 and 5.55 above!)

(5) Show that a union of a *finite* number of closed sets is closed.

(6) Give an example to show that the union of an *infinite* number of closed sets need not be closed.

Exercise 5.57. Let
$$M = \begin{pmatrix} \lambda & 1 \\ 0 & \lambda \end{pmatrix},$$
where λ is some fixed complex number. Determine explicitly the set of all matrices $N \in M_2(\mathbb{C})$ such that $MN = NM$. (You should be able to describe the elements of the desired set by a simple rule.)

Exercise 5.58. Let M be an element of $M_2(\mathbb{C})$. You have seen in Example 5.44 that M multiplies vectors $\overline{v}$ in $\mathbb{C}^2$, via the rule described in that example. Show that this multiplication is *linear,* that is,
$$M \cdot (\lambda_1 \cdot \overline{v_1} + \lambda_2 \cdot \overline{v_2}) = \lambda_1 \cdot (M \cdot \overline{v_1}) + \lambda_2 \cdot (M \cdot \overline{v_2})$$
for all vectors $\overline{v_1}$ and $\overline{v_2}$ in $\mathbb{C}^2$ and all scalars λ_1 and λ_2 in $\mathbb{C}$.

Exercise 5.59. We will turn the situation of Exercise 5.58 around. We will show here why the formula for a matrix acting on a vector that we considered at the end of Example 5.44 is a very natural one, if we *start with the assumption* that we have some function on $\mathbb{C}^2$ that is linear! Accordingly, suppose $T: \mathbb{C}^2 \to \mathbb{C}^2$ is some function that is known to be linear. What this means, exactly as described in Exercise 5.58, is that

$$T\left(\lambda_1 \cdot \overline{v_1} + \lambda_2 \cdot \overline{v_2}\right) = \lambda_1 \cdot T\left(\overline{v_1}\right) + \lambda_2 \cdot T\left(\overline{v_2}\right)$$

for all vectors $\overline{v_1}$ and $\overline{v_2}$ in $\mathbb{C}^2$ and all scalars λ_1 and λ_2 in $\mathbb{C}$. Now suppose you know how T acts on two special vectors

$$\overline{e_1} = \begin{pmatrix} 1 \\ 0 \end{pmatrix} \quad \text{and} \quad \overline{e_2} = \begin{pmatrix} 0 \\ 1 \end{pmatrix}$$

as follows:

$$T(\overline{e_1}) = \begin{pmatrix} a \\ c \end{pmatrix} \quad \text{and} \quad T(\overline{e_2}) = \begin{pmatrix} b \\ d \end{pmatrix}.$$

Show that the action of T on a general vector

$$\overline{v} = \begin{pmatrix} z_1 \\ z_2 \end{pmatrix}$$

is given precisely by the action of the 2×2 matrix

$$\begin{pmatrix} a & b \\ c & d \end{pmatrix}$$

on the vector $\overline{v}$ as defined at the end of Example 5.44. That is,

$$T(\overline{v}) = \begin{pmatrix} a & b \\ c & d \end{pmatrix} \cdot \overline{v} = \begin{pmatrix} az_1 + bz_2 \\ cz_1 + dz_2 \end{pmatrix}.$$

(Hint: Write the vector $\begin{pmatrix} z_1 \\ z_2 \end{pmatrix}$ as $z_1\overline{e_1} + z_2\overline{e_2}$ and invoke the linearity of T.)

Remark 5.59.1. Note that the columns of the 2×2 matrix $\begin{pmatrix} a & b \\ c & d \end{pmatrix}$ produced above are just the images under T of the two special vectors $\overline{e_1}$ and $\overline{e_1}$. Conversely, it is easy to see that for any matrix $\begin{pmatrix} p & q \\ r & s \end{pmatrix}$ the columns represent the "action" of this matrix on the special vectors $\overline{e_1}$ and $\overline{e_1}$.

Exercise 5.60. In this exercise we will discover why matrix multiplication is defined by such a "weird" formula! We will build on the result of Exercise 5.59 above.

Let $S: \mathbb{C}^2 \to \mathbb{C}^2$ and $T: \mathbb{C}^2 \to \mathbb{C}^2$ be two linear functions. We have seen in Exercise 5.59 above that to each of these we can associate a matrix whose columns are the images of the special vectors $\overline{e_1}$ and $\overline{e_2}$. If

$$M_S = \begin{pmatrix} a & b \\ c & d \end{pmatrix} \quad \text{and} \quad M_T = \begin{pmatrix} e & f \\ g & h \end{pmatrix}$$

are the matrices associated to S and T respectively, show that the matrix associated to the composite function $S \circ T$ is the following:

$$\begin{pmatrix} ae + bg & af + bh \\ ce + dg & cf + dh \end{pmatrix}.$$

Looking at the formula for multiplying M_S and M_T in Example 5.44, we see that the matrix for $S \circ T$ is precisely what we would get if we multiplied M_S and M_T by that formula in Example 5.44. It is for this reason that mathematicians have chosen that formula for matrix multiplication: it corresponds exactly to the composition of linear functions on $\mathbb{C}^2$!

(Hint: You need to consider $S \circ T(\overline{e_1}) = S(T(\overline{e_1}))$ and $S \circ T(\overline{e_2}) = S(T(\overline{e_2}))$. Use the results of Exercise 5.59 to interpret these quantities in terms of the matrices M_S and M_T.)

Exercise 5.61. Let M be the 2×2 matrix of Exericise 5.57 above. Determine all vectors

$$\overline{v} = \begin{pmatrix} z_1 \\ z_2 \end{pmatrix}$$

and suitable complex numbers r (that depend on $\overline{v}$) such that $M \cdot \overline{v} = r\overline{v}$.

Exercise 5.62. Given the matrix $M = \begin{pmatrix} a & b \\ c & d \end{pmatrix}$ in $M_2(\mathbb{C})$, we define the *trace* of M, denoted $\text{Tr}(M)$, to be the real number $a + d$, and the *determinant* of M, denoted $\det(M)$, to be the real number $ad - bc$. Given matrices M and N in $M_2(\mathbb{C})$, show that:

(1) $\text{Tr}(M + N) = \text{Tr}(M) + \text{Tr}(N)$.

(2) $\det(MN) = \det(M) \cdot \det(N)$.

(3) $M^2 - \text{Tr}(M) \cdot M + \det(M) \cdot \mathbf{I}_2 = \mathbf{0}_2$, where $\mathbf{I}_2 = \begin{pmatrix} 1 & 0 \\ 0 & 1 \end{pmatrix}$ and $\mathbf{0}_2 = \begin{pmatrix} 0 & 0 \\ 0 & 0 \end{pmatrix}$.

(Here, given the real number r, the product $r \cdot \begin{pmatrix} a & b \\ c & d \end{pmatrix}$ is defined to be the matrix $\begin{pmatrix} ra & rb \\ rc & rd \end{pmatrix}$.)

Exercise 5.63. Let $M = \begin{pmatrix} a & b \\ c & d \end{pmatrix}$ be a 2x2 matrix with entries in $\mathbb{C}$, and suppose, as in Example 5.45, we wish to solve the equation

$$\begin{pmatrix} a & b \\ c & d \end{pmatrix} \cdot \begin{pmatrix} x \\ y \end{pmatrix} = \begin{pmatrix} p \\ q \end{pmatrix} \tag{5.2}$$

for x and y. We will study in this exercise one method of solving this equation using matrix operations, under a certain assumption about M:

(1) Assume that $\det(M) \neq 0$. Show that the matrix

$$N = \begin{pmatrix} d/\det(M) & -b/\det(M) \\ -c/\det(M) & a/\det(M) \end{pmatrix}$$

satisfies $MN = NM = I_2 \stackrel{\text{def}}{=} \begin{pmatrix} 1 & 0 \\ 0 & 1 \end{pmatrix}$.

(2) Use part (1) to solve for x and y by multiplying both sides of Equation 5.2 on the left by N. (You may assume that if v is a 2-dimensional vector, and A and B are two 2×2 matrices, then $A(B \cdot v) = (AB) \cdot v$. Of course, after you have solved this exercise, it would be instructive to prove this fact too!)

6

Interlude: So, How to Prove It? An Essay

By now you have read through several chapters of the book and have absorbed many ideas and arguments contained there. But perhaps you are still stumped every time you have to work on a problem on your own. "How can I prove it?" you ask in frustration, about something you are asked to establish in a problem. This chapter is an essay on how to do mathematics, and in particular, how you may go about finding a proof.

As is stressed repeatedly in this book, mathematics, first and foremost, is about ideas. Proofs are simply vehicles for expressing those ideas in a water-tight fashion. Thus, to write a proof, you must first understand the key idea undergirding the problem you are working on. Once you understand the idea, the proof simply codifies the idea and makes it explicit in a sequence of logical steps, taking care that all i's are dotted and all t's crossed.

So, when students say that they find it difficult to prove something in a problem, what they are really saying is that they do not understand the idea behind the problem! Thus, mathematics is not about proof *techniques,* and it is not about *procedure,* but about ideas. (All the same, we will review some proof techniques later in this chapter.)

Now, how does one generate ideas that will help solve a problem? This comes from practice, by first inculcating what may be called the mathematical attitude. One must study mathematics *actively*, working through every statement in detail, probing for patterns, and looking for generalizations. One must constantly ask questions like: "Where is this hypothesis used?," "Why are these constraints placed on these parameters?," "How will the result be modified if we tweak the hypotheses just a bit?," "How does this fit in with these previous theorems that I just studied?," and very importantly, "What are some examples that illustrate this phenomenon?" (In fact, this last aspect of the attitude cannot be overemphasized: to understand mathematics is to understand examples on which theorems are based!)

As part of this mathematical attitude, one must also keep a lot of mathematics in the active parts of one's mind: ideas only come by building on previous ideas, so these previous ideas must be readily accessible! Thus, as part of this attitude, one must think about mathematics during a significant amount of one's time, churning ideas in one's head even when engaged in other more routine activities!

As you develop such an attitude, you will find yourself naturally becoming more and more creative, and ideas will flow! With greater frequency you will be able to see the specific ideas undergirding the problems you are working on, and will be able to translate those ideas into a logically water-tight proof.

There is by now a classic essay on how to solve problems (much of which applies more generally to the question of how to do mathematics) written by George Polya. It is called, appropriately, *How To Solve It.* Summaries of this essay are widely available on the internet. It would be a good idea for you to read the original essay, or at least to read some of the widely available summaries.

Polya suggests a four-step approach to problem solving: Understand the problem, Devise a plan, Carry out the plan, and Look back at the solution. These steps are not procedures guaranteed to solve your problem, but guideposts for how you might generate ideas that will help you solve problems!

What is particularly relevant to us here is his systematic listing of various heuristics[1] aimed at generating ideas (they form part of his "Devise a plan" step). We cannot do better than simply quote from his essay, which we do below. Note that Polya mainly addresses the solving of problems like "Calculate the diagonal given these sides." Some of his language needs to be interpreted slightly differently for the sorts of problems we solve in this book, problems like "Prove that Statement B is true given that Statement A is true" — a fact that Polya himself acknowledges. For instance, both "data" and "condition" below may roughly be interpreted as "hypotheses" in our context.

> (Polya:) If you cannot solve the proposed problem try to solve first some related problem. Could you imagine a more accessible related problem? A more general problem? A more special problem? An analogous problem? Could you solve a part of the problem? Keep only a part of the condition, drop the other part; how far is the unknown then determined, how can it vary? Could you derive something useful from the data? Could you think of other data appropriate to determine the unknown? Could you change the unknown or the data, or both if necessary, so that the new unknown and the new data are nearer to each other? Did you use all the data? Did you use the whole condition? Have you taken into account all essential notions involved in the problem?

The paragraph above thus suggests the heuristics below to help us solve our problems of the "Statement $A \implies$ Statement B" kind. Notice that these heuristics do not themselves solve the original problem. However, very often, they will allow you to see enough of what is going on to help you then solve the original problem. In fact, very often, they will allow you to see *more* than what is needed to help you solve the original problem: very often you will come away with a deeper understanding of the situation than if you had solved the problem directly!

[1] Heuristics: exploratory techniques, designed to probe into a problem, often by by focusing on parts of the problem or disregarding parts of the data, and using trial and error methods or guesses or reasoning by analogy to generate insights and even partial solutions. Such insights and partial solutions can potentially lead to deep understanding.

Polya reinterpreted:

- If you cannot solve the proposed problem, can you find a simpler problem to solve? For instance, can you work in a special case, perhaps even within a concrete example rather than in any general situation, and see if you can prove it in that case?
- Can you solve a seemingly more difficult problem, one with fewer hypotheses perhaps? (Sometimes, the details of the hypotheses in Statement A may cloud what is really happening, and it may be helpful to work in even more general situations, in other words, with fewer restrictions than what is in Statement A. This is analogous to the situation where it is sometimes easier to see what is on the ground by being up in the air! The three-dimensional perspective sheds light on the two-dimensional situation.)
- Can you find an analogous problem: the same sort of question in perhaps a different context, and solve that?
- What other conclusions, possibly as intermediate steps, can you arrive at from Statement A?
- Can you find extra hypotheses to those in Statement A that you know will make Statement B true?
- Can you modify Statement A slightly to Statement A' and Statement B to Statement B', and prove that $A' \implies B'$?
- Are you certain you are taking all the hypotheses into account?
- Are you certain you are bringing to bear everything you know about the context of the problem: theorems you have previously learned, concrete examples you have previously studied?

Much as Polya systematized heuristic approaches to problem solving, one may catalog various types of proofs. In what follows, we attempt such a catalog. We assume that we are given two statements A and B, and we are asked to prove that $A \implies B$.

- *Direct Proof.* This is as the name suggests: You start with the hypotheses in A, and combine those with results from previously established theorems, and by a series of logical arguments, show that B must be true. For instance, both proofs in Example 1.2 in Chapter 1, where we showed that the sum of the squares of two odd integers is even, are direct proofs.
- *Proof of Contrapositive.* Here, you prove instead that $\neg B \implies \neg A$. For instance, if you were asked to show that (Statement A) the equation $ax^2+bx+c=0$ has equal roots implies that (Statement B) $b^2-4ac=0$, you show instead that if $b^2-4ac \neq 0$, then $ax^2+bx+c=0$ cannot have equal roots. We have already considered a proof of contrapositive in Proposition 5.28 in Chapter 5.
- *Proof by contradiction.* This is a very widely used technique: you assume that B is not true, and you deduce, using a series of logical arguments, that some particular statement (call it Statement C) ought to be true. But independently, you know that C is false! So you conclude from this that B must be true, since the alternative leads to a logical fallacy!

We have already considered proofs by contradiction in Proposition 5.32 in Chapter 5, in the portion where we showed that if $f : A \to B$ is invertible, then it is injective.

A classic example of proof by contradiction is the proof that $\sqrt{2}$ is irrational, which we will present in Example 10.28 in Chapter 10.

- *Proof by Cases.* (Or Proof by Exhaustion.) Here, you partition the universe of possibilities into subsets (not necessarily disjoint), called *cases*, and you show that the theorem is true for each of these cases. For instance, if you are asked to prove that $|x| \geq x$ for real numbers x, you may divide the proof into two cases: $x \geq 0$ and $x < 0$. Together these cover all possibilities for x. You then prove the assertion for each case separately. In this example, you divide your proof into these two cases because $|x|$ has different definitions for $x \geq 0$ and for $x < 0$, but in other situations you may wish to prove by cases simply because different cases yield to different arguments.

  A noteworthy family of proof by cases is where the number of cases is infinite. For instance suppose you are asked to prove that a function $f : \mathbb{C} \to \mathbb{C}$ is just a constant, that is, that $f(z) = c$ for all $z \in C$, where c is some fixed complex number. Perhaps this function is initially defined in some convoluted manner, and it is not at all obvious from this definition that the function is just a constant. Now suppose that you are able to prove after some detailed analysis that for any subset $S_r = \{z \in \mathbb{C} \mid |z| < r\}$, the function $f|_{S_r} : S_r \to \mathbb{C}$, which is f restricted to just the elements in S_r, has this property. Then necessarily $f(z) = c$ for all $z \in \mathbb{C}$, because any one z lies, for instance, in $S_{|z|+1}$, and we know that the statement is true for f restricted to $S_{|z|+1}$. (Effectively, here, we are considering the cases $\{z \mid |z| < r\}$ for larger and larger values of r, and these cases of course exhaust all possibilities for z.)

- *Inductive Proof.* You will learn about these in Chapter 7 ahead. Induction is a process which, in its simplest form, works as follows: you are given a family of statements $S(n)$, one for each $n \in \mathbb{N}$, and you are asked to show that $S(n)$ is true for all $n \in \mathbb{N}$. You show first by explicit methods that $S(1)$ is true. This is called the base step. Next, you show that the following auxiliary statement is true: "If $S(k)$ is true for some $k \in \mathbb{N}$, then necessarily $S(k + 1)$ is also true." This is known as the induction step. Mind you, you are not yet proving in the induction step that $S(k+1)$ is true: you are only showing that it is true *under the additional hypothesis* that $S(k)$ is true! The principle of induction then allows you to conclude from the combination of the base step and the induction step that $S(n)$ is true for all $n \in \mathbb{N}$. Chapter 7 gives several examples of inductive proofs and also describes some variants of the steps outlined above.

- *Constructive Proof.* This sort of proof is used to show the existence of an object with specific properties by explicitly showing how to construct such an object. A simple instance is in Euclidean geometry: If you are asked to prove that every angle can be bisected, you do that by proposing a method for constructing the half-angle, and showing in a series of logical steps that the proposed method indeed produces the correct half-angle. Here is another instance, from later in this book: In Chapter 13, you will learn about groups and cyclic groups; for now, just think of them as some objects. If you are asked to prove that a certain group is cyclic, you do this by explicitly producing a generator g in the group, that is, an element g such that all elements of the group are just powers of g (this is the definition of a cyclic group: a set in which all elements are "powers" — suitably interpreted — of a single element).

A constructive proof is in contrast to:

- *Nonconstructive Proof.* In this sort of proof, you show the existence of an object with specific properties not by an explicit construction, but by an indirect argument. For instance, you may show that if such an object does not exist, then some known theorem is violated! Or perhaps the hypotheses are violated. Here is an example: Suppose somebody gives you some infinite set S of real numbers that all lie in the closed interval $[0, 1]$. They then subdivide this interval into n disjoint subintervals $[0, 1/n)$, $[1/n, 2/n), \ldots, [(n-1)/n, 1]$ (for some integer $n \geq 2$). They now ask you to show that there exists at least one subinterval from these n that contains infinitely many numbers from S. Instead of trying to produce an actual subinterval that contains infinitely many numbers from S, you argue as follows: if none of these n subintervals contained infinitely many numbers from S, then each must contain only finitely many numbers. Say there are m_1 numbers from S in $[0, 1/n)$, m_2 numbers in $[1/n, 2/n)$, ..., and m_n numbers in $[(n-1)/n, 1]$. But this means that in the union $[0, 1] = [0, 1/n) \cup [1/n, 2/n) \cup \cdots \cup [(n-1)/n, 1]$, there are only $m_1 + m_2 + \cdots + m_n$ numbers from S. This is a finite number. But all of S is contained in $[0, 1]$, and S is infinite! This contradiction proves the existence of at least one subinterval that contains infinitely many numbers from S.

  (Note that this proof is also a proof by contradiction.)

Naturally, given that the ideas that occur in mathematics are so rich, this cannot be a complete classification. Also, as seen above in the example of a nonconstructive proof, the various elements in this list are not mutually exclusive!

Being aware of such a catalog may help in generating some ideas: you start tentatively along one path, say the way of direct proof, searching for ideas, and then you find yourself stuck, so you say to yourself, "maybe I should see instead what can happen if the conclusion were false" and hope that you can find a contradiction to your hypotheses, and maybe that leads to a proof by contradiction. Or maybe that does not work either, so you say to yourself "maybe I should see what happens in special cases," and perhaps that leads to a proof by cases. Thus, recalling and attempting various types of proofs is itself a part of heuristics!

7

Induction

In this chapter we will consider an extremely useful principle, the *Principle of Induction*, that comes in very handy when proving statements that are indexed by the positive integers. First, remember from Remark 3.4 in Chapter 3 that to "prove a statement" is to show that the given statement, which up front could either be true or false, is actually true. Next, recall from Remark 3.20 in Chapter 3 what "statements that are indexed by the positive integers" means: it is a family of statements, one for each positive integer. For instance, the statement "For all positive integers n, we have $n^2+n-1>0$" is a family of statements, which we may label $S(n)$, $n=1,2,\ldots$. Here, $S(1)$ is the statement "$1^2+1-1>0$," $S(2)$ is the statement "$2^2+2-1>0$," and so on. Now suppose that while trying to prove some such family of statements $S(n)$, we are able to accomplish two things: *Step* 1, *known as the "base step",* where we are able to prove that $S(1)$ is a true statement, and *Step* 2, *known as the "induction step",* where we are also able to prove the following, which at first may appear somewhat weird: "if $S(k)$ happens to be true for some positive integer k, then $S(k+1)$ must also be true." What can we do with these two things that we have proved? Well, start with $S(1)$, which we have proved in the base step (Step 1) to be true. Now apply the induction step (Step 2) to this fact: Taking $k=1$ in what we have proved in the induction step, we find that since $S(1)$ has been proven to be true, "$S(k+1)$," i.e., $S(2)$, must also be true! Now apply what we have proved in the induction step to this new fact: taking $k=2$ in the induction step, we find that since $S(2)$ must be true, $S(3)$ must also be true! Now once again apply what we have proved in the induction step to this latest new fact, where we find that since $S(3)$ must be true, $S(4)$ must be true. And so on! Repeating as often as necessary, we find that for any positive integer n, $S(n)$ must be a true statement.

7.1 Principle of Induction

This technique of proof can be visualized by a set of tiles (or bricks, or dominoes), each representing $S(n)$ for $n=1,2,\ldots$ (see Figure 7.1). At first, these tiles are standing upright in a row, minding their own business. But then, someone comes along and proves

Figure 7.1. Illustration of proof by induction: The nth tile falls because the $(n - 1)$th tile falls on it and pushes it forward, and it then pushes the $(n + 1)$th tile forward, causing that tile also to fall, and so on (this is due to the induction step of the proof). This process needs the first tile to be pushed forward to get started (this is the base step of the proof). From `http://www.texample.net/tikz/examples/dominoes/`, Author: Mark Wibrow, Creative Commons Licence CC By 2.5.

that $S(1)$ is a true statement (the base step of the proof). The first tile, the one representing $S(1)$, falls forward—here, falling forward symbolizes being a true statement. But since the first tile falls on the second tile, the second tile in turn falls forward—this symbolizes the fact that since $S(1)$ is true, $S(2)$ must also be true by the induction step above. In turn, the second tile falls on the third tile, so the third tile falls forward—this symbolizes the fact that since $S(2)$ is true, $S(3)$ must also be true by the induction step above, and so on.

A different way to think of this technique of proof is to imagine the statements $S(n)$, $n = 1, 2, \ldots$, as rungs of an infinite ladder, with the nth statement representing the nth rung. The induction step consists of showing that if you are already on the nth rung, then you have the capacity to climb to the $(n+1)$st rung. The base step consists of showing that you have the capacity to get on the first rung. Thus, the proof that shows that you can get on to the first rung (this is due to the base step of the proof), and then because you are already on the first rung you can climb to the second, and then because you are already on the second rung you can climb to the third, and so on (these are due to the induction step of the proof).

We will illustrate these ideas with several examples shortly, but first, we should be aware that while the approach described above to show that $S(n)$ is true for all $n \in \mathbb{N}$ appears reasonable, it nevertheless needs to be stated as a formal assumption (or axiom) in axiomatic set theory. Of course, as we discussed in Chapter 5, we are adopting an intuitive approach to set theory and need not be concerned with the finer points of set theory at this early juncture, but all the same, let us state this first as a principle (or axiom) we will adopt:

Definition 7.1. *Principle (or Axiom) of Induction:* Let $S(n)$, $n \in \mathbb{N}$ be a family of statements. Assume that

(1) $S(1)$ is true, and

(2) If $S(k)$ is true for some $k \in \mathbb{N}$, then $S(k+1)$ is also true.

Then $S(n)$ is true for all $n \in \mathbb{N}$.

Let us begin with the following example:

Example 7.2. Show that

$$1+2+3+\cdots+n = \frac{n(n+1)}{2} \tag{7.1}$$

for all $n \in \mathbb{N}$.

The expression in (7.1) above represents a family of statements $S(n)$. For any one value of n, we plug that value for n in $S(n)$ to determine what the statement reads for that value. For instance, when $n=1$, we obtain $S(1)$ by plugging $n=1$ to obtain the following:

$$1+2+3+\cdots+\underbrace{1}_{n=1} = \underbrace{\frac{1(1+1)}{2}}_{n=1}.$$

We interpret the left side as follows: we start at 1 and keep adding to it one by one the next integer, and stop after we have added the last integer on the left. In this case, the last integer is 1. This means that there is no addition to be done, we stop at the first step itself! Thus, for $n=1$ we find $S(1)$ is the statement

$$1 = \frac{1 \cdot 2}{2}.$$

Similarly, $S(2)$ is obtained by plugging $n=2$ in $S(n)$, to obtain the following:

$$1+2+3+\cdots+\underbrace{2}_{n=2} = \underbrace{\frac{2(2+1)}{2}}_{n=2}.$$

As discussed above in the $n=1$ situation on how to interpret the left side of this expression, the left side just becomes $1+2$, so $S(2)$ is the statement

$$1+2 = \frac{2 \cdot 3}{2}.$$

And so on, for each value of n. Remember that at this point, these are just statements, with the potential of being either true or false, and we need to show that these are actually true.

Although this family of statements can be proved by other means too, let us use the principle of induction to prove them.

Base Step: We need to verify that $S(1)$ is true. This is patently clear: the left side of $S(1)$ is 1, while the right side is $\frac{1 \cdot 2}{2}$, which is just 1. Since the left and right sides are equal, $S(1)$ is true.

Induction Step: Let us assume that $S(k)$ is true for some $k \in \mathbb{N}$, and under this assumption, we need to show that $S(k+1)$ is true. Observe that $S(k)$ is the statement

$$1 + 2 + 3 + \cdots + \underbrace{k}_{n=k} = \underbrace{\frac{k(k+1)}{2}}_{n=k}$$

and $S(k+1)$ is the statement

$$1 + 2 + 3 + \cdots + k + \underbrace{k+1}_{n=k+1} = \underbrace{\frac{(k+1)[(k+1)+1]}{2}}_{n=k+1}.$$

Now the left side of $S(k+1)$ can be grouped as follows:

$$\text{Left side of } S(k+1) = \underbrace{1 + 2 + 3 + \cdots + k}_{\text{Left side of } S(k)} + (k+1)$$

Now, since $S(k)$ is assumed true, we may replace the left side of $S(k)$ with the right side of $S(k)$, so the stuff in the underbrace can be written as $\frac{k(k+1)}{2}$. We thus find

$$\begin{aligned}\text{Left side of } S(k+1) &= \frac{k(k+1)}{2} + (k+1) = (k+1)\left(\frac{k}{2} + 1\right)\\ &= (k+1)\frac{(k+2)}{2} = \text{Right side of } S(k+1)\end{aligned}$$

Thus, under the assumption that $S(k)$ is true, we have shown, using simple algebra, that the left side of $S(k+1)$ equals the right side of $S(k+1)$, that is, that $S(k+1)$ is a true statement. Thus, we have accomplished the induction step.

By the principle of induction (Definition 7.1 above), $S(n)$ is a true statement for all $n \in \mathbb{N}$.

Here is another family of statements: the sum of a (finite) geometric series:

Example 7.3. An expression such as $G_{a,r}(n) = a + ar + ar^2 + \cdots + ar^{n-1}$, for fixed real numbers a and r, and for a fixed integer $n \geq 1$ is known as a *(finite) geometric series.* (The notation is to suggest that a and r are parameters, and for fixed a and r, the sum depends on the number of summands n.) When $r = 1$, the value of $G_{a,1}(n)$ is clear: it is just an. So assume that $r \neq 1$. We will show by induction on n that

$$G_{a,r}(n) = \frac{a(1-r^n)}{1-r}, \quad r \neq 1. \tag{7.2}$$

for all $n \in \mathbb{N}$. (Note that division by $1 - r$ is allowed as $r \neq 1$.)

As before, write $S(n)$ for the statement in (7.2).

Base Step: We'll verify $S(1)$ is true. The left side of $S(1)$ is the expresstion $G_{a,r}(1)$, which is a. The right side is $\frac{a(1-r)}{1-r}$, which is also a. Thus, $S(1)$ is a true statement.

Induction Step: Let us assume that $S(k)$ is true for some $k \in \mathbb{N}$. Under this assumption, we need to show that $S(k+1)$ is true. The left side of $S(k+1)$ is the expression $G_{a,r}(k+1)$, which can be grouped as follows:

$$\text{Left side of } S(k+1) = G_{a,r}(k+1) = \underbrace{a + ar + \cdots + ar^{k-1}}_{\text{Left side of } S(k)} + ar^k.$$

Now, since $S(k)$ is assumed true, we may replace the left side of $S(k)$ with the right side, so the stuff in the underbrace can be written as $\frac{a(1-r^k)}{1-r}$. We thus find

$$\text{Left side of } S(k+1) = \underbrace{\frac{a(1-r^k)}{1-r} + ar^k = \frac{a - ar^k + ar^k - rar^k}{1-r}}_{\text{open parentheses, add fractions}}$$

$$= \frac{a(1-r^{k+1})}{1-r} = \text{Right side of } S(k+1).$$

Thus, under the assumption that $S(k)$ is true, we have shown, using simple algebra, that the left side of $S(k+1)$ equals the right side of $S(k+1)$, that is, that $S(k+1)$ is a true statement. Thus, we have accomplished the induction step.

By the principle of induction (Definition 7.1 above), $S(n)$ is a true statement for all $n \in \mathbb{N}$.

We will consider infinite geometric series in Section 11.6 in Chapter 11.

Let us now consider a mild modification of the induction process we have considered so far. Our basic principle (Definition 7.1) stated that if $S(n)$ is a family of statements for which (1) *Base case:* $S(1)$ is true, and (2) *Induction Step:* if for some $k \in \mathbb{N}$ $S(k)$ is true then $S(k+1)$ is also true, then $S(n)$ must be true for all $n \in \mathbb{N}$. Now very often in mathematics, we run across statements $S(n)$ that are not necessarily true for all $n \geq 1$, but only for all $n \geq n_0$ (for some n_0). Consider, for example, the following: $n! > 2^n$. This is not true when you put in $n = 1$, not true for $n = 2$, and not true for $n = 3$. However, as we shall see soon, it is true for any $n \geq 4$. To prove this result using induction, it turns out that all we have to do is establish the truth of the case $n = 4$ numerically, and then show that if $S(k)$ is true for some $k \geq 4$ (note, *four,* not one!), then $S(k+1)$ is true. It will then follow that the statements are true for all $n \geq 4$. This is the content of the following:

Proposition 7.4. *Let n_0 be a fixed integer, and let $S(n)$, $n \geq n_0$, be a family of statements. If $S(n_0)$ is true, and if $S(k)$ is true for some $k \geq n_0$ implies $S(k+1)$ is true, then $S(n)$ is true for all $n \geq n_0$.*

We will prove this shortly, but it is instructive to see how we use this proposition while proving the statement we just discussed:

Example 7.5. For integers $n \geq 4$, $n! > 2^n$.

Let $S(n)$ denote the statement $n! > 2^n$. We will use Proposition 7.4. Thus, we will prove that (*base case*) $S(4)$ is true, and then (*induction step*) that if $S(k)$ is true for some $k \geq 4$, $S(k+1)$ must be true. That $S(4)$ is true is clear: the left side of $S(4)$ is $4! = 24$, and the right side is $2^4 = 16$, and of course $24 > 16$. Now assume that $S(k)$ is true for some $k \geq 4$. We wish to show that $S(k+1)$ must then be true. Now

$$\begin{aligned}
\text{Left side of } S(k+1) &= (k+1)! \\
&= \underbrace{k!}_{\text{Left side of } S(k)} \cdot (k+1) \\
&> \underbrace{(2^k)}_{\text{Right side of } S(k)} \cdot (k+1) \\
&> \underbrace{2^k \cdot 2}_{k+1>2 \text{ since } k\geq 4} = 2^{k+1} \\
&= \text{Right side of } S(k+1).
\end{aligned}$$

Putting these inequalities and equalities together, we find that the left side of $S(k+1)$ is greater than the right side, i.e., $S(k+1)$ is true! By Proposition 7.4, $S(n)$ is true for all integers $n \geq 4$.

Remark 7.6. You should be able to work through the proof of the induction step in Example 7.5 above and convince yourself that the proof would have gone through even if only assumed that $k \geq 1$: we did not need the assumption $k \geq 4$. However, $S(1)$ *is not true!* Thus, we could not have proved that $S(n)$ is true for all $n \geq 1$ just on the strength of the fact that the induction step would have worked! For the entirety of the induction argument to work, both the base step and the induction step need to work!

Continuing with this remark, there was at least one redeeming feature in the example above: as long as we started at the appropriate base step of $n = 4$, the result turned out to be true for all $n \geq 4$. Here is a more egregious example: a statement $S(n)$ that is false for all values of n, but for which the induction step works! Consider the following assertion:

$$1 + 3 + 5 + \cdots + 2n + 3 = n^2 + 4n.$$

Let $S(n)$ denote this statement above. (The sum on the left is meaningless if $2n+3 < 1$, or in other words if $n < -1$, so implicitly, we are assuming that $n \geq -1$.) It easy to see that whenever $S(k)$ is true for some integer k, $S(k+1)$ must also be true, exactly as in the examples above:

$$\begin{aligned}
\text{Left side of } S(k+1) &= 1 + 3 + 5 + \cdots + 2(k+1) + 3 \\
&= \underbrace{1 + 3 + 5 + \cdots + 2k + 3}_{\text{Left side of } S(k)} + \underbrace{2(k+1) + 3}_{=2k+5} \\
&= \underbrace{(k^2 + 4k)}_{\text{Right side of } S(k)} + (2k+5) \\
&= k^2 + 2k + 1 + 4k + 4 = (k+1)^2 + 4(k+1) \\
&= \text{Right side of } S(k+1).
\end{aligned}$$

Yet, for no integer $n \geq -1$ is $S(n)$ true! In fact, writing $2n + 3$ as $2(n + 2) - 1$, the result Exercise 7.10.1 of 7.10 shows that for all $n \geq 1$, the left side of $S(n)$ above should equal $(n+2)^2 = n^2 + 4n + 4$. So the right side of $S(n)$ is off from the correct answer by 4 for $n \geq 1$! (For $n = -1$ and $n = 0$ too, the result is false, as can be easily checked.)

The moral is that both the base step and the induction step are vital ingredients in any proof by induction!

Proof of Proposition 7.4. Let us define the statements $T(m)$, $m \geq 1$, by the rule $T(m) = S(m + n_0 - 1)$. Thus, $T(1) = S(1 + n_0 - 1) = S(n_0)$, $T(2) = S(2 + n_0 - 1) = S(n_0 + 1)$, $T(3) = S(3 + n_0 - 1) = S(n_0 + 2)$, and so on. Essentially, the statements S and T are "the same" except that the index for T is shifted $n_0 - 1$ steps to the left of the index of S. The advantage is that the index for the statements $T(m)$ start at $m = 1$, so we can apply the principle of induction, as stated in Definition 7.1 to the statements T. So, the fact that $S(n_0)$ is true translates to $T(1)$ being true, since $S(n_0)$ and $T(1)$ are the same statement. The induction step "for some $k \geq n_0$, $S(k)$ is true, implies $S(k + 1)$ is true" translates to the corresponding induction step "for some $l \geq 1$, $T(l)$ is true, implies that $T(l + 1)$ is true." By the principle of induction applied to the family of statements T, $T(m)$ is true for all $m \geq 1$. It follows that $S(n)$ is true for all $n \geq n_0$, once again because the statements $S(n)$ for $n \geq n_0$ and $T(m)$ for $m \geq 1$ are the same except for the shifted indexing. □

7.2 Another Form of the Induction Principle

Often, the induction principle is stated in a slightly different form:

Definition 7.7. *Principle of Induction, "Strong" Form:* Let $S(n)$, $n \in \mathbb{N}$, be a family of statements. Assume that

(1) $S(1)$ is true, and

(2) If $S(1), S(2), \ldots, S(k)$ are all true for some $k \in \mathbb{N}$, then $S(k + 1)$ is also true.

Then $S(n)$ is true for all $n \in \mathbb{N}$.

Remark 7.8. It can be proved that the two forms of the induction principle that we have stated here, in Definitions 7.1 and 7.7, are logically equivalent, that is, one form of the principle holds if and only if the other form of the principle holds. A proof of this equivalence is detailed in the Notes section at the end of this chapter.

Note though that, as we have said several times before, we are adopting an intuitive approach to set theory, and since this principle in Definition 7.7 seems just as obvious as the earlier principle in Definition 7.1, we would be well advised at this juncture to just accept these as intuitively clear, not worry too much about the logical relationship between the two, and focus on absorbing deeper mathematical ideas.

The word "strong" in the form of the induction principle given above in Definition 7.7 refers to the fact that the principle seems to accept a much stronger set of conditions as input in the induction stage (that is, all of $S(1)$ through $S(k)$ need to be true, instead of just $S(k)$). It should be stressed that the word "strong" does not indicate that this principle is somehow stronger than the original form that was presented in Definition 7.1; indeed, the two are equivalent!

We will consider an example where the principle of induction is invoked in the form given in Definition 7.7 above.

Example 7.9. The Fibonacci numbers are defined by the following rule: $F_1 = 1$, $F_2 = 1$, and for $n \geq 3$, F_n is defined to be $F_{n-2} + F_{n-1}$. Prove that $F_n < \left(\frac{7}{4}\right)^n$ for all $n \in \mathbb{N}$.

(Such a definition of objects indexed by $\mathbb{N}$, or $\mathbb{Z}_{\geq 0}$, where the n-th object is defined in terms of the previously defined $(n-1)$-th object, the $(n-2)$-th object, and so on, is known as a *recursive definition.*)

We will prove this using the form of the induction principle given in Definition 7.7 (Strong Form). As always, let $S(n)$, for $n \in \mathbb{N}$, denote the statement $F_n < \left(\frac{7}{4}\right)^n$.

Base case: Looking ahead at the induction step, we find that we need to establish the truth of both $S(1)$ and $S(2)$ to start our induction process—this arises from the way F_n is defined, as the sum of the previous *two* terms. Now $F_1 = 1$, which is clearly less than $\left(\frac{7}{4}\right)^1$, and F_2 is also 1, which is also less than $\left(\frac{7}{4}\right)^2$.

Induction Step: We have already shown that $S(1)$ *and* $S(2)$ are true. Therefore, we only need to show the truth of $S(n)$ from $n = 3$ onwards. So assume that for some $k \geq 2$, F_j has been proven to be less than $\left(\frac{7}{4}\right)^j$, for $1 \leq j \leq k$. We wish to show that F_{k+1} is less than $\left(\frac{7}{4}\right)^{k+1}$. Now,

$$
\begin{aligned}
F_{k+1} = \underbrace{F_{k-1} + F_k}_{\text{Definition of } F_{k+1}} &< \underbrace{\left(\frac{7}{4}\right)^{k-1} + \left(\frac{7}{4}\right)^{k}}_{\text{induction hypothesis, note } k-1 \geq 1} \\
&= \left(\frac{7}{4}\right)^{k-1} \cdot \left(1 + \frac{7}{4}\right) = \left(\frac{7}{4}\right)^{k-1} \cdot \frac{11}{4} = \left(\frac{7}{4}\right)^{k-1} \cdot \frac{44}{16} \\
&< \left(\frac{7}{4}\right)^{k-1} \cdot \underbrace{\frac{49}{16}}_{\left(\frac{7}{4}\right)^2} = \left(\frac{7}{4}\right)^{k+1}.
\end{aligned}
$$

Tracing the inequalities and equalities above, we find $F_{k+1} < \left(\frac{7}{4}\right)^{k+1}$, as desired. Hence, $S(n)$ is true for all $n \in \mathbb{N}$.

7.3 Further Exercises

In the following exercises, you are required to use induction to prove the desired result. Of course, many of these can be proved by several other techniques, and you are encouraged to find other proofs as well!

Exercise 7.10. Practice Exercises:

Exercise 7.10.1. Show that $1 + 3 + 5 + \cdots + 2n - 1 = n^2$, for all $n \geq 1$.

Exercise 7.10.2. Show that $1 + 4 + 7 + \cdots + (3n - 2) = \dfrac{3n^2 - n}{2}$, for all $n \geq 1$.

Exercise 7.10.3. Show that $1 + 5 + 9 + \cdots + (4n - 3) = 2n^2 - n$, for all $n \geq 1$.

Exercise 7.10.4. Show that $1 \cdot 1! + 2 \cdot 2! + \cdots + n \cdot n! = (n + 1)! - 1$, for all $n \geq 1$.

Exercise 7.11. Let x be a real number, $x \geq 0$. Show that $(1 + x)^n \geq 1 + nx$ for all integers $n \geq 0$. Show that equality holds for all n if $x = 0$, and for $x > 0$, equality holds if and only if $n = 0$ or $n = 1$.

Exercise 7.12. Show the following:

(1) Show that $3^n > n^2$, for all $n \geq 1$. (Hint: Show that the statement is true for both $n = 1$ and $n = 2$ first, so at the induction step you only need to assume $n \geq 2$.)

(2) $n^n > n!$, for all $n \geq 2$.

Exercise 7.13. Show the following:

(1) $n^3 + 5n$ is divisible by 3 for all $n \geq 0$.

(2) $5^{2n} - 1$ is divisible by 3 for all $n \geq 0$.

Exercise 7.14. Show that $n^5 - n$ is a multiple of 5 for all integers $n \geq 0$. (By contrast, is $n^4 - n$ a multiple of 4 for all integers $n \geq 0$? See Exercise 7.25 ahead as well!)

Exercise 7.15. Show that $\dfrac{n^5}{5} + \dfrac{n^3}{3} + \dfrac{7n}{15}$ is an integer for all $n \geq 0$.

Exercise 7.16. The following contains both Example 7.2 and Exercises 7.10.1 through 7.10.3 as special cases. An expression such as $A_{a,d}(n) = a + (a + d) + (a + 2d) + \cdots + (a + (n - 1)d)$, where a and d are real numbers and n is an integer,

$n \geq 1$, is known as an *Arithmetic Series.* (Compare the pattern here with that of geometric series we looked at in Example 7.2. The notation is to suggest that a and d are parameters, and for fixed a and d, the sum depends on the number of summands n.)

Show that $A_{a,d}(n) = \dfrac{(2a + (n-1)d)\,n}{2}$.

(Notice that this works out to the average of the first and last terms, multiplied by n. It is as if we are simply summing the average of the first and last terms n times!)

Exercise 7.17. For integers n, r with $n > r > 0$, show

$$\sum_{i=r}^{n} \binom{i}{r} = \binom{n+1}{r+1}.$$

(Hint: Proposition 4.15(2) (Pascal's Identity) in Chapter 4 may be of help.)

Exercise 7.18. Show that $|sin(nx)| \leq n\ sinx$ for all integers $n \geq 0$, for any x in the range $0 \leq x \leq \pi$. (Hint: you would need the addition formula for the sine function.)

Exercise 7.19. (For those who have seen some calculus.) Let $f_1(x), \ldots, f_n(x)$ ($n \geq 2$) be real-valued functions that are *differentiable* (that is, possess derivatives) for all x in some interval (a, b). Recall the notation f' for the derivative of $f(x)$ with respect to x. Show that the following holds for all $x \in (a, b)$:

(1)

$$(f_1 \cdot f_2 \cdots f_n)' = \sum_{i=1}^{n} f_1 \cdots f_{i-1} \cdot f_i' \cdot f_{i+1} \cdots f_n,$$

(2)

$$\frac{(f_1 \cdot f_2 \cdots f_n)'}{f.f_2 \cdots f_n} = \sum_{i=1}^{n} \frac{f_i'}{f_i}.$$

Exercise 7.20. The following exercise generalizes the familiar $(a+b)^2 = a^2 + b^2 + 2ab$ identity for n summands: Prove using induction that if $x_1, \ldots, x_n$, $n \geq 2$, are complex numbers, then $(x_1 + \cdots + x_n)^2 = x_1^2 + \cdots + x_n^2 + 2x_1x_2 + \cdots + 2x_1x_n + 2x_2x_3 + \cdots + 2x_2x_n + \cdots + 2x_{n-1}x_n$.

Exercise 7.21. Show the following:

(1) If x and y are two *distinct* integers, then for all $n \geq 1$, $x^n - y^n$ is divisible by $x - y$. (Why is the hypothesis that x and y are distinct necessary?)

(2) If x and y are two integers, then for all *odd* integers $n \geq 1$, $x^n + y^n$ is divisible by $x + y$. Is the result true if n is allowed to be even? (Hint: Write n as $2k + 1$ for suitable k and perform induction on k.)

Exercise 7.22. Use the principle of induction to prove the Binomial Theorem (Theorem 4.16 in Chapter 4).

(Hint: At the induction step, write $(x + y)^{k+1}$ as $(x + y)(x + y)^k$. Expand $(x + y)^k$ using the induction assumption, multiply by $(x + y)$. Collect like terms and use Proposition 4.15(2) in Chapter 4.)

Exercise 7.23. Show that for all $n \in \mathbb{Z}$, $(\cos(\theta) + i\sin(\theta))^n = \cos(n\theta) + i\sin(n\theta)$. (Hint: First prove this for $n \in \mathbb{Z}_{\geq 0}$ using induction starting at $n = 0$ as well as addition formulas for the sine and the cosine. For negative n, say $n = -m$ for $m > 0$, write $(\cos(\theta) + i\sin(\theta))^n$ as $\dfrac{1}{(\cos(\theta) + i\sin(\theta))^m}$ and use the fact that you have already proved the theorem for positive exponents.)

Exercise 7.24. Establish the following formula that directly describes the nth Fibonacci number, introduced in Example 7.9. (This is in contrast to the recursive definition introduced in that example!)

$$F_n = \frac{1}{\sqrt{5}}\left(\left(\frac{1+\sqrt{5}}{2}\right)^n - \left(\frac{1-\sqrt{5}}{2}\right)^n\right), \quad n \geq 1.$$

(Hint: Write G_n for the right-hand side of the equation above. Show that $G_1 = F_1$, $G_2 = F_2$, and then using induction, that $G_{n+1} = G_n + G_{n-1}$ for $n \geq 2$. Since $G_n + G_{n-1} = F_n + F_{n-1}$ by the induction step, this will establish the result. It may be helpful first to find a quadratic polynomial $x^2 + ax + b$, whose roots are precisely $\phi = \frac{1+\sqrt{5}}{2}$ and $\psi = \frac{1-\sqrt{5}}{2}$.)

Remark 7.24.1. The amazing thing is that the right side of the equation above, which appears so messy, turns out be just an integer!)

Exercise 7.25. This exercise generalizes Exercise 7.14 above. Prove the following:

(1) Prove that if p is a prime, then for all integers i with $0 < i < p$, the quantity $\binom{p}{i}$ (which is an *integer* as it represents a *count* of something) is a multiple of p. (Note: you do not need the principle of induction for this step.)

(2) Use part (1) above and induction to prove *Fermat's Little Theorem:* If p is a prime, then $n^p - n$ is a multiple of p for all integers $n \geq 0$.

Exercise 7.26. Use Fermat's Little Theorem in Exercise 7.25 above to prove the statements in Exercises 7.13(1), and 7.15.

Exercise 7.27. We will use induction to derive a result known as the *Inclusion Exclusion Principle* in this exercise. It is a very useful tool for counting the number of elements of a union of sets. (One example is in Exercise 10.47 of Chapter 10 ahead.) So, let U be a set, and let $A_1, A_2, \ldots, A_n$, $n \geq 2$, be finite subsets of U.

(1) We will do the base case where $n = 2$ here: Prove that $|A_1 \cup A_2| = |A_1| + |A_2| - |A_1 \cap A_2|$. (Hint: draw a diagram of the two sets and stare at it!)

(2) Now use induction to establish the general case of the Inclusion Exclusion Principle: For $n \geq 2$, show that

$$\begin{aligned}|A_1 \cup A_2 \cup \cdots \cup A_n| = &\sum_{1\leq i\leq n} |A_i| - \sum_{1\leq i<j\leq n} |A_i \cap A_j| \\ &+ \sum_{1\leq i<j<k\leq n} |A_i \cap A_j \cap A_k| \\ &- \sum_{1\leq i<j<k<l\leq n} |A_i \cap A_j \cap A_k \cap A_l| \\ &\pm \cdots + (-1)^{n-1}|A_1 \cap A_2 \cap \cdots A_n|.\end{aligned}$$

(Hint: Write $A_1 \cup A_2 \cup \cdots \cup A_{n+1}$ as $[A_1 \cup A_2 \cup \cdots \cup A_n] \cup A_{n+1}$ and use part (1). Note too that $(A_1 \cup A_2 \cup \cdots \cup A_n) \cap A_{n+1}$ can be written as $(A_1 \cap A_{n+1}) \cup (A_2 \cap A_{n+1}) \cup \cdots \cup (A_n \cap A_{n+1})$ to which the induction hypothesis can be applied.)

7.4 Notes

We will prove here that indeed the two forms of the principle of induction we have considered in the chapter, namely the one in Definition 7.1 (which we called the *Principle of Induction* there, but will now additionally refer to as the *Standard Form* of the principle of induction) and the one in Definition 7.7 (which we called the *Principle of Induction, "Strong" Form*), are logically equivalent. Just to have them both in front of us, let us recall the two forms:

Principle of Induction (Standard Form): Let $S(n)$, $n \in \mathbb{N}$ be a family of statements.

(1) *Standard Hypothesis* 1: If $S(1)$ is true, and

(2) *Standard Hypothesis* 2: If, for some $k \in \mathbb{N}$, $S(k)$ is true, then $S(k+1)$ is also true,

(3) *Standard Conclusion*: Then $S(n)$ is true for all $n \in \mathbb{N}$.

Principle of Induction (Strong Form): Let $S(n)$, $n \in \mathbb{N}$ be a family of statements.

(1) *Strong Hypothesis* 1: If $S(1)$ is true, and

(2) *Strong Hypothesis* 2: If, for some $k \in \mathbb{N}$, $S(1), S(2), \dots, S(k)$ are all true, then $S(k+1)$ is also true,

(3) *Strong Conclusion*: Then $S(n)$ is true for all $n \in \mathbb{N}$.

First, let us assume that the standard form of the principle of induction (see above) holds. We need to show that the strong form also holds.

Let $T(m)$, $m = 1, 2, \dots,$ stand for the statement "$S(1), \dots, S(m)$ are true." Thus, Strong Hypothesis 1 says that $T(1)$ is true (note that "$T(1)$ is true" is just the statement "$S(1)$ is true!"), while Strong Hypothesis 2 says that if for some $k \geq 1$, $T(k)$ is true, then $S(k+1)$ is also true. But let's see how to bring $T(k+1)$ into the picture. Notice that when $T(k)$ and $S(k+1)$ are both true, it means that $S(1), \dots, S(k)$ *and* $S(k+1)$ are true! Thus, Strong Hypothesis 2 really says if $T(k)$ is true then $T(k+1)$ is true.

Just to summarize, this is how the two strong hypotheses read in terms of $T(m)$:

(1) *Strong Hypothesis* 1 *Rewritten*: If $T(1)$ is true, and

(2) *Strong Hypothesis* 2 *Rewritten*: If for some $k \geq 1$, $T(k)$ is true, then $T(k+1)$ is also true,

But these are now precisely the hypotheses that go into the standard form of the principle of induction for the family of statements $T(k)$ (see above). Thus, by Standard Conclusion (remember we are assuming that the standard form of the principle of induction holds), $T(n)$ must be true for all n. But what does this mean? This means $S(1)$, $\dots$, $S(n)$ are true for all n. This is just another way to say that $S(n)$ is true for all n! Hence, the strong form of the principle of induction holds!

Now we go the other way: we assume that the strong form of the principle of induction holds, and we will show that the standard form must also hold. So suppose that we have a family of statements $S(n)$, $n \geq 1$, for which the two standard hypotheses hold: $S(1)$ is true, and if $S(k)$ is true for some $k \geq 1$, then $S(k+1)$ is also true. We need to be able to conclude that $S(n)$ is true for all $n \in \mathbb{N}$. Now, given these hypotheses, we can first conclude that the statement "if $S(1), S(2), \ldots, S(k)$ are true for some $k \geq 1$, then $S(k+1)$ must also be true" must be true. This is because Standard Hypothesis 2 says, under even the weaker assumption that just $S(k)$ is true, that $S(k+1)$ must be true. Thus, these statements $S(n)$ satisfy the two strong hypotheses: $S(1)$ is true, and if $S(1), S(2), \ldots, S(k)$ are true for some $k \geq 1$, then $S(k+1)$ is also true. Since we have assumed that the strong form of the induction principle holds, we can say (Strong Conclusion) that $S(n)$ is true for all $n \geq 1$. Note that the strong conclusion is the same as the standard conclusion! Thus, under the assumption of the strong form of the principle, we have shown that the standard form of the principle of induction holds!

As it turns out, each of these forms of the principle of induction is equivalent to the Well-Ordering Principle, which we will cover in Chapter 8 ahead. (See Definition 8.12 in that chapter.)

8

Cardinality of Sets

In this chapter, we will consider a remarkable sequence of ideas that enable us to measure the *size* of sets. What this ought to be is intuitively clear for a finite set (it ought to be the number of elements in the set), but this is a little harder to fathom when the set is infinite. Accordingly, we need some new concepts to help us measure the size of a set precisely; in fact, we even need to pin down what "finite" and "infinite" mean! We will consider these concepts below. As an offshoot of these concepts, we will find that there are "as many" integers as there are rational numbers! We will even find that there are different types of "infinities." For instance, viewed under the lens of these ideas, we will see that the infinitude of the integers is of a different kind than the infinitude of the real numbers!

These ideas are due to Georg Cantor, who also introduced the concept of sets to mathematics. When they were first proposed, even established mathematicians had a difficult time accepting them! However, with experience, these ideas turned out to be very natural and very fruitful. Moreover, certain questions concerning these ideas led to further research, which led to several foundational results in that field, including the development of various axioms for set theory. (We have alluded to the existence of these axioms already in Chapter 5.)

8.1 Finite and Infinite Sets, Countability, Uncountability

Let us go back and look at how we count objects: If there are, say three objects lying on a table, we count them by pointing to each in turn, and saying "one, two, three." If you look at the mathematics that is happening behind the scenes, what we are really doing is creating a bijection between the set of objects on the table and the set $\{1, 2, 3\}$. Motivated by this, we are tempted to say that a set S has n elements in it if there is a bijection between S and the subset $\{1, 2, \dots, n\}$ of $\mathbb{N}$. The advantage here is that we understand sets like $\{1, 2, \dots, n\}$ very well, where we already have an intuitive idea of size.

But already, there is a potential problem: that of uniqueness. Can S admit a bijection between itself and $\{1, 2, \ldots, n\}$ and also between itself and $\{1, 2, \ldots, m\}$ for different positive integers n and m? The answer, given our long experience with finite sets, ought to be "no!" And indeed this is correct, although a formal proof of this very intuitive result needs to be given. We will merely state this uniqueness result as a proposition here, and push the formal proof out to the exercises so that we can quickly move to deeper results in the subject:

Proposition 8.1. *Suppose a set S admits a bijection between itself and $\{1, 2, \ldots, n\}$ and also between itself and $\{1, 2, \ldots, m\}$, for positive integers m and n. Then $n = m$.*

Proof. See Exercise 8.40 ahead. □

Accordingly, we make the following definition:

Definition 8.2. A set S is said to be *finite* if there exists a bijection between S and the set $\{1, 2, \ldots, n\}$ for some positive integer n. The integer n (guaranteed to be unique by Proposition 8.1) is called the *cardinality* of S. In addition, we will adopt the convention that the empty set is a finite set of cardinality 0. The notation $|S|$ will be used to denote the cardinality of a finite set S.

If S is not finite, we say that S is *infinite.* Thus, if S is infinite, then S is nonempty, and for no positive integer n is there a bijection between S and $\{1, 2, \ldots, n\}$.

Remark 8.3. So, for finite sets, cardinality measures what we would intuitively consider to be the size of the set: the number of elements in it.

But we are not done. How do we measure the size of sets like $\mathbb{N}$ or $\mathbb{Z}_{\geq 0}$ or $\mathbb{Z}$? Intuitively it should be clear that these are not finite sets, but how do we prove that? In fact, how do we even characterize infinite sets other than by saying that they are nonempty and do not admit a bijection with $\{1, 2, \ldots, n\}$ for any positive integer n? Infinite sets are indeed tricky objects! We will introduce further concepts below that will help us deal with them.

Let us start a more useful characterization of infinite sets than the one described above in Definition 8.2. We have the following:

Proposition 8.4. *A set S is infinite iff it is nonempty and there exists an injective function $f : \mathbb{N} \to S$.*

Proof. Suppose S is infinite. In particular, S is not empty (see Definition 8.2). Pick an element of S, call it s_1. Then $S \setminus \{s_1\}$ is not empty, else, $S = \{s_1\}$, which clearly is a finite set of cardinality 1 (it is in bijection with $\{1\}$), and this contradicts the fact that S is infinite. Since $S \setminus \{s_1\}$ is not empty we can pick an element $s_2 \in S \setminus \{s_1\}$. Once again, $S \setminus \{s_1, s_2\}$ is not empty, else, $S = \{s_1, s_2\}$, which clearly is a finite set of cardinality 2 (it is in bijection with $\{1, 2\}$). Proceeding, we can create a subset of distinct elements $s_1, s_2, \ldots,$ of S, and this process cannot stop, since if it stops at some positive integer k, then $S = \{s_1, s_2, \ldots, s_k\}$, which is a finite set of cardinality k (it is in bijection with $\{1, 2, \ldots, k\}$): a contradiction. We define the function $f : \mathbb{N} \to S$ to be one that takes 1 to s_1, 2 to $s_2, \ldots$; this function is then injective as the s_i are all distinct.

Now assume that S is nonempty, and that we have an injective function $f : \mathbb{N} \to S$. Assume to the contrary that S is finite. Thus, there is a bijection $g : S \to \{1, 2, \ldots, n\}$ for some positive integer n. Composing, we find that $h = g \circ f : \mathbb{N} \to \{1, 2, \ldots, n\}$ is an injective function (see Proposition 5.28). But this is absurd: if we view the $n + 1$ elements $h(1)$, $h(2)$, ..., $h(n + 1)$ as $n + 1$ letters going into the n pigeon holes 1, 2, ..., n, we'll find by the Pigeon Hole Principle that $h(i) = h(j)$ for two distinct i and j, $1 \leq i < j \leq n + 1$, contradicting the injectivity of h. Thus, S cannot be finite. □

This leads to the following definition:

Definition 8.5. Let S be an infinite set. If, further, there exists a bijection $f : \mathbb{N} \to S$, then S is said to be *countably infinite.* An infinite set that is not countably infinite is said to *uncountable.*

The term *countable* is used to denote either a finite set or a countably infinite set.

Remark 8.6. By Proposition 8.4, a set that is not finite already admits an injective map $f : \mathbb{N} \to S$. The key point around which Definition 8.5 is constructed is that such a map need not be surjective. However, if among all such injective maps one can be found that is also bijective, the set is called countably infinite.

Remark 8.7. Figure 8.1 below depicts the classification of sets based on the Definitions 8.2 and 8.5. The examples shown will be developed as we go along, both in the text and the exercises.

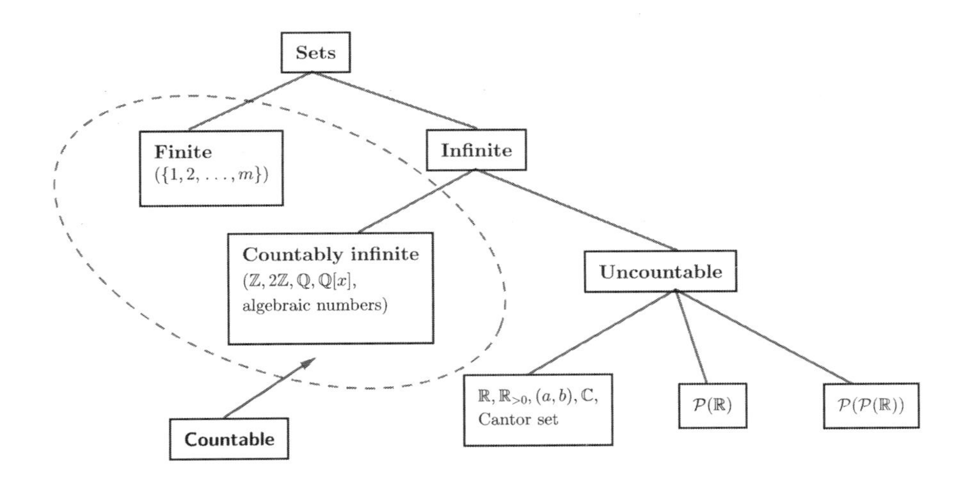

Figure 8.1. Classification of sets. The examples shown will be developed in the text and exercises.

Example 8.8. The following are some examples of infinite sets:

(1) The set $\mathbb{N}$ is infinite. The identity map $\mathbb{1}_{\mathbb{N}}$ that sends any $n \in \mathbb{N}$ to itself is clearly an injection, in fact a bijection. Thus, $\mathbb{N}$ is in fact countably infinite.

(2) $\mathbb{Z}_{\geq 0}$, $\mathbb{Z}$, $\mathbb{Q}$, $\mathbb{R}$, and $\mathbb{C}$ are all infinite: $\mathbb{N}$ sits inside each of these as a subset, so the inclusion map (see Remark 5.22 in Chapter 5) of $\mathbb{N}$ into each of $\mathbb{Z}_{\geq 0}$, $\mathbb{Z}$, $\mathbb{Q}$, $\mathbb{R}$, and $\mathbb{C}$ shows that each is an infinite set.

(3) The set $\mathbb{Z}$ is countably infinite. We can create a bijective map f from $\mathbb{N}$ to $\mathbb{Z}$ as follows: we first arrange $\mathbb{Z}$ in the order $\{0, 1, -1, 2, -2, 3, -3, \dots\}$, and simply "count out" this list one by one. In other words, to create f, we send 1 to 0, 2 to 1, 3 to -1, 4 to 2, 5 to -2, and so on. (You can check the general formula: a positive even integer of the form $2k$ goes to k, while a positive odd integer of the form $2k+1$ goes to $-k$). This map is clearly bijective.

(4) The set $2\mathbb{Z} = \{n \in \mathbb{Z} \mid n \text{ is even}\}$ is also countably infinite. You should be able to provide the bijection between $\mathbb{N}$ and $2\mathbb{Z}$ yourself, along the same lines as in part (3).

This example gives us an interesting phenomenon. Consider the function $g : \mathbb{N} \to 2\mathbb{Z}$ that you have constructed, and also consider the function f from $\mathbb{N}$ to $\mathbb{Z}$ in part (3) above. The function $g \circ f^{-1}$ is then a bijection between $\mathbb{Z}$ and $2\mathbb{Z}$. (Alternatively, the map $h : \mathbb{Z} \to 2\mathbb{Z}$ that sends m to $2m$ provides a clear bijection between $\mathbb{Z}$ and $2\mathbb{Z}$.) But $2\mathbb{Z}$ is a proper subset of $\mathbb{Z}$, and yet is in bijection with $\mathbb{Z}$! As we will see below in Theorem 8.11, this is not an accident: every infinite subset of $\mathbb{Z}$ is in bijection with $\mathbb{Z}$!

(5) What is interesting is that $\mathbb{Q}$ is also countably infinite! We will prove this in Theorem 8.16 below.

(6) On the other hand, $\mathbb{R}$ is not countably infinite. The proof of this uses the famous Cantor Diagonalization argument, which we will explain in Theorem 8.17. Thus, $\mathbb{R}$ is an example of an uncountable set.

The following will help motivate the definition of cardinality for arbitrary sets (Definition 8.14 ahead):

Proposition 8.9. *Let A and B be sets. Assume A and B are in bijection. Then,*

(1) *A is finite, of cardinality n, if and only if B is finite, of cardinality n.*

(2) *A is countably infinite if and only if B is countably infinite.*

Now assume that A and B are either both finite of cardinality n, or both countably infinite. Then A and B are in bijection.

Proof. If $f : A \to B$ is a bijection, and if B is finite of cardinality n, we can see that A is also finite of cardinality n as follows: By definition, there is a bijection $g : B \to \{1, 2, \dots, n\}$ for some $n \in \mathbb{N}$, and it follows from this that the composition $g \circ f : A \to \{1, 2, \dots, n\}$ is a bijection. Moreover, applying this same reasoning to $f^{-1} : B \to A$, which is a bijection, we can conclude that if A is finite of cardinality n, then B is also finite of cardinality n. Replacing the maps to $\{1, 2, \dots, n\}$ above with maps to $\mathbb{N}$, we can conclude with the same arguments that one of them is countably infinite if and only the other is also countably infinite.

In the reverse situation, if two sets A and B are both finite of cardinality n, then we can conclude that A and B are in bijection as follows: if $\phi : A \to \{1, 2, \ldots, n\}$ and $\psi : B \to \{1, 2, \ldots, n\}$ are the bijections that establish the cardinalities of A and B, then $\psi^{-1} \circ \phi : A \to B$ is the desired bijection. Likewise, replacing maps to $\{1, 2, \ldots, n\}$ with maps to $\mathbb{N}$, the same arguments show that if A and B are both countably infinite, then A and B are in bijection. □

Proposition 8.9 allows us to derive the following characterization of a countability infinite set:

Proposition 8.10. *A set is a countably infinite set if and only if it is in bijection with the set of integers.*

Proof. Example 8.8(3) shows that $\mathbb{Z}$ is countably infinite. The proof of the proposition now follows from Proposition 8.9. □

We will now show that every subset of $\mathbb{Z}$ is countable, that is, it is either finite, or else countably infinite. As a consequence, Proposition 8.10 shows that if $S \subsetneq \mathbb{Z}$ is an infinite subset, then S—a *proper subset* of $\mathbb{Z}$—is in bijection with $\mathbb{Z}$! Thus, the example of $2\mathbb{Z}$ being in bijection with $\mathbb{Z}$ that we considered in Example 8.8(4) above is part of a wider phenomenon.

Theorem 8.11. *Any subset of the set of integers is countable. In particular, any infinite subset of the integers is countably infinite and in bijection with the integers.*

Proof. We will prove the first statement. As already described in the paragraph above the theorem, the second statement is a consequence of the first and Proposition 8.10.

Our proof will use the intuitively clear statement that every nonempty subset of $\mathbb{N}$ has a least element. In actuality, this is a formal axiom of logic, known as the well-ordering principle, that is equivalent to another axiom of logic, the axiom of induction that we have already considered in Chapter 7 (see Exercise 8.43 at the end of this chapter). We will just state it here as a tool that we will use henceforth without losing sleep over its soundness:

Definition 8.12. *Well-Ordering Principle:* Every nonempty subset of the set of positive integers has a least element.

Let S be a subset of $\mathbb{Z}$. Assume that S is not finite (if it is, there is nothing to prove). Write g for the map $\mathbb{Z} \to \mathbb{N}$ that is the inverse of the map $f : \mathbb{N} \to \mathbb{Z}$ of Example 8.8(3). Writing i for the inclusion map $S \to \mathbb{Z}$ (see Remark 5.22, Chapter 5), the composition $g \circ i : S \to \mathbb{N}$ is then an injective map (see Proposition 5.28, Chapter 5). Since S is in bijection with its image $g \circ i(S)$, and since $g \circ i(S)$ is infinite (why?), it is sufficient to instead prove the following statement: *Any infinite subset of* $\mathbb{N}$ *is countably infinite.* The sufficiency follows from Proposition 8.9. The same remark (or more formally, Proposition 8.10) will also show that S is then in bijection with $\mathbb{Z}$.

So, let T be any infinite subset of $\mathbb{N}$; we will prove that T is countably infinite. By the well-ordering principle, T has a least element, call it t_1. Notice that $t_1 \geq 1$. Notice, too, that $T \setminus \{t_1\}$ is nonempty, else T would equal $\{t_1\}$, a finite set, which contradicts the fact that T is infinite. Hence, by the well-ordering principle, $T \setminus \{t_1\}$ has a least element, call it t_2. Now $t_2 > t_1 \geq 1$, so $t_2 \geq 2$. Once again, $T \setminus \{t_1, t_2\}$ is nonempty, else T would equal $\{t_1, t_2\}$, a finite set, while we know T is infinite. Hence $T \setminus \{t_1, t_2\}$ has a least element, call it t_3. Proceeding thus, we create a list of elements of T, $t_1 < t_2 < t_3 < \dots$ (with $t_i \geq i$). This process cannot stop, since if it stops at some t_k for some integer $k \geq 1$, then we would find $T = \{t_1, t_2, \dots, t_k\}$, a finite set, which is a contradiction. We thus now have an injective function $\phi : \mathbb{N} \to T$, that sends 1 to t_1, 2 to t_2, and so on. It is surjective as well, since given any $m \in T$, we know $m \leq t_m$, so m must have appeared as either the least element in T (if so m would equal t_1), or else as the least element in $T \setminus \{t_1\}$ (if so m would equal t_2), and so on. Thus, m must be one of $t_1, t_2, \dots, t_m$, and hence must be in the image of ϕ. □

We get the following:

Corollary 8.13. *Any subset of a countable set is countable.*

Proof. Let $A \subseteq B$ be the two sets, with B countable. We wish to show that A is countable. If A is already finite, then there is nothing to prove, so we will assume that A is infinite. It follows that B is also infinite (Exercise 8.25). Thus, B is countably infinite, so there is a bijection $f : B \to \mathbb{N}$. In particular, the restriction $f|_A : A \to f(A)$ is injective, and by the definition of $f(A)$, it is surjective as well. Thus, $f|_A : A \to f(A)$ is a bijection. (Here, $f(A) \subseteq \mathbb{N}$.) Since A is assumed to be infinite, $f(A)$ must also be infinite (Proposition 8.9). As we have seen in the proof of Theorem 8.11 above, $f(A)$ must be countably infinite. Since A is in bijection with a countably infinite set, A is itself countably infinite. □

For finite sets, we used the word cardinality (see Definition 8.2) for the number of elements in the set. What should be the corresponding notion for infinite sets? Well, the whole point about being infinite is that there is no bijection between the given set and $\{1, 2, \dots, n\}$ for any positive integer n. So, we cannot measure the size of infinite sets using the numbers we have at our disposal so far. But at this stage, we can work around the problem by simply recognizing when two sets ought to have the same size. Proposition 8.9 laid bare the connection between two sets being in bijection and being either both finite of cardinality n or countably infinite: for *countable* sets, these two are equivalent. This connection suggests that even when sets are infinite and *not* countable, we should simply consider any two sets that are in bijection with each other to be of the same "size." Continuing to use the word cardinality, we therefore have the following:

Definition 8.14. Two sets A and B are said to be of the *same cardinality* if there exists a bijection between them.

At this stage, we will be content with this characterization of "size." We simply recognize when sets are of the same size, without giving independent meaning to this size. As it turns out, the actual cardinality, or size, of infinite sets is measured by a new set of "numbers" called cardinal numbers. We will not go into these "numbers" — this is an area that is a bit advanced for where we are in our mathematical studies (but read about the *continuum hypothesis* in your physical or e-library).

There is a special case where conventionally a name is given to the size of an infinite set: the size of $\mathbb{N}$ is denoted $\aleph_0$ (read "aleph null").

Example 8.15. We have already seen some examples of infinite sets of equal cardinality: for instance, Theorem 8.11 shows that any infinite subset of $\mathbb{Z}$ has the same cardinality as $\mathbb{Z}$. Here are a few further examples:

(1) $\mathbb{R}$ and $\mathbb{R}_{>0}$. One example of a function $f : \mathbb{R} \to \mathbb{R}_{>0}$ that is both injective and surjective is the function $f(x) = e^x$ already considered in Example 5.37, Part (1) in Chapter 5. (See particularly, the graph of the function displayed in that example.)

(2) $(-\pi/2, \pi/2)$ and $\mathbb{R}$. The function $\tan : (-\pi/2, \pi/2) \to \mathbb{R}$ provides the bijection. You would of course be familiar with the trigonometric functions from your studies in school, and in particular with the tangent function of an angle x, denoted $\tan(x)$. See the graph below in Figure 8.2.

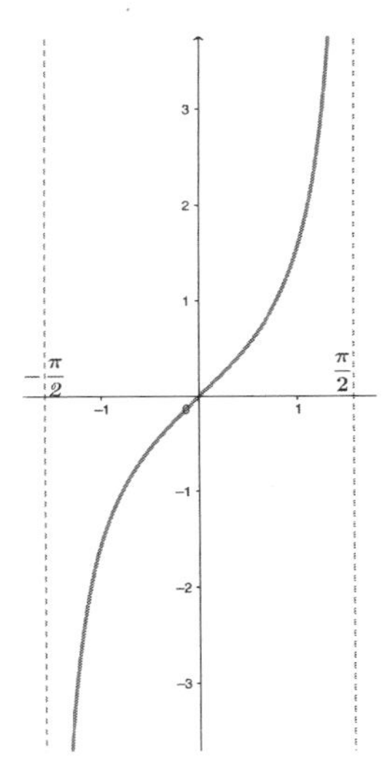

Figure 8.2. The tangent function, shown in the interval $(-\pi/2, \pi/2)$.

(3) For any real numbers $a < b$ and $c < d$, the closed intervals $[a, b]$ and $[c, d]$. For a function that provides the bijection, all we have to do is to choose the function $y\,(= f(x)) = Mx + B$ which represents the equation of the line joining (a, c) and (b, d). Here, M is the slope of the line, namely $\frac{d-c}{b-a}$, and B is the y-intercept. This is clearly a bijection; see the graph in Figure 8.3 below.

(4) For any real numbers $a < b$ and $c < d$, the open intervals (a, b) and (c, d). The same function as in part (3) above, but restricted to the open interval (a, b), will provide the bijection.

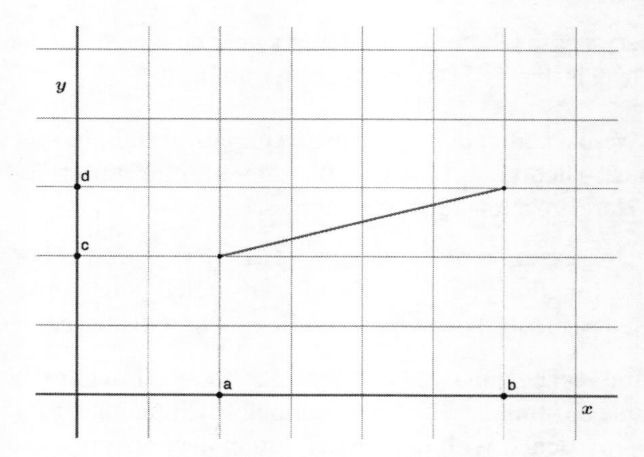

Figure 8.3. Bijection between two closed intervals $[a, b]$ and $[c, d]$.

8.2 Cardinalities of $\mathbb{Q}$ and $\mathbb{R}$

We now prove:

Theorem 8.16. *The set of rational numbers is countably infinite.*

Proof. We have already noted in Example 8.8(2) that $\mathbb{Q}$ is infinite (that is, there exists an injective function $\mathbb{N} \to \mathbb{Q}$). Now we need to prove that there exists a *bijective* function $\mathbb{N} \to \mathbb{Q}$. In other words, we need to find a way of pairing each rational number with a unique positive integer, so we can say "here is the first rational number, here is the second rational number," and so on, and exhaust all the rational numbers in the process.

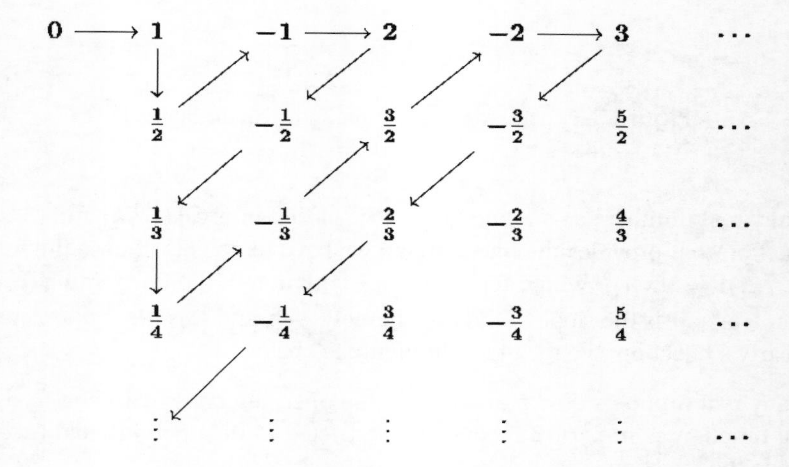

Figure 8.4. We can count the elements of $\mathbb{Q}$ as shown, following the arrows.

For this, we order the rational numbers as shown in Figure 8.4: we write the rational numbers in reduced form, and in the first row, we write those with denominator 1 (i.e., the integers), and we do so in the order shown: 0, 1, -1, 2, -2, and so on. In the second row, we write those with denominator 2 with the same alternating sign pattern, remembering that we are working with *reduced* fractions, so for instance, $\frac{2}{2}$ does not show up as it has already been considered. In general, in the i-th row, we write down the *reduced* fractions with denominator i. And so on. We then proceed to "count" these rational numbers in the order shown in the figure, proceeding along diagonals: 0, the first; 1, the second; $\frac{1}{2}$, the third; -1, the fourth; 2, the fifth; $-\frac{1}{2}$, the sixth; $\frac{1}{3}$, the seventh; $\frac{1}{4}$, the eighth; $-\frac{1}{3}$, the ninth; $\frac{3}{2}$, the tenth; .... This provides a clear bijection between $\mathbb{N}$ and $\mathbb{Q}$. □

Next, we show that $\mathbb{R}$ is uncountable. As mentioned in Example 8.8(6), the proof involves Cantor's diagonalization argument.

Theorem 8.17. *The set of real numbers is uncountable.*

Before we give the proof, let us describe the strategy behind it. Since we know that $\mathbb{R}$ contains $\mathbb{N}$, for instance, it is already infinite, so we need to show that it is not countably infinite (see Definition 8.5). In other words, we need to show that no bijection between $\mathbb{N}$ and $\mathbb{R}$ can exist. Ordinarily, we would expect that it should take a bit of cleverness to prove that something *cannot* exist! And indeed, Cantor's argument is remarkably clever! It is a proof by contradiction. It assumes that there is such a bijection $f : \mathbb{N} \to \mathbb{R}$. Such an f must therefore be surjective, so the set $\{f(1), f(2), f(3), \dots\}$ must be all of $\mathbb{R}$. The argument then looks at the decimal expansions of $f(1)$, $f(2)$, $f(3)$, etc., and cleverly constructs from these decimal expansions a real number that cannot possibly be $f(n)$ for any positive integer n. This contradicts the assumed surjectivity of f!

Proof of Theorem 8.17. We have already seen in Example 8.8(2) that $\mathbb{R}$ is infinite; now we need to show that $\mathbb{R}$ is not countably infinite, that there is no bijection $f : \mathbb{N} \to \mathbb{R}$. For this, we first write each real number as an infinite decimal, agreeing that (1) we will write a finite decimal such as 1.25 as 1.2500000... (ending in an infinite string of zeros), and (2) we will rewrite any infinite decimal ending with an infinite string of 9s as one ending in an infinite string of zeros: for example, we will rewrite 3.5799999... as 3.58000.... (Recall that 3.5799999... and 3.58000... are the same numbers. By rewriting numbers like 3.5799999... consistently as 3.58000..., we guarantee that each real number is represented in a *unique* fashion as an infinite decimal.)

We now assume to the contrary that there exists a bijection $f : \mathbb{N} \to \mathbb{R}$. As explained above the start of the proof, we will arrive at a contradiction by using the decimal expansions of $f(1), f(2), f(3), \dots$ to find an element of $\mathbb{R}$ that is not in the image of f! This contradicts the fact that f is surjective, so our assumption that such a bijection exists must be flawed!

With f as described, we create a table such as Table 8.1, with the elements 1, 2, 3, ... in the first column, and the elements $f(1), f(2), f(3), \dots$ in the second column, written as described above as infinite decimals. (Table 8.1 depicts $f(1)$ through $f(7)$ for a *sample* f.) Now let d_1 be the first digit after the decimal point in $f(1)$, d_2 be the second digit after the decimal in $f(2), \dots$, and d_n the n-th digit after the decimal point in $f(n)$, for any integer $n \geq 1$. (Thus, the digits d_i occur along the diagonal in the column on the right side, see the figure, hence the term "diagonalization argument.")

Table 8.1. Cantor's diagonalization: The digits d_i in ith decimal places, $i = 1, 2, \ldots$ run down the diagonal. We create a real number that differs in the i-th place with d_i, and so cannot equal any real number in a countable listing.

n	$f(n)$	d_n	e_n
1	51.**2**5793569...	$d_1 = 2$	$e_1 = 0$
2	03.1**7**159260...	$d_2 = 7$	$e_2 = 0$
3	00.56**5**79153...	$d_3 = 5$	$e_3 = 0$
4	43.123**5**5555...	$d_4 = 5$	$e_4 = 0$
5	19.0000**0**000...	$d_5 = 0$	$e_5 = 1$
6	27.53291**2**65...	$d_6 = 2$	$e_6 = 0$
7	03.590320**0**5...	$d_7 = 0$	$e_7 = 1$
$\vdots$	$\vdots$	$\vdots$	$\vdots$

$$r = 0.0000101\ldots$$

Now consider any real number r whose decimal expansion is of the form $0.e_1e_2e_3\ldots$ with the property that $e_1 \neq d_1, e_2 \neq d_2, \ldots, e_n \neq d_n$ for all integers $n \geq 1$. (For instance, choose the e_i as follows: if $d_i = 0$, let $e_i = 1$, else, let $e_i = 0$.) Then r cannot appear anywhere in the right-hand column, in other words, r cannot equal $f(n)$ for any $n \in \mathbb{N}$. This is because if $r = f(n)$ for some integer n, then the n-th digit of r after the decimal place must be that of $f(n)$, so it must be d_n. But the n-th digit after the decimal place of r is e_n, *which has explicitly been selected to not equal d_n!* Hence f cannot be surjective, as r is not in its range. This contradiction shows that such an f cannot exist, proving the theorem! □

8.3 The Schröder-Bernstein Theorem

In Example 8.15 above, we saw various examples of sets of equal cardinality; in all of these, we explicitly constructed a function that provides the desired bijection between the respective pairs of sets. In many situations where we are attempting to show that two sets A and B have the same cardinality, however, it can be difficult to construct an explicit function $f : A \to B$ that is both injective and surjective. What is often the case is that we can construct a function $f : A \to B$ that is *injective*, and a similar function in the *reverse* direction $g : B \to A$ that is also *injective*. After identifying A with $f(A)$, we can think of A as "a subset" of B, and equally, after identifying B with $g(B)$, we can think of B as "a subset" of A. Intuition therefore suggests that if each can be viewed as a subset of the other, then A and B must then have the same cardinality. As it turns out, our intuition is indeed correct. But as always, a formal proof is in order! We state the following as a theorem, and provide a proof:

Theorem 8.18 (Schröder-Bernstein[1]). *Let A and B be sets. Suppose that we have a function $f : A \to B$ that is injective, and another function $g : B \to A$ that is also injective. Then, there exists a bijection $h : A \to B$, and in particular, A and B are of the same cardinality.*

Proof. If either f or g were itself surjective, then f or g^{-1} would already be the desired bijective function from A to B, so we may assume that neither is surjective, that is, $B \supsetneq f(A)$ and $A \supsetneq g(B)$. The heart of the proof of the theorem is the following easy result:

Lemma 8.19. *Suppose X, Y, and Z are sets, and suppose that X is a proper subset of Y. If $\psi : Y \to Z$ is any injective function, then*

(1) *$\psi(X)$ is a proper subset of $\psi(Y)$, and*

(2) *The subset $Y \setminus X$ is sent under ψ to $\psi(Y) \setminus \psi(X)$, and the function $\psi|_{Y \setminus X} : Y \setminus X \to \psi(Y) \setminus \psi(X)$ is a bijection.*

Proof. We will prove part (2). Part (1) can then be proved immediately, as follows: We start with the fact that $Y \setminus X$ is nonempty because X is a proper subset of Y. Now $\psi|_{Y \setminus X}$ maps $Y \setminus X$ bijectively to $\psi(Y) \setminus \psi(X)$ by part (2). Therefore $\psi(Y) \setminus \psi(X)$ will also be nonempty. But this is just saying that $\psi(X)$ will be a proper subset of $\psi(Y)$.

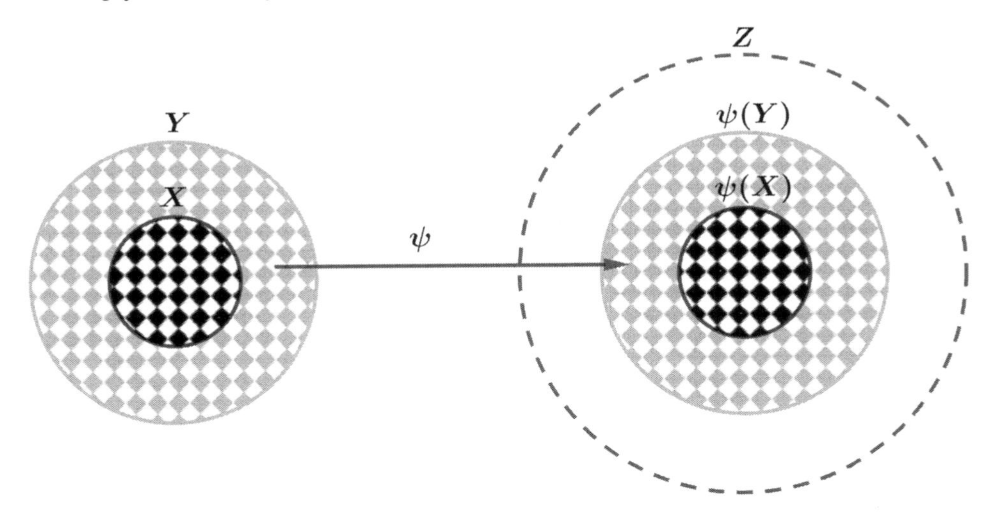

Figure 8.5. Lemma 8.19. The map ψ is a bijection between the blue region in Y and the blue region in $\psi(Y) \subseteq Z$.

See Figure 8.5. We need to show that the regions $Y \setminus X$ and $\psi(Y) \setminus \psi(X)$ are in bijection under ψ.

As noted above, $Y \setminus X$ is nonempty because X is given to be a proper subset of Y. Take $y \in Y \setminus X$. Clearly, $\psi(y) \in \psi(Y)$, so what we first need to show is that $\psi(y) \notin \psi(X)$.

[1]Sometimes referred to as the Cantor-Schröder-Bernstein theorem or also as the Cantor-Bernstein theorem.

Assume to the contrary that $\psi(y) \in \psi(X)$, so $\psi(y) = \psi(x)$ for some $x \in X$. Since ψ is injective, we find $y = x$, which contradicts the fact $y \notin X$. Thus, $\psi|_{Y\setminus X}$ sends $Y \setminus X$ to $\psi(Y) \setminus \psi(X)$. The restricted function $\psi|_{Y\setminus X}$ is injective because ψ is injective on all of Y. For surjectivity, note that just by the very definition of $\psi(Y)$, any $z \in \psi(Y) \setminus \psi(X)$ is of the form $\psi(y)$ for some $y \in Y$. This y cannot be in X, for if so, $z = \psi(y)$ would be in $\psi(X)$, contradicting the choice of z.

It follows that the function $\psi|_{Y\setminus X} : Y \setminus X \to \psi(Y) \setminus \psi(X)$ is a bijection. □

Proof of Theorem 8.18 *continued:* We start with the pair of strict containments in A and B (which, we recall, arose from our assumption that neither f nor g is surjective):

$$\begin{aligned} A &\supsetneq g(B), \\ B &\supsetneq f(A). \end{aligned} \tag{8.1}$$

We apply the injective function $f : A \to B$ to the first of equations (8.1) and invoke Lemma 8.19(1): taking $Y = A$ and $X = g(B)$, we get $f(A) \supsetneq fg(B)$, where we have written fg for $f \circ g$. Similarly, we apply the injective function $g : B \to A$ to the second of equations (8.1) and invoke Lemma 8.19(1): taking $Y = B$ and $X = f(A)$, we get $g(B) \supsetneq gf(A)$. Chaining these two new relations to the right of the containments in equations 8.1, we get the pair of (chains of) strict containments in A and B (see Figure 8.6 below):

$$\begin{aligned} A &\supsetneq g(B) \supsetneq gf(A), \\ B &\supsetneq f(A) \supsetneq fg(B). \end{aligned} \tag{8.2}$$

If we apply f to the relation $g(B) \supsetneq gf(A)$, we find, by Lemma 8.19(1) again, $fg(B) \supsetneq fgf(A)$. Similarly, applying g to the relation $f(A) \supsetneq fg(B)$ we find $gf(A) \supsetneq gfg(B)$. Chaining these two to the right of the containments in equations (8.2), we get a new pair of chains of strict containments in A and B (see Figure 8.6 below):

$$\begin{aligned} A &\supsetneq g(B) \supsetneq gf(A) \supsetneq gfg(B), \\ B &\supsetneq f(A) \supsetneq fg(B) \supsetneq fgf(A). \end{aligned} \tag{8.3}$$

Proceeding thus, we get the following pair of chains of strict containments in A and B going on indefinitely (see Figure 8.6, which depicts the first four sets in each pair):

$$\begin{aligned} A \supsetneq g(B) &\supsetneq gf(A) \supsetneq gfg(B) \supsetneq \cdots \\ &\supsetneq (gf)^k(A) \supsetneq g(fg)^k(B) \supsetneq (gf)^{k+1}(A) \supsetneq \cdots \\ B \supsetneq f(A) &\supsetneq fg(B) \supsetneq fgf(A) \supsetneq \cdots \\ &\supsetneq (fg)^k(B) \supsetneq f(gf)^k(A) \supsetneq (fg)^{k+1}(B) \supsetneq \cdots \end{aligned} \tag{8.4}$$

The chain of containments in equation (8.4), along with Lemma 8.19(2), essentially finishes the proof of the theorem! See Figure 8.6: Each light region in A is in bijection with the corresponding light region in B that is obtained by applying f to it, and

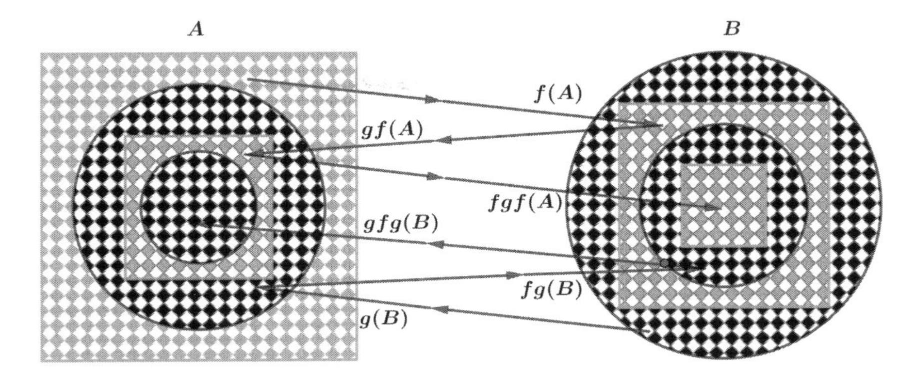

Figure 8.6. The first four steps of the containments $A \supsetneq g(B) \supsetneq gf(A) \supsetneq gfg(B) \supsetneq \cdots$ and $B \supsetneq f(A) \supsetneq fg(B) \supsetneq fgf(A) \supsetneq \cdots$ in the proof of the Schröder-Bernstein theorem. This nested pattern continues indefinitely: in A, a light region will correspond to some $(gf)^k(A) \setminus g(fg)^k(B)$, and a dark region will correspond to some $g(fg)^k(B) \setminus (gf)^{k+1}(A)$. In B, a dark region will correspond to some $(fg)^k(B) \setminus f(gf)^k(A)$, and a light region will correspond to some $f(gf)^k(A) \setminus (fg)^{k+1}(B)$.

similarly, each dark region in B is in bijection with the corresponding dark region in B that is obtained by applying g to it! This is because each light region in A is of the form $(gf)^k(A)\setminus g(fg)^k(B)$ for $k = 0, 1, \ldots$. By Lemma 8.19(2), this region, after applying f, is in bijection with the region $f(gf)^k(A) \setminus fg(fg)^k(B)$—but $fg(fg)^k(B) = (fg)^{k+1}(B)$, so this is precisely the corresponding light region in B! Similarly, given a dark region in B, it is of the form $(fg)^k(B) \setminus f(gf)^k(A)$. By Lemma 8.19(2), this region, after applying g, is in bijection with the region $g(fg)^k(B) \setminus gf(gf)^k(A)$—but $gf(gf)^k(A) = (gf)^{k+1}(A)$, so this is precisely the corresponding dark region in B!

So, to create one function h from A to B that will be injective and surjective, all we have to do is to send the light regions in A to the corresponding light regions in B via the map f, and send the dark regions in A back to the dark regions in B where they came from, i.e., via the map g^{-1}. Thus, we "follow the f arrows" in the light regions in A, and "reverse the g arrows" in the dark regions in A.

Formally, we define (see Figure 8.7 for the first four regions in A):

$$h(a) = \begin{cases} f(a) & \text{if } a \in (gf)^k(A) \setminus g(fg)^k(B), \quad k = 0, 1, \ldots, \\ g^{-1}(a) & \text{if } a \in g(fg)^k(B) \setminus (gf)^{k+1}(A), \quad k = 0, 1, \ldots. \end{cases} \tag{8.5}$$

This provides the desired bijection and proves the theorem. □

Example 8.20. We can use the Schröder-Bernstein theorem (Theorem 8.18) to show that various sets are in bijection:

(1) For real numbers $a < b$ the intervals $C = [a, b]$ and $O = (a, b)$ are in bijection. We have an obvious injection of O into C. For the other direction, note that we can use Example 8.15(3): we find a closed interval $[c, d] \subseteq O$; for example, we can take $c = a + (b - a)/4$ and $d = a + 3(b - a)/4$, and then, as shown in that example, there is a bijection $f : C \to [c, d]$. In particular, f maps C injectively into O. By

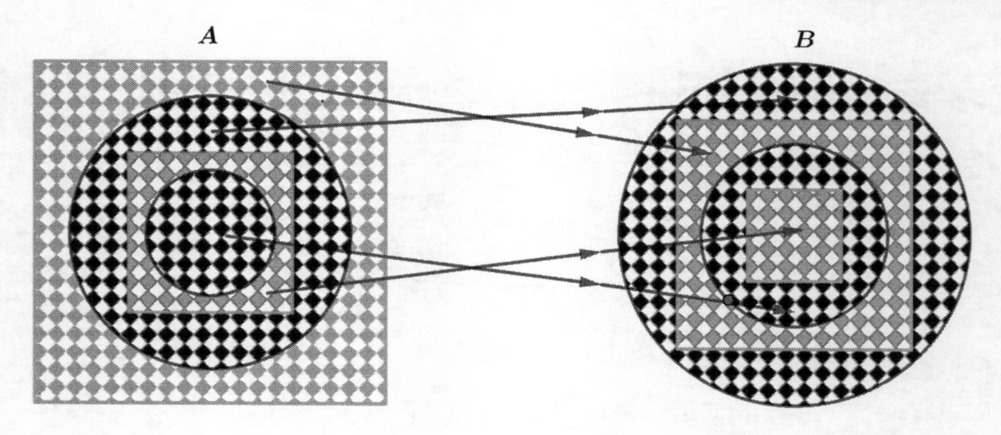

Figure 8.7. The map h defined on the first four regions $A \supsetneq g(B) \supsetneq gf(A) \supsetneq gfg(B) \supsetneq \cdots$ in the proof of the Schröder-Bernstein theorem.

the Schröder-Bernstein theorem, O and C are in bijection. (See also Exercise 8.33 ahead.)

(2) Essentially the same proof as in part (1) above shows that for real numbers $a < b$ the half open interval $H = [a, b)$ (as also $H' = (a, b]$) and $O = (a, b)$ are in bijection. The bijection created in part (1) above between $[a, b]$ and $[c, d]$ restricts to a bijection between $[a, b)$ and $[c, d)$, and hence yields an injective map from $[a, b)$ into (a, b).

(3) The set $\mathbb{Z}\times\mathbb{R}$ is in bijection with $\mathbb{R}$. Here is one way to see this, using the Schröder-Bernstein theorem: We know that $\mathbb{R}$ is in bijection with the open interval $(0, 1)$. This follows from Example 8.15, by combining the bijections between $\mathbb{R}$ and $(-\pi/2, \pi/2)$ (part (2)) and between $(-\pi/2, \pi/2)$ and $(0, 1)$ (part (4)). Write f for the function $\mathbb{R} \to (0, 1)$ that provides this bijection. Then the map h from $\mathbb{Z} \times \mathbb{R}$ to $\mathbb{R}$ that sends (n, r) to $n + f(r)$ is easily seen to be injective as follows: Suppose $h(n, r) = h(m, s)$. This means that $n + f(r) = m + f(s)$. Since $0 < f(r), f(s) < 1$, this must means that $m = n$. It follows that $f(r) = f(s)$. But f is injective, so $r = s$. Thus, $(n, r) = (m, s)$. In the other direction we have the map $\mathbb{R} \to \mathbb{Z} \times \mathbb{R}$ that sends r to $(r, 0)$, which is clearly an injection. By the theorem, $\mathbb{Z} \times \mathbb{R}$ and $\mathbb{R}$ are in bijection.

(4) The set $\mathbb{R}\times\mathbb{R}$ is in bijection with $\mathbb{R}$. We have an injection of $\mathbb{R}$ into $\mathbb{R}\times\mathbb{R}$ that sends r to $(r, 0)$. For the other direction, let $f : \mathbb{R} \to (0, 1)$ be the map in part (3) above that provides a bijection between $\mathbb{R}$ and $(0, 1)$. It is clear that the map (which we will suggestively label "$f\times f$" — the $\times$ symbol here serves only as a part of the label and has no other connotation!) that sends $\mathbb{R}\times\mathbb{R}$ to $(0, 1)\times(0, 1)$ by sending (r, s) to $(f(r), f(s))$ is a bijection (check!), and in particular is injective. We now compose $f\times f$ with the following map g from $(0, 1)\times(0, 1)$ to $\mathbb{R}$: given (u, v) in $(0, 1)\times(0, 1)$, we represent u by its decimal expansion $0.d_1d_2d_3\ldots$, and v by its decimal expansion $0.e_1e_2e_3\ldots$, with the same understanding as in the proof that $\mathbb{R}$ is uncountable (Theorem 8.17): (1) we will write a finite decimal as one ending in an infinite string of zeros, and (2) we will rewrite any infinite decimal ending with an infinite string

of 9s as one ending in an infinite string of zeros. As discussed in that proof, this guarantees that each real number in $(0, 1)$ is represented in a *unique* fashion as an infinite decimal. We now define $g(u, v)$ to be the decimal $0.d_1e_1d_2e_2d_3e_3\ldots$, alternating the digits of u with the digits of v. It is easy to check that g is injective, so it follows (see Proposition 5.28 in Chapter 5) that the composite $g \circ (f \times f)$ that sends $\mathbb{R}\times\mathbb{R}$ to $\mathbb{R}$ is also injective. The Schröder-Bernstein theorem now shows that indeed $\mathbb{R} \times \mathbb{R}$ and $\mathbb{R}$ are in bijection!

We can immediately conclude from this that the set of complex numbers is of the same cardinality as $\mathbb{R}$. For, we have the obvious bijection $f : \mathbb{C} \to \mathbb{R} \times \mathbb{R}$ that sends the complex number $a + ib$ to the ordered pair (a, b); its inverse is the map $\mathbb{R} \times \mathbb{R} \to \mathbb{C}$ that sends (u, v) to the complex number $u + iv$.

8.4 Cantor Set

We end the chapter with a discussion of a set introduced by Cantor himself, known appropriately as the Cantor set. It has many properties of much interest in topology and analysis, and although a full study of all its properties will have to wait until further concepts are introduced in future courses, it is instructive to consider the set at this stage. We will consider one of its properties here: it is an uncountable set!

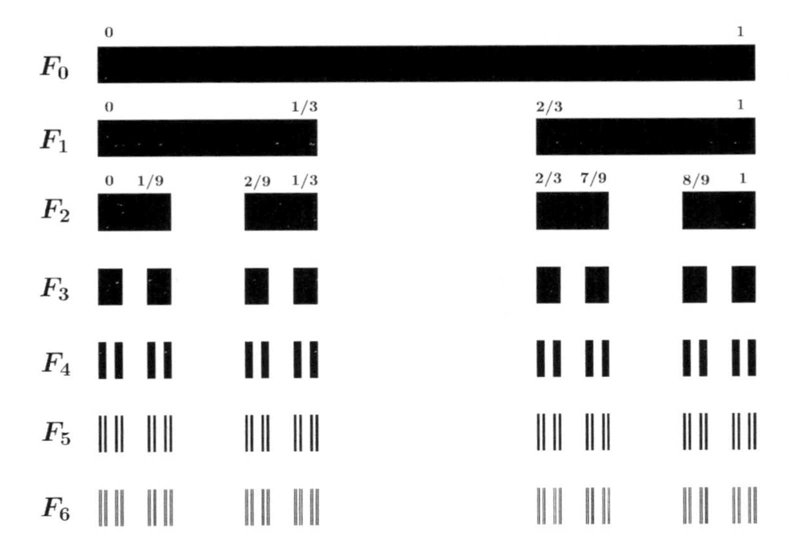

Figure 8.8. The Cantor set. At each stage, the middle third open interval is deleted to produce the next stage. (Adapted from work of Thefrettinghand at English Wikibooks. Used under the terms of the GNU Free Documentation License, licensed under Creative Commons Attribution-Share Alike 3.0 Unported `https://creativecommons.org/licenses/by-sa/3.0/deed.en` License.

We will construct the Cantor set by first constructing a sequence of auxiliary sets F_i, $i = 0, 1, \ldots$. Let F_0 denote the closed interval $[0, 1]$. We remove the open interval that represents the "middle third" to create our next set F_1. Thus, F_1 is defined as $F_0 \setminus (1/3, 2/3)$. Alternatively, $F_1 = [0, 1/3] \cup [2/3, 1]$, a disjoint union of two closed intervals, each of length 1/3. (It is worth emphasizing that the intervals are non-contiguous, that is, there is a gap between successive intervals: the upper limit of one interval is less than the lower limit of the next interval.) Next, we remove the open interval that corresponds to the "middle third" of each of the two intervals that comprise F_1. Thus, F_2 is defined as $F_1 \setminus ((1/9, 2/9) \cup (7/9, 8/9))$. Alternatively, $F_2 = [0, 1/9] \cup [2/9, 1/3] \cup [2/3, 7/9] \cup [8/9, 1]$, a disjoint union of four (non-contiguous) closed intervals, each of length 1/9. We proceed thusly: Each F_n, $n = 1, 2, \ldots$, is a disjoint union of 2^n (non-contiguous) closed intervals of length $1/3^n$ each, and we obtain F_{n+1} by removing the middle third open interval from each of the constituent intervals of F_n.

See Figure 8.8.

The *Cantor set* C is defined as what is left of $[0, 1]$ after we keep taking away the middle third open intervals in this fashion: more precisely,

$$C \stackrel{\text{def}}{=} \bigcap_{i=1}^{\infty} F_i.$$

(There are other versions of the Cantor set obtained by removing other open intervals, for instance, the middle fifths, and technically, we should call this version the *ternary* Cantor set. This version is the most common, however, and it is harmless to refer to it as *the* Cantor set, and the other versions as variants of this.)

Notice something about the set defined above: it cannot contain any open interval (a, b) with $a < b$. This is because if C contains one such interval I, then, by the definition of C, $I \subseteq F_n$ for all n. But F_n is composed of disjoint closed intervals of length $1/3^n$ as we have seen. For n large enough, we have $1/3^n < (b - a)$, and I will not fit inside any of the constituent non-contiguous intervals of F_n, giving a contradiction.

Remark 8.21. Subsets of $\mathbb{R}$ that contain no open interval are called *totally disconnected*. We have thus shown that the Cantor set is totally disconnected.

It is easy to see that C is an infinite set. For instance, since we are deleting the *middle* third of the closed intervals that constitute F_n, the end points of the closed intervals that constitute F_n never get deleted in F_j for $j > n$, and this is true for all F_n. So, for instance, the right end points of the very first constituent intervals of the various F_n are in C: $1, 1/3, 1/9, \ldots, 1/3^n, \ldots$ are all in C.

We will now show that the Cantor set is uncountable. We will have to use a theorem that we will prove in Chapter 12 ahead, the nested intervals theorem. (See Theorem 12.33 in that chapter.) In turn, this theorem invokes the notion of a limit. This is something you would have seen before if you have taken calculus, and is something we deal with extensively in Chapter 11, but at this point an intuitive understanding is sufficient. We will discuss these notions in intuitive terms now:

Suppose that $I_0 \supset I_1 \supset \cdots$ are closed intervals in $\mathbb{R}$. We call such intervals *nested*. See Figure 8.9, where an example of nested intervals has been shown, vertically offset from one another for clarity. If $I_k = [a_k, b_k]$, let $w_k = b_k - a_k$; this is known as the *width* of I_k. Since $I_0 \supset I_1 \supset \ldots$, it is clear that $w_0 \geq w_1 \geq \cdots$. In fact, in the context that we are considering, our inclusions $I_k \supset I_{k+1}$ will be strict, and we will have $w_0 > w_1 > \cdots$. Suppose that the limit of the w_k is 0 as k goes to infinity. What

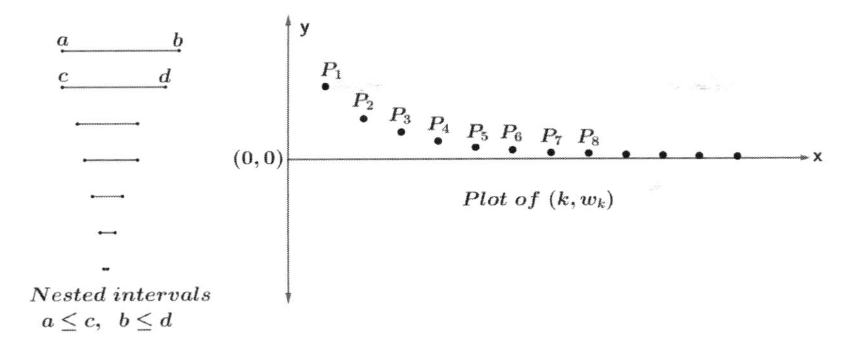

Figure 8.9. Nested closed intervals shown (left) vertically offset from one another for clarity, and an example plot shown (right) of (k, w_k) . It is clear from the graph that the w_k, which are the y-coordinates of the plotted points, has limit 0.

this effectively means in our context of decreasing w_k is that the w_k come closer and closer to 0 as k gets larger, and in fact, they come arbitrarily close to 0 and *stay close to* 0. So, for instance, if we want the w_k to be less than 10^{-3}, there will be an integer K_3, say, such that for all $k \geq K_3$, $w_k < 10^{-3}$. If we want the w_k to be less than 10^{-10}, there will be an integer K_{10}, say, such that for all $k \geq K_{10}$, $w_k < 10^{-10}$. And so on. See Figure 8.9, where an example plot of (k, w_k) has been shown. The w_k in that figure, which are the y-coordinates of the plotted points, clearly are approaching arbitrarily close to 0.

The nested intervals theorem says that if the I_j, $j = 0, 1, \ldots$, are closed nested intervals as above, then $\cap I_j$ is nonempty (this is not obvious at all, and is a false statement if the I_j are changed to be open nested intervals, for instance). Moreover (and this will be crucial to us), when the widths w_k defined above have limit 0, then $\cap I_j$ consists of *precisely* one point.

There is one more ingredient we need to show our result. Let L and R be two symbols, without further mathematical meaning. Using just these two symbols, we can create infinite sequences of the form

$$LRRRLLRLRLLRRLR\ldots, \text{ or}$$
$$LRRLLLLLLRLRLLRRRRLLLRLRLRL\ldots,$$

and so on.

Lemma 8.22. *Let L and R be two symbols. The set consisting of all infinite sequences made up from L and R is uncountable.*

Proof. The proof is just an application of the same Cantor's diagonalization argument that we used to show that $\mathbb{R}$ is uncountable, and we will sketch it. (If we think about it, real numbers are just infinite sequences that we make from ten symbols $0, 1, \ldots, 9$, except that we include a decimal point somewhere in the sequence!) We assume that there is an injective map f from $\mathbb{N}$ to this set, and write down a sequence that differs in the i-th place from the symbol that is in the i-th place of $f(i)$. This sequence cannot be in the image of f, so f cannot be surjective. □

We now prove:

Theorem 8.23. *The Cantor set is uncountable.*

Proof. Let S denote the set of all infinite sequences made from the symbols L and R. We construct a function f from the Cantor set C to S as follows: An element $x \in C$ is by definition in all the F_i. So, to start, x is either in the left interval $[0, 1/3]$ or the right interval $[2/3, 1]$ of F_1. Our image sequence $f(x)$ will have an L in the first place if x is in the left interval and an R if it is in the right interval. Thus, if we write $f(x)_i$ for the ith symbol in the desired sequence $f(x)$, this process uniquely determines $f(x)_1$.

Now assume that $f(x)_1 = L$, so $x \in [0, 1/3]$. In F_2, the interval $[0, 1/3]$ subdivides into $[0, 1/9] \cup [2/9, 1/3]$ (with the middle third open interval $(1/9, 2/9)$ deleted). View $[0, 1/9]$ as the new left interval and $[2/9, 1/3]$ as the new right interval. Since $x \in F_2$, and since $x \in [0, 1/3] \subseteq F_1$, x must either be in the new left interval $[0, 1/9]$ or the new right interval $[2/9, 1/3]$. If the former, define $f(x)_2 = L$, and if the latter, define $f(x)_2 = R$. If on the other hand $f(x)_1 = R$, so $x \in [2/3, 1]$, then in F_2, x must be in the left interval $[2/3, 7/9]$ or the new right interval $[8/9, 1]$. Once again, if the former, define $f(x)_2 = L$, and if the latter, define $f(x)_2 = R$.

We continue this process. Once we have determined $f(x)_1, \ldots, f(x)_n$, we would have determined explicitly which of the constituent closed intervals of $F_1, \ldots, F_n$ our x lies in. The interval in F_n that x lies in further subdivides into a left interval and a right interval in F_{n+1} (with the middle third open interval deleted). If x is now in the left interval, we define $f(x)_{n+1} = L$, and if in the right, we define $f(x)_{n+1} = R$. Proceeding, we determine $f(x) \in S$.

The nested intervals theorem guarantees that this map f is both injective and surjective. To show injectivity, suppose $f(x) = f(y)$. Since $f(x)_1 = f(y)_1$, our construction shows that x and y are in the same constituent interval of F_1. Denote this interval by I_1. Since $f(x)_2 = f(y)_2$, x and y are in the same constituent interval of F_2 obtained by splitting I_1. Call this I_2. Proceeding thus, we find that x and y are in the same constituent interval of F_n for all n. Thus $x \in \cap_n I_n$ and $y \in \cap I_n$. But the I_n are nested closed intervals, whose widths $1/3^n$ has limit 0 as n goes to infinity. By the nested intervals theorem, $\cap_n I_n$ consists of a single point. Both x and y have to be this point. Hence $x = y$.

For the surjectivity, we just run this argument backwards. An infinite sequence (s_n) of two symbols L and R determines a nested set of closed intervals whose widths have limit 0. If $s_1 = L$ it determines the left interval $I_1 = [0, 1/3]$ of F_1, and if $s_1 = R$ it determines the right interval $I_1 = [2/3, 1]$ of F_1. In general, having determined $I_1, \ldots,$ I_n, the next interval I_{n+1} is determined as the left interval in F_{n+1} that I_n splits into if $s_{n+1} = L$, and as the right interval that I_n splits into if $s_{n+1} = R$. These I_n are nested closed intervals whose width goes down to zero, so $\cap_n I_n = x$ for some $x \in [0, 1]$. Clearly $f(x) = (s_n)$. Thus f is surjective.

C is hence in bijection with an uncountable set (Lemma 8.22 above), so C is uncountable. □

Some features of the Cantor set. We should pause to admire those depths of the Cantor set that can be revealed just with the concepts at our disposal now; it is a truly amazing set! Many of its properties are counterintuitive. For instance, we have seen that it is uncountable (Theorem 8.23). We have also seen that it is totally disconnected (Remark 8.21). Because it is totally disconnected, we may imagine the individual elements of the Cantor set as being separated from one another, or more colloquially, as mere *dust* lying in the interval $[0, 1]$! Now we intuitively think of dust as "discrete," or in other words, countable. (For instance, if we examine the surface of a desk under a microscope, we will be able see the individual particles of dust and therefore count them.) Yet, the dust of the Cantor set is uncountable!

Notice that the Cantor set contains the endpoints of the intervals in the sets F_n we have considered. Now the set of these endpoints is countable (see Exercise 8.41 ahead), while the Cantor set is uncountable. It follows that the Cantor set contains lots more points besides just the endpoints, in fact, uncountably many of them!

Here is another feature of the Cantor set we can appreciate. We have described the Cantor set as the set of all infinite sequences formed from the symbols L and R. We may consider the subset C_1 of sequences that start with L: these correspond to those $x \in C$ that lie in the left interval $[0, 1/3]$ of F_1. We may similarly consider the subset C_2 of sequences that start with R: the set of $x \in C$ that lie in the right interval $[2/3, 1]$ of F_1. Then C_2 is just C_1 shifted by $2/3$. But more is true: each of C_1 and C_2 is just the original Cantor set C shrunk by a factor of three! (The reason for this should be intuitively clear: We may directly construct C_1, for instance, by starting with the interval F_1 instead of F_0 and proceeding exactly as in the definition of the Cantor set. But F_1 is only one-third the length of F_0. This sort of behavior is known as *self-similarity,* making the Cantor set a *fractal.*) Note that we can carry this on: the various subsets consisting of $x \in C$ that lie not just in the left interval of F_1 but also in the interval $[0, 1/9]$, or in the interval $[2/9, 1/3]$, or in the interval $[2/3, 7/9]$, or in the interval $[8/9, 1]$ of F_2, are all just translates of one another, and also are all just the original Cantor set, scaled down now by a factor of 9!

There is a very pretty characterization of the Cantor set in terms of the *ternary expansion* of real numbers in $[0, 1]$. To understand what this means, let us first consider the usual decimal expansion. What a decimal such as $x = 0.46598203\ldots$ means is the following: you first divide $[0, 1]$ into ten parts. The fact that x starts with 4 in the first place after the decimal means that it lies in the interval $[4/10, 5/10)$. Now, you further divide $[4/10, 5/10)$ into ten intervals $[4/10, 4/10 + 1/100)$, $[4/10 + 1/100, 4/10 + 2/100)$, $\cdots$, $[4/10 + 9/100, 5/10)$. The fact that x has 6 in the second place means that it lies in the interval $[4/10 + 6/100, 4/10 + 7/100)$. Now you further subdivide the interval $[4/10 + 6/100, 4/10 + 7/100)$ into ten subintervals of the form $[4/10 + 6/100 + i/1000, 4/10 + 6/100 + (i + 1)/1000)$, $i = 0, \ldots, 9$. Then x lies in the interval $[4/10 + 6/100 + 5/1000, 4/10 + 6/100 + 6/1000)$. And so on. Now instead of subdividing by a tenth each time, we could have chosen to subdivide *a third* each time! Thus, given an $x \in [0, 1]$, we would subdivide $[0, 1]$ into $[0, 1/3)$, $[1/3, 2/3)$, and $[2/3, 1)$. We would consider where x lies. If x is in the first interval, it's ternary expansion would start with $0.0\ldots$, if it is in the second it would start with $0.1\ldots$, and if in the third, it would start with $0.2\ldots$. Say it is in the third, so x starts with $0.2\ldots$. Next you would subdivide $[2/3, 1)$ into a further third: $[2/3, 2/3 + 1/9)$, $[2/3 + 1/9, 2/3 + 2/9)$, and $[2/3 + 2/9, 1]$. If x lies in, say the second interval, x would start with $0.21\ldots$. And so on. Thus, x would be represented as an infinite sequence consisting of just the digits 0, 1 and 2. Now this process of repeatedly dividing into a third is exactly what is involved in the construction of the Cantor set. Except, the Cantor set leaves out all numbers which at any stage falls into the middle interval! Thus, at no stage will the ternary expansion of an $x \in C$ contain 1. In fact, it can be seen by using the arguments we used to show that C is in bijection with the set of sequences made from L and R, that the Cantor set is *precisely* the set of real numbers in $[0, 1]$ whose ternary expansion only consists of 0 and 2! This is striking!

8.5 Further Exericses

Exercise 8.24. Practice Exercises:

> **Exercise 8.24.1.** Show that $2\mathbb{Z}$ and $3\mathbb{Z}$ have the same cardinality. (Recall that "$n\mathbb{Z}$" stands for the set $\{kn \mid k \in \mathbb{Z}\}$, the set of integer multiples of n.)
>
> **Exercise 8.24.2.** Show that for any integer $n \geq 2$, $n\mathbb{Z}$ is in bijection with $\mathbb{Z}$ and is therefore countably infinite (Proposition 8.10).
>
> **Exercise 8.24.3.** Show that the open intervals $(0, 1)$ and $(-\pi/2, \pi/2)$ have the same cardinality.
>
> **Exercise 8.24.4.** Show that the open interval $(0, 1)$ and $\mathbb{R}$ have the same cardinality.
>
> **Exercise 8.24.5.** Show that $\mathbb{Q}_{>0}$ is countably infinite by modifying the proof of Theorem 8.16 to include only the positive rational numbers in Figure 8.4.
>
> **Exercise 8.24.6.** Let S denote the set of real numbers in the interval $(0.3, 0.4)$. Modify the proof of Theorem 8.17 to show that S is uncountable.

Exercise 8.25. Suppose A and B are sets with $A \subseteq B$. If B is finite show that A is also finite. (Hint: Assume A is not finite and apply Proposition 8.4.)

Exercise 8.26. Let $A \subset B$ be sets, and assume that A is uncountable. Show that B is also uncountable.

Exercise 8.27. Let S_1 and S_2 be two countable sets. Show that $S_1 \times S_2$ is also countable. Conclude using induction that the product $S_1 \times S_2 \times \cdots \times S_n$ of n countable sets is countable, for any $n \geq 2$. (Hint: For $n = 2$, if S_1 and S_2 are both finite, the proof should be clear. In the case where S_1 or S_2 is countably infinite, mimic the proof of Theorem 8.16. Exercise 8.37 furnishes a proof along different lines, but look at that only *after* you have solved this exercise on your own!)

Exercise 8.28. Let S and T be two countable sets. Show that $S \cup T$ is also countable. (Hint: $S \cup T = S \cup (T \setminus S)$, and S and $T \setminus S$ are disjoint. If S is finite, count the elements of $S \cup T$ by shifting the count of $T \setminus S$ forward by $|S|$ and counting the elements of S first. Proceed similarly if $T \setminus S$ is finite. Otherwise, both S and $T \setminus S$ are in bijection with $\mathbb{N}$. Mimic the argument in Example 8.8(3).). See also Exercise 8.30 ahead.

Exercise 8.29. By modifying the proof of Theorem 8.17 (or otherwise), show that $\mathbb{R} \times S$ is uncountable for any nonempty set S.

Exercise 8.30. Let A_i, $i = 1, 2, \ldots$, be countable sets. Let A be the set $\cup_{i=1}^{\infty} A_i$, i.e. the union of all the sets A_i. We will show that A is also countable in this exercise. Since A_i is countable for each i, there exist one-to-one and onto functions ϕ_i from A_i to either $\mathbb{N}$ (if A_i is countably infinite) or to some finite subset $\{1, 2, \ldots, n_i\}$ of $\mathbb{N}$ (if A_i is finite). Define a function from f from A to $\mathbb{N} \times \mathbb{N}$ as follows: for any $x \in A$, let i be least such as $x \in A_i$ (note that the well-ordering principle—see Definition 8.12—guarantees such an i exists, although this fact is intuitively obvious). Define $f(x)$ to be the pair $(i, \phi_i(x))$.

(1) Show that f is injective.

(2) Use f and Exercise 8.27 and Corollary 8.13 to show that A is countable.

Exercise 8.31. In this exercise we will show that the set $\mathbb{Q}[x]$, the set of all polynomials with coefficients in $\mathbb{Q}$, is countable.

(1) For each i, $i = 0, 1, 2, \ldots$, let P_i denote the set of all polynomials in $\mathbb{Q}[x]$ of degree exactly i. Thus, P_0 consists of all degree zero polynomials, namely the constants, that is, $\mathbb{Q}$, while P_1 consists of all linear polynomials, i.e., polynomials of the form $ax + b$, $a, b \in \mathbb{Q}$, $a \neq 0$, P_2 consists of all quadratic polynomials, i.e., polynomials of the form $ax^2 + bx + c$, $a, b, c \in \mathbb{Q}$, $a \neq 0$, and so on. Show that each P_i is countable (and infinite).

(2) Use Exercise 8.30 to show that $\mathbb{Q}[x]$ is countable (and infinite).

Exercise 8.32. We call a number α (real or complex) *algebraic* if it satisfies a polynomial with coefficients in $\mathbb{Q}$. We will show that the set of all algebraic numbers is countable (and infinite). We will need one result that you may have seen earlier: any polynomial with coefficients in $\mathbb{Q}$, of degree n, has at most n distinct roots in the complex numbers. (In fact, more is true: every such polynomial has *exactly* n roots in the complex numbers, if we count each root as many times as it appears in a factorization of the polynomial into linear factors; for instance, the polynomial $x^2 - 2x + 1$ has the root 1 counted twice, corresponding to the fact that $x^2 - 2x + 1 = (x - 1)^2$. Note that given a polynomial $f \in \mathbb{Q}[x]$, there is no guarantee that f has any root whatsoever in $\mathbb{Q}$, but it is guaranteed to have all its roots in $\mathbb{C}$: this is the celebrated *Fundamental Theorem of Algebra*.)

(1) For each $d \in \mathbb{N}$, let P'_d denote the set of all polynomials of degree exactly d in $\mathbb{Q}[x]$ whose highest coefficient is 1. (Such a polynomial is called a *monic polynomial of degree* d.) Show that P'_d is countable (and infinite).

(2) For each nonzero algebraic number α, show that there exists a monic polynomial in $\mathbb{Q}[x]$ satisfied by α. Now let d be the least degree of all monic polynomials in $\mathbb{Q}[x]$ satisfied by α. Show that there is exactly one monic polynomial of degree d that alpha satisfies. This polynomial is called the *minimum polynomial of* α, and α is said to be *of degree* d. (Hint: if there are two such monic polynomials of degree d satisfied by α, play with them to find a contradiction to how d is defined.)

(3) Write A_d for the set of algebraic numbers of degree d. By the result on roots of polynomials discussed in the introduction to this problem, each polynomial $p \in P_d'$ has at most d distinct roots. Once and for all, for each $p \in P_d'$, select an ordering of the (at most d) roots of p. Use this ordering and the minimal polynomial from part (2) above to construct an injective map from A_d to the set $\mathbb{N} \times P_d'$. Combine Exercise 8.27 and part (1) to show that A_d is countable.

(4) Use Exercise 8.30 to conclude that the set of algebraic numbers is countable. Exhibit infinitely many distinct algebraic numbers to conclude that the set of algebraic numbers is countably infinite.

Exercise 8.33. We saw in Example 8.20(1) that the closed interval $[a,b]$ and the open interval (a,b) are in bijection: the proof used the Schröder-Bernstein theorem. This exercise asks you to construct *directly* a bijection $f : [a,b] \to (a,b)$. (Here is a hint: if x is a symbol, then $\{x\} \cup \mathbb{N}$ is countable, and in fact, the map that sends x to 1, 1 to 2, 2 to 3, ..., provides a bijection with $\mathbb{N}$. Find two disjoint sets S and T in (a,b) that are each in bijection with $\mathbb{N}$ and form the bijections between $\{a\} \cup S$ and S as also $\{b\} \cup T$ and T described above. Extend to the desired bijection between $[a,b]$ and (a,b).)

Exercise 8.34. Let S be any subset of the plane (i.e., $\mathbb{R} \times \mathbb{R}$; see Example 5.42 in Chapter 5) that contains a region of the form $(a,b) \times (c,d)$ for some real numbers $a < b$ and $c < d$. (Such a region is also known as an *open rectangle*.) Show that S is in bijection with $\mathbb{R}$.

Exercise 8.35. Recall from Definition 5.11 in Chapter 5 that if S is a set, the power set $\mathcal{P}(S)$ of S is the set of all subsets of S. Show that if S is any set, there cannot exist a bijection between S and $\mathcal{P}(S)$. (Hint: If S is finite, then this result falls out of Exercise 5.53 in Chapter 5, since the cardinality of S is clearly less than the cardinality of $\mathcal{P}(S)$ in this case. Now let S be an arbitrary set, which we can assume to be infinite as we have established the result for finite sets anyway. Let $f : S \to \mathcal{P}(S)$ be a bijection. Consider the subset A of S defined as $A = \{s \in S \mid s \notin f(s)\}$. Play with it.)

The significance of this result is the following: it is possible to order the cardinalities of sets (including infinite sets) in a manner that generalizes the ordering of cardinalities of finite sets (which is just the natural ordering of the nonnegative integers). In this generalized scheme of ordering, we say $|X| \leq |Y|$ if there exists an injective map $f : X \to Y$, and we say $X < Y$ if there exists an injective map $f : X \to Y$ but no injective map $Y \to X$. It follows from the Schröder-Bernstein theorem (Theorem 8.18) that we may equivalently define $X < Y$ if there exists an injective map $f : X \to Y$ but no bijection between X and Y. Since for any set S the map $f : S \to \mathcal{P}(S)$ defined by $s \mapsto \{s\}$ is clearly an injection, we find $|S| \leq |\mathcal{P}(S)|$ for any set S. The result of this exercise therefore shows that something stronger is afoot: we actually have a *strict* inequality, $|S| < |\mathcal{P}(S)|$.

See Figure 8.1 earlier in the chapter: $\mathcal{P}(\mathbb{R})$ is of a different cardinality from $\mathbb{R}$, and $\mathcal{P}(\mathcal{P}(\mathbb{R}))$ is of a different cardinality from $\mathcal{P}(\mathbb{R})$.

Exercise 8.36. If X and Y are sets, let Y^X denote the set of all functions from X to Y.

(1) If X and Y are both finite sets, what is the cardinality of Y^X?

(2) Let $Y = \{0, 1\}$. Show that $|Y^X| = |\mathcal{P}(X)|$ for any set X.

Exercise 8.37. Although we will formally consider the theorem on the unique prime factorization of integers ahead in Chapter 10, you are doubtless familiar with the result already: Every integer (> 1) factors into a product of primes, and this factorization is *unique* except for the order in which the primes are written. (The uniqueness here means that if an integer n has the factorization $p_1^{k_1} p_2^{k_2} \cdots p_l^{k_l}$, where the p_i are primes and the k_i positive integers, and if n appears to also have the factorization $q_1^{r_1} q_2^{r_2} \cdots q_s^{r_s}$, where the q_i are primes and the r_i positive integers, then $l = s$, and after reordering, say, the primes q_i if necessary, we have $p_i = q_i$ and $k_i = r_i$ for all i.) We can use the uniqueness of the prime factorization to give an alternative proof of the result in Exercise 8.27 above. It is sufficient to prove that $\mathbb{N} \times \mathbb{N}$ is countable (why?). We define a function $f : \mathbb{N} \times \mathbb{N} \to \mathbb{N}$ by sending (m, n) to $2^m 3^n$. The uniqueness part of the prime factorization theorem shows that this function is injective, so $\mathbb{N} \times \mathbb{N}$ is in bijection with its image under f (see Remark 5.21 in Chapter 5), and Theorem 8.11 then shows that $\mathbb{N} \times \mathbb{N}$ is countable.

Using the ideas above (or otherwise!), do the following exercises:

(1) Let $\mathcal{PF}(\mathbb{N})$ denote the set of *finite* subsets of $\mathbb{N}$. Show that $\mathcal{PF}(\mathbb{N})$ is countable.

(2) For any $n \in \mathbb{N}$, let S_n denote the set of all finite subsets of $\mathbb{N}$ whose least element is n. Using the result in part (1), show that S_n is countable.

(3) Use part (2) to show that $\mathcal{PF}(\mathbb{N})$ can be partitioned into infinitely many (disjoint) sets, each of the same cardinality as $\mathcal{PF}(\mathbb{N})$.

(4) Combine parts (3) and (1) above to derive a partition of $\mathbb{N}$ into infinitely many (disjoint) subsets of the same cardinality as $\mathbb{N}$. (See also Exercise 8.38 below.)

Exercise 8.38. In this exercise we will partition $\mathbb{N}$ into infinitely many (disjoint) sets, each of the same cardinality as $\mathbb{N}$ in a manner different from that of Exercise 8.37(4) above.

(1) Let n be a positive integer. Show that the set M_n of all positive multiples of n is in bijection with $\mathbb{N}$. (This is really just an observation!)

(2) Let p be a prime, and let $q_1, q_2, \ldots, q_t$ be primes distinct from p. Show that the set $M_p \setminus (M_{q_1} \cup M_{q_2} \cup \cdots \cup M_{q_t})$ is in bijection with $\mathbb{N}$. (Hint: Exhibit infinitely many elements in M_p that are not in any of the M_{q_i}.)

(3) Use parts (1) and (2) to partition $\mathbb{N}$ into infinitely many (disjoint) sets, each of the same cardinality as $\mathbb{N}$. (Do not forget to include 1 in your partition somewhere!)

Exercise 8.39. Exercise 8.35 above shows us that $\mathcal{P}(\mathbb{N})$ is not countable. This exercise will further show that the cardinality of $\mathcal{P}(\mathbb{N})$ is the same as that of $\mathbb{R}$.

(1) Suppose sets A and B have the same cardinality. Establish a bijection between $\mathcal{P}(A)$ and $\mathcal{P}(B)$. (Hint: if $f : A \to B$ is a bijection, use f to define a suitable map from $\mathcal{P}(A)$ to $\mathcal{P}(B)$.)

(2) Find an injective map from $\mathcal{P}(\mathbb{N})$ to the interval $(0, 1)$. (Hint: find a way of associating to a subset of the positive integers an infinite decimal in $(0, 1)$.)

(3) Find an injective map from the interval $(0, 1)$ to $\mathcal{P}(\mathbb{Z}_{\geq 0} \times \mathbb{Z}_{\geq 0})$. (Hint: find a way of associating to the i-th digit of an element r in $(0, 1)$, an element (a_i, b_i) in $\mathbb{Z}_{\geq 0} \times \mathbb{Z}_{\geq 0}$, and consider the collection of all such (a_i, b_i) over all the digits of r.)

(4) Use parts (3) and (1) above to show that there is an injective map from the interval $(0, 1)$ to $\mathcal{P}(\mathbb{N})$, and finish the problem by appealing to the Schröder-Bernstein theorem as well as to parts (2) and (4) in Example 8.15.

Exercise 8.40. We will prove Proposition 8.1. Let $f : S \to \{1, 2, \dots, n\}$ be a bijection between S and $\{1, 2, \dots, n\}$, and likewise, let $g : S \to \{1, 2, \dots, m\}$ be a bijection between S and $\{1, 2, \dots, m\}$. We will show that $m = n$. Note that by Exercises 5.36 and 5.29 in Chapter 5, $h \stackrel{\text{def}}{=} g \circ f^{-1}$ is a bijection between $\{1, 2, \dots, n\}$ and $\{1, 2, \dots, m\}$. We will prove here: if there exists an injective map from $\{1, 2, \dots, n\}$ to $\{1, 2, \dots, m\}$, then $n \leq m$. Applying this first to h and then h^{-1} (which are both bijective, and in particular injective), we find $n \leq m$ and $m \leq n$, as desired.

We will fix n and prove this by induction on m. Thus, we will let $P(m)$ be the statement: "If there exists an injective map from $\{1, 2, \dots, n\}$ to $\{1, 2, \dots, m\}$, then $n \leq m$," and we will show that $P(m)$ is true for all $m \in \mathbb{N}$.

(1) Show that $P(1)$ is true. (Hint: Let ϕ be the injective map from $\{1, 2, \dots, n\}$ to $\{1\}$. If $n > 1$, what can you say about $\phi(2)$?)

(2) Now we will assume that $P(k)$ is true for some $k \geq 1$, and we will show that $P(k + 1)$ must also be true. Let $\phi : \{1, 2, \dots, n\} \to \{1, 2, \dots, k + 1\}$ be an injective map. Assume that $n > k + 1$.

  (a) Assume that the range of ϕ (see Definition 1 in Chapter 5) does not include $k + 1$. Show that $n > k + 1$ is impossible in this case. (Hint: The induction hypothesis.)

  (b) If $k + 1$ is in the range of ϕ and if $\phi(n) = k + 1$, show by considering the restriction of ϕ to the subset $\{1, 2, \dots, n - 1\}$ that $n > k + 1$ is impossible in this case too.

  (c) Finally, continue to assume that $k + 1$ is in the range of ϕ, but now assume that $\phi(s) = k + 1$ for some $s < n$. Consider an auxiliary map $\psi : \{1, 2, \dots, n\} \to \{1, 2, \dots, n\}$ that sends s to n, n to s, and acts as the identity map on all other elements. Combine the maps ψ and ϕ and apply part (2b) to show that $n > k + 1$ is impossible in this case as well.

Remark 8.40.1. Observe that the statement that $P(m)$ is true for all $m \in \mathbb{N}$ is nothing but the pigeonhole principle (see Principle 2.2 in Chapter 2). This is clearer if we consider $P(m)$ in its contrapositive form: If $n > m$, then there cannot exist an injective map from $\{1, 2, \ldots, n\}$ to $\{1, 2, \ldots, m\}$, or what is the same thing, any map ϕ from $\{1, 2, \ldots, n\}$ to $\{1, 2, \ldots, m\}$ must satisfy $\phi(i) = \phi(j)$ for distinct i and j in $\{1, 2, \ldots, n\}$. Now, we can view a placement of n letters into m pigeonholes as a map ϕ from $\{1, 2, \ldots, n\}$ to $\{1, 2, \ldots, m\}$ as follows: if we place the i-th letter in the p-th pigeonhole, then we define $\phi(i) = p$. The truth of $P(m)$ now says that ϕ is not injective, that is, for distinct i and j, the i-th and j-th letters will end up in the same pigeonhole.

Thus, the pigeonhole principle, which really is an "obvious" principle, has a formal proof of its validity based on the principle of induction, and therefore (see Exercise 8.43 below) on the well-ordering principle.

Exercise 8.41. Show that the subset of the Cantor set consisting of the set of endpoints of the intervals F_n is a countable set. Conclude that the Cantor set contains uncountably many elements that are not endpoints of the intervals of the various F_n. (Hint: Set up the set of endpoints of the F_n as a subset of $\mathbb{Z}_{\geq 0} \times \mathbb{Z}_{\geq 0}$.)

Exercise 8.42. We saw in Theorem 8.23 that the Cantor set is uncountable. Extend the ideas in the proof of that theorem to show that the Cantor set has the same cardinality as $\mathbb{R}$. (Hint: It is sufficient to show that the Cantor set has the same cardinality as $[0, 1]$ — why? Connect the sequences considered in that theeorem with elements of $[0, 1]$. It may help you to think in terms of bifurcating intervals such as $[a, b]$ or $[a, b)$ into $[a, (a + b)/2)$ and $[(a + b)/2, b]$ or $[a, (a + b)/2)$ and $[(a + b)/2, b)$.)

Exercise 8.43. We will show here that the well-ordering principle (Definition 8.12) is equivalent to the principle of induction. (See Definition 7.1 in Chapter 7 for the standard form of the principle of induction, and Definition 7.7 in the same chapter for the strong form. Note that the standard form and the strong form are equivalent to one another, as is shown in the Notes at the end of that chapter.)

(1) Assume the well-ordering principle holds. We will show that the strong form of the principle of induction must also hold. Let $S(n)$, $n \in \mathbb{N}$, be a family of statements. Assume the two hypotheses of the strong form of the principle of induction hold about the statements $S(n)$. Assume to the contrary that $S(n)$ is not true for all n. Let Σ be the set of all $n \in \mathbb{N}$ such that $S(n)$ is not true. By assumption it is nonempty. By the well-ordering principle, it has a least element k. What can you say about k and $S(1), \ldots,$ $S(k-1)$? Use your observations and the hypotheses of the strong form of induction to arrive at a contradiction.

(2) Now assume that the strong form of the principle of induction holds. We will show that the well-ordering principle must also hold. Let Σ be

a nonempty subset of $\mathbb{N}$. Assume it does not have a least element. For $n \in \mathbb{N}$, write $S(n)$ for the statement that n is not in Σ. Show that the two hypotheses of the strong form of the principle of induction must hold for the statements $S(n)$. Use this to show that the well-ordering principle must hold.

9

Equivalence Relations

Much of mathematics consists of studying objects *governed by some loose notion of equality.* The meaning of this will become clearer as we study more examples later in the chapter (and in other courses as you proceed further in your study of mathematics), but here is an example from high school geometry: when we study triangles in the plane, we typically do not differentiate between *congruent* triangles—all that matters to us is the lengths of the three sides, not where it is located in the plane or how it is aligned. Here is another example, this time from trigonometry: when considering trigonometric ratios in right triangles, we do not differentiate between *similar* right triangles—all that matters to us is the angles in that right triangle, not the lengths of the sides. And here is an example that you have already considered in Exercise 1.3.1, Chapter 1: we create a new number system $\mathbb{Z}/2\mathbb{Z}$ by considering all even integers to be a new number, and all odd integers to be another new number, that is, we do not differentiate between two integers that leave the same remainder on dividing by 2—all that matters to us is whether the remainder is 0 or 1.

We will develop in this chapter the machinery to describe this concept of objects being the same up to some loose notion of "sameness." The basic idea that we will consider is that of a relation on a set (and more generally, of a relation from a set A to a set B). We will specialize this notion of relation to the notion of an *equivalence relation,* which is precisely the concept used in mathematics to specify an appropriate notion of sameness.

9.1 Relations, Equivalence Relations, Equivalence Classes

To motivate the definition of relations, let us consider the following:

Example 9.1. Take A to be the set of all adults in a room, and take B to be the set of all children. We wish to create a mathematical object that will represent the fact that a certain adult, call her Maria, is the parent of a certain child, call him Jose. The simplest way to designate this relation is to simply write the two names in order, as an ordered pair: (Maria, Jose), with the understanding that the entry in the first slot of this ordered pair is the parent, and the entry in the second slot is the child. Thus, another ordered pair such as (Hamid, Fatima) would represent the fact that Hamid is the parent of Fatima. If we were to consider the totality of parent-child relationships, then we would consider the set, call it R, of all ordered pairs of the form (x, y), where $x \in A$ is the parent of $y \in B$. This set R is thus a subset of $A \times B$. (Note that not all ordered pairs in $A \times B$ will represent a parent-child relationship. If an adult a is not the parent of a child b, then the element (a, b) will not be in R.)

For example, if the adults in the room are Maria, Hamid, Blake, and Julie, and if the children in the room are Jose, Fatima, Syliva, Abdul, David, Ravi, and Laura, then

$$A = \{\text{Maria, Hamid, Blake, Julie}\}$$

and

$$B = \{\text{Jose, Fatima, Syliva, Abdul, David, Ravi, Laura}\}.$$

Now suppose that Jose and Sylvia are both Maria's children, and Fatima is Hamid's daughter. Also suppose that Blake does not have children and Julie's children are not in the room, and that the parents of Abdul, David, Ravi, and Laura are also not in the room. Then, R is the subset

$$\{(\text{Maria, Jose}), (\text{Maria, Sylvia}), (\text{Hamid, Fatima})\}$$

of $A \times B$.

Here is another example, a slight twist on the one before:

Example 9.2. Let us assume that we are in a different room at a different time, and once again take A to be the set of all adults in this room, and B to be the set of all children in this room. This time, let us consider the relation from A to B represented by "is the grandparent of." We will use the same machinery of an ordered pair to represent this relation; thus, an ordered pair such as (x, y) will signify the fact that the person x is the grandparent of the child y. As before, the totality of grandparent-grandchild relationships will be represented by the set, call it S, of *all* ordered pairs of the form (x, y), where $x \in A$ is the grandparent of $y \in B$. This set S is of course a subset of $A \times B$.

For example, suppose that Maria, Hamid, Blake, and Julie have moved over to this room, and there are no other adults in this room, so

$$A = \{\text{Maria, Hamid, Blake, Julie}\}$$

as before, but

$$B = \{\text{Ernesto, Felicita, Adrian, Abbie, Laura}\}.$$

Maria has an older daughter Guadalupe, whose children are Ernesto and Felicita, and Abbie is Julie's daughter's daughter. Hamid and Blake have no grandchildren, and Adrian and Laura's grandparents are elsewhere. Then

$$S = \{(\text{Maria, Ernesto}), (\text{Maria, Felicita}), (\text{Julie, Abbie})\}.$$

Motivated by these examples, we have the following:

Definition 9.3. Let A and B be sets. A *relation from A to B* is a subset R of $A \times B$.

Well that was simple! Almost deceptively so. Lurking in this definition is the possibilty that relations are allowed to be quite wild! In Examples 9.1 and 9.2, we considered only certain natural subsets of $A \times B$—natural in the sense that they represented meaningful relationships such as parent-child and grandparent-grandchild. In Definition 9.3 however, we are allowing *arbitrary* subsets of $A \times B$ to represent a relation! But this is just a typical example of how mathematics works: since what is a "meaningful" relation and what is not varies from person to person and from situation to situation and cannot be pinned down exactly, mathematics simply *decrees* that all possible subsets of $A \times B$ can represent a relation from A to B! But then how do we give practical meaning to a relation defined by an arbitrary subset of $A \times B$? Well, we don't bother! In general we don't try to imbue an arbitrary subset of $A \times B$ with practical significance (although we'd certainly be delighted if someone discovers some significance one day!), and simply say that the relation is defined by the subset! Here is an example:

Example 9.4. For simplicity, consider

$$A = \{\text{Apple, Banana, Orange}\}$$

and B to be $\{1, 2, 3\}$. Take R to be the subset

$$\{(\text{Banana}, 2), (\text{Banana}, 3), (\text{Orange}, 2)\}$$

of $A \times B$. This is indeed a relation from A to B. But what possible meaning can we give to this relation? Don't even try! Just accept that, mathematically speaking, it is a relation!

Remark 9.5. If R is a relation from A to B and if $(a, b) \in R$, we say "a is related to b." Sometimes, we write aRb to indicate that a is related to b. Another symbol that is often used for relations is the "tilde:" instead of writing $(a, b) \in R$ or aRb, we often write $a \sim b$.

While the previous example should convince you that relations can be quite wild beasts, we will now consider a more restricted type of a relation that captures the notion of "sameness" that we often consider in mathematics, including those considered in the examples at the very beginning of this chapter. But to begin with, some supporting notions:

First, when A and B are the same set, a relation R from A to B (in other words a subset R of $A \times A$) is simply known as a *relation on A*. We will now consider various kinds of relations on a set A:

Definition 9.6. A relation R on A is said to be *reflexive* if $(a, a) \in R$ for all $a \in A$.

Example 9.7. Take A to be $\mathbb{Z}$, and R to be the relation "less than or equal to ($\leq$)." Thus, $(a, b) \in R$, for $a, b \in \mathbb{Z}$, if and only if $a \leq b$. Then $(n, n) \in R$ for all $n \in \mathbb{Z}$, because clearly $n \leq n$ for all $n \in \mathbb{Z}$. So the relation $\leq$ is a reflexive relation.

By contrast, if the relation had been " less than ($<$)" (so $(a, b) \in R$, for $a, b \in \mathbb{Z}$, if and only if $a < b$), then this relation would not have been reflexive. In fact, for *no* $n \in \mathbb{Z}$ would you find $(n, n) \in R$, since it is never true that $n < n$.

Definition 9.8. A relation R on A is said to be *symmetric* if $(a, b) \in R$ implies that (b, a) is also in R.

Example 9.9. Let A be the set of all adults in a city, and let R be the relation on A defined by "is married to." Thus, $(x, y) \in R$ means that x is married to y. But if x is married to y, then clearly y is also married to x. There is thus symmetry in this relation: (y, x) is also in R. So, R is a symmetric relation.

By constrast, if for the same set A above we considered the relation "is a brother of", then this relation need not be symmetric. You could have a brother-sister pair in the city, so if x is the brother and y the sister, then $(x, y) \in R$, but clearly $(y, x) \notin R$! (Of course, if by some odd chance, your city is such that there are no brother-sister pairs in it, then "is a brother of" would actually be a symmetric relation! This is because any $(x, y) \in R$ would have to represent a brother-brother pair, so y would be a brother of x, and (y, x) would be in R.)

Definition 9.10. A relation R on A is said to be *anti-symmetric* if $(a, b) \in R$ and $(b, a) \in R$ implies $a = b$.

Example 9.11. The relation "$\leq$" on $\mathbb{Z}$ that we considered in Example 9.7 above is clearly anti-symmetric: if $a \leq b$ and $b \leq a$, then we must have $a = b$.

By contrast, the relation "is married to" in Example 9.9 is not anti-symmetric. If (x, y) is in this relation, then as we have seen, (y, x) is also in this relation (the relation is symmetric), and then, the fact that (x, y) and (y, x) are both in the relation certainly does not mean that $x = y$! (If William is married to Audrey, then Audrey is of course married to William; we most certainly cannot conclude from this that William and Audrey are the same person!)

Definition 9.12. A relation R on A is said to be *transitive* if $(a, b) \in R$ and $(b, c) \in R$ implies $(a, c) \in R$.

Example 9.13. The relation "$\leq$" on $\mathbb{Z}$ that we considered in Example 9.7 above is clearly transitive: if $a \leq b$ and $b \leq c$, then we must have $a \leq c$.

By contrast, if A is the set of all human beings, then the relation "is a friend of" is not transitive. If (a, b) is in this relation, this means a is a friend of b. If further (b, c) is in this relation, then this means b is a friend of c. Now of course, a friend of a friend need not be a friend at all—so a need not be a friend of c in general. (Moreover, this is not just a possibility—we can actually exhibit many such instances where a is a friend of b, b is a friend of c, but a is not a friend of c.) Thus, this relation is not transitive.

With these concepts under our belt, let us consider the first example at the beginning of this chapter: we consider triangles "up to congruence" in geometry. Here, the set A is the set of all triangles in the plane, and the relation R is "congruence:" if T_1 and T_2 are triangles, we say (T_1, T_2) is in R if T_1 is conguent to T_2. Now let us consider the properties of this relation. It is clearly reflexive: every triangle is congruent to itself. It is symmetric as well: if T_1 is congruent to T_2, then T_2 is congruent to T_1 too. Finally, the relation is transitive: if T_1 is congruent to T_2 and T_2 is congruent to T_3, then T_1 is congruent to T_3.

This example is not an exception: it turns out that these three properties of R above are precisely those that capture the notion of "sameness" one often encounters in mathematics. As another instance, consider the notion of integers up to being odd or even. The set here is $\mathbb{Z}$, and the relation R is the "same remainder on dividing by 2," that is,

$(m, n) \in R$ if m and n either both leave a remainder of 0 (in which case we call them both even) or they both leave a remainder of 1 (in which case we call them both odd). We check that this relation is (1) reflexive: yes, m leaves the same remainder as m on dividing by 2 (a rather silly statement, but something we need to consider to test reflexivity), so $(m, m) \in R$ for all integers m; (2) symmetric: yes, if m leaves the same remainder as n, then clearly n leaves the same remainder as m; and (3) transitive: yes, because if m and n leave the same remainder, and n and s leave the same remainder, then of course, m and s leave the same remainder.

We give this a name:

Definition 9.14. A relation R on a set A is said to be an *equivalence relation* if R is (1) reflexive, (2) symmetric, and (3) transitive.

Remark 9.15. If R is an equivalence relation on a set A, then by definition it is symmetric, so for $a, b, \in A$, if a is related to b, b is necessarily related to a. We will describe this by the symmetric expression "a and b are related."

Before proceeding to study other examples of equivalence relations, let us consider one other notion:

Definition 9.16. Given an equivalence relation R on a set A, the *equivalence class* of an element $a \in A$, denoted $[a]_R$ or simply $[a]$ if the context is clear, is the set consisting of all elements of A that a is related to. Thus, $[a]_R = \{b \in A \mid (a, b) \in R\}$. We will refer to any $b \in [a]_R$ as a *representative* of the equivalence class $[a]_R$.

We will prove Theorem 9.19 below, which is a set of results that is quite critical to how equivalence classes are used in mathematics. A piece of terminology first:

Definition 9.17. If A_α, $\alpha \in I$ are nonempty subsets of a set A with the property that (1) $A_\alpha \cap A_\beta = \emptyset$ if $\alpha \neq \beta$, and (2) $\cup_{\alpha \in I} A_\alpha = A$, we say that A is the *disjoint union* of the A_α, and we write this as $A = \sqcup_{\alpha \in I} A_\alpha$. We also describe the collection of these A_α as a *partition of A into disjoint nonempty subsets.*

If I is finite in the situation above, say $I = \{1, 2, \ldots, n\}$, we often write out the disjoint union explicitly as $A = A_1 \sqcup A_2 \sqcup \cdots \sqcup A_n$.

Example 9.18. We consider some examples of the partition of a set into disjoint subsets:

(1) Let $A = \{a, b, c, d, e\}$, and let $A_1 = \{a\}$ and $A_2 = \{b, c, d, e\}$. Then A_1 and A_2 partition A into disjoint subsets.

(2) Let $A = \mathbb{R}^2$, and for each $\alpha \in \mathbb{R}$, let $A_\alpha = \{(\alpha, y) \mid y \in \mathbb{R}\}$. Then the subsets A_α are the vertical lines passing through $(\alpha, 0)$, $\alpha \in \mathbb{R}$, and $\mathbb{R}^2$ is the disjoint union of these A_α.

(3) Let A be the set of all triangles in the plane. Let A_a be the subset of all triangles all of whose angles are acute (less than $\pi/2$), let A_r be the set of right triangles, and let A_o be the subset of all triangles one of whose angles is obtuse (greater than $\pi/2$). Then A is the disjoint union of A_a, A_r, and A_o.

Theorem 9.19. *Let R be an equivalence relation on a set A, and let a and b be arbitrary elements of A. Then (see Figure 9.1):*

(1) $a \in [a]_R$.

(2) *If a and b are related, the equivalence classes of a and b are the same sets, that is, $[a]_R = [b]_R$.*

(3) *If a and b are not related, the equivalence classes of a and b are disjoint, that is, $[a]_R \cap [b]_R = \emptyset$.*

(4) *Given a and b, either $[a]_R = [b]_R$, and this happens when a and b are related, or else $[a]_R \cap [b]_R = \emptyset$, and this happens when a and b are not related.*

(5) *A is the disjoint union of its equivalence classes, that is, $A = \sqcup_{\substack{\text{one } a \text{ from each} \\ \text{equivalence class}}} [a]_R$.*

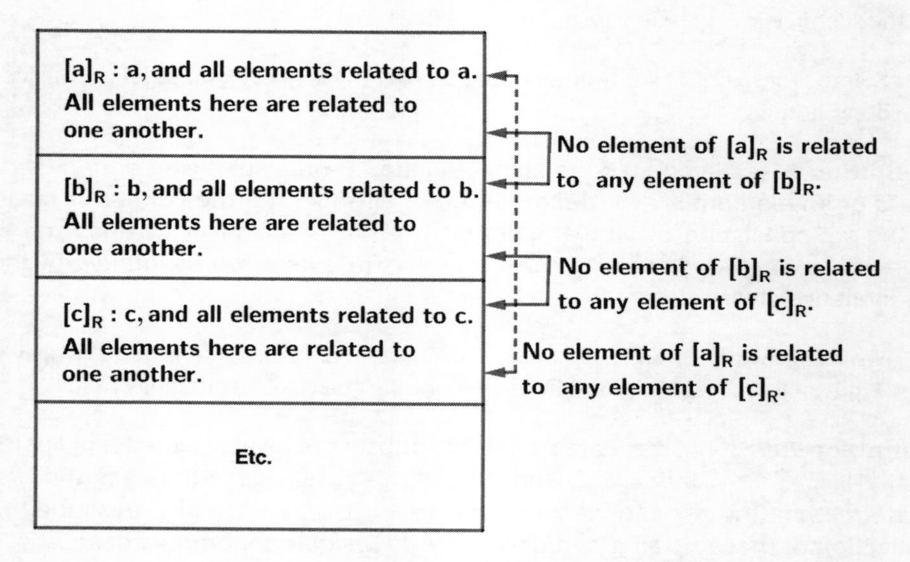

Figure 9.1. The partitioning of a set into its various equivalence classes.

Proof. We consider each statement one by one:

(1) This follows from the reflexivity property of equivalence relations: $(a, a) \in R$, so by definition of $[a]_R$, $a \in [a]_R$.

(2) Note that $[a]_R$ and $[b]_R$ are sets. So, to prove that these two sets are equal, we will show that each is contained in the other. Consider any element of $[a]_R$. By definition, if $x \in [a]_R$, then $(a, x) \in R$. By the symmetry property of equivalence relations, $(x, a) \in R$ as well. We thus have $(x, a) \in R$, and then $(a, b) \in R$ as a is related to b. By the transitivity property of equivalence relations, we find $(x, b) \in R$. Invoking symmetry again, we find $(b, x) \in R$. This means, just by definition of $[b]_R$, that $x \in [b]_R$. Since x was an arbitrary member of $[a]_R$, we have shown that $[a]_R \subseteq [b]_R$. Working in the other direction, consider an arbitrary $y \in [b]_R$; by

definition this means $(b, y) \in R$. We thus have $(a, b) \in R$ and $(b, y) \in R$, which implies, by transitivity, $(a, y) \in R$. But this says that $y \in [a]_R$. Thus, $[b]_R \subseteq [a]_R$. We conclude now that $[a]_r = [b]_R$.

(3) Suppose to the contrary that $[a]_R \cap [b]_R$ is nonempty, and let x be an element of this intersection. By definitions of the set $[a]_R$ and $[b]_R$, we find $(a, x) \in R$ and $(b, x) \in R$. Applying the property of symmetry of equivalence relations to this second containment, we find $(x, b) \in R$ as well. Stringing together $(a, x) \in R$ and $(x, b) \in R$ using the transitivity property of equivalence relations, we find $(a, b) \in R$, that is, a is related to b. But we are given that a and b are not related! Hence our original assumption must have been flawed. We have thus shown that $[a]_r \cap [b]_R = \emptyset$.

(4) This is just an amalgam of parts (2) and (3). We have two mutually exclusive possibilities: either $(a, b) \in R$ or $(a, b) \notin R$. If $(a, b) \in R$, then $[a]_R = [b]_R$ by part (2), and if $(a, b) \notin R$, then $[a]_R \cap [b]_R = \emptyset$ by part (3).

(5) Since any equivalence class $[x]_R$ can be written as $[a]_R$ for any a in that class (part (2) above), we may once and for all select one representative a from each equivalence class, and we can write the union of the equivalence classes as the union of these $[a]_R$, as a ranges through the representatives that we have selected. The various sets $[a]_R$, as a ranges through this representative set, are disjoint: this is just part (4) above. We thus find that the union $\cup_{\substack{\text{one } a \text{ from each} \\ \text{equivalence class}}} [a]_R$ is a *disjoint* union. This disjoint union is clearly a subset of A, since each $[a]_R$ is a subset of A. To show that it equals A, we only need to show the reverse containment, that is, that $A \subseteq \sqcup_{\substack{\text{one } a \text{ from each} \\ \text{equivalence class}}} [a]_R$. But for this, we have part (1) at our disposal: any $x \in A$ belongs to the equivalence class $[x]_R$, and if a_x (say) is the representative we have chosen in $[x]_R$ (so $[x]_R$ is the same as $[a_x]_R$), we find that x appears in the set $[a_x]_R$ and hence in the set $\sqcup_{\substack{\text{one } a \text{ from each} \\ \text{equivalence class}}} [a]_R$.

□

To review (see Figure 9.1 again): we find that an equivalence relation R on a set A gives rise to sets ("equivalence classes") of the form $[a]_R$—the set of all elements of A which a is related to. Each $a \in A$ is a member of its own equivalence class $[a]_R$. Any two elements in the same equivalence class are related to each other, and determine the same equivalence class. Two *distinct* equivalence classes are *disjoint*, and the union of these (disjoint) equivalence classes then fills out A.

Thus, Theorem 9.19 shows that the various sets of the form $[a]_R$ we obtain by selecting one a from each equivalence class partitions A into disjoint nonempty subsets. But here is an interesting fact—we can turn this situation around! Given any partition of a set A into disjoint nonempty subsets, we can use this partition to define an equivalence relation! We leave this as the following:

Exercise 9.20. Let $A = \sqcup_{\alpha \in I} A_\alpha$ be a partition of A into disjoint nonempty subsets. Define a relation R on A by the rule $(a, b) \in R$ if a and b belong to the same partition, i.e., $a, b \in A_\alpha$ for some $\alpha \in I$. Show that R is an equivalence relation.

As discussed at the beginning of this chapter, much of mathematics is the study of various objects "up to" some appropriate notion of sameness. For instance, as we explained, the study of triangles in geometry is really the study of triangles "up to" congruence. We have seen earlier in the chapter that the relation on the set of triangles on the plane that declares $T_1 \sim T_2$ if T_1 is congruent to T_2 is an equivalence relation. In the language of equivalence relations, when we study triangles, we choose not to distinguish between members of the same equivalence class (that is, between congruent triangles). Instead, we choose one representative from each equivalence class (that is, one triangle from all the triangles congruent to this triangle) and study the set of such representatives.

In the general context of an equivalence relation R on a set A, we similarly choose to not distinguish between two elements that are in the same equivalence class. Instead we pick one representative from each equivalence class, and study the set of such representatives. It is useful therefore to introduce notation for this. We will write A/R (read "A mod R") for the set of equivalence classes of A under R. (Thus, each member of A/R is itself a set, an equivalence class to be precise.)

9.2 Examples

Let us now consider several examples of equivalence relations, and the equivalence classes determined by them:

Example 9.21. Let S be a set, and write S^2 for the set $S \times S$. Thus, S^2 consists of ordered pairs of elements of S (see Definition 5.12, Chapter 5). We wish to consider *unordered pairs* of elements of S, that is, we do not wish to distinguish a pair (x, y) from the pair (y, x). The easiest way to do this is to define $R \subseteq S^2 \times S^2$ (remember that the elements of $S^2 \times S^2$ are *ordered pairs of ordered pairs* of elements of S!) to consist of all pairs $((x, y), (s, t))$ such that either $x = s$ and $y = t$ or else $x = t$ and $y = s$. Verify that R is an equivalence relation. The equivalence class of any (x, y) under R is the set $\{(x, y), (y, x)\}$ (of course if $x = y$, then (x, y) and (y, x) are the same), and S^2/R captures pairs of elements of S *up to order*. Thus, S^2/R is the set of unordered pairs of elements of S!

Figure 9.2 shows the case where $S = \mathbb{R}$. Recall from your high school coordinate geometry that pairs of points (a, b) and (b, a) are situated symmetrically across the $y = x$ line (that runs at 45° to the positive x-axis). Of course, if $a = b$, then these two points are the same, and are situated *on* the $y = x$ line. The equivalence classes come in two kinds: classes that contain only one point, of the form (a, a), as a varies in $\mathbb{R}$, and classes that contain only the pair of points (a, b) and (b, a), as a and b vary in $\mathbb{R}$, with $a \neq b$.

Example 9.22. This is an example we considered at the beginning of the chapter; it arises while defining trigonometric ratios: Let A be the set of right triangles in the plane, and R the relation "is similar to." Thus, if T_1 and T_2 are right triangles, then $(T_1, T_2) \in R$ if T_1 is similar to T_2. Verify that R is an equivalence relation. For any right triangle T, we can describe the equivalence class of T in terms of its interior angles: if the set of interior angles of T is $\{\alpha, \beta, \pi/2\}$ $(0 < \alpha \leq \beta < \pi/2,\ \alpha + \beta = \pi/2)$ (where we have listed the angles in order of increasing magnitude), the equivalence class $[T]_R$

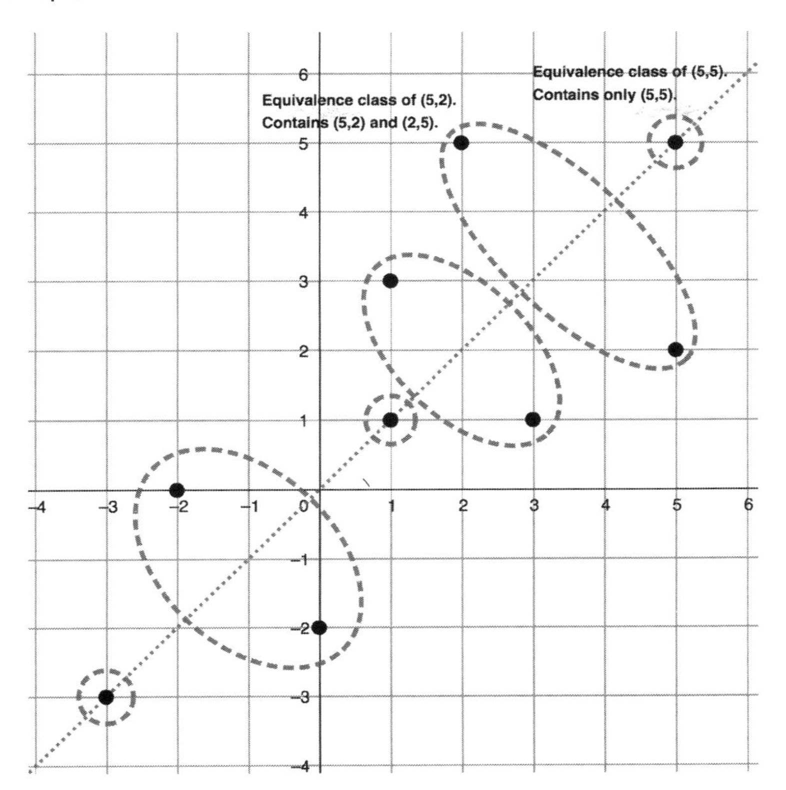

Figure 9.2. Equivalence classes in $\mathbb{R}^2$ that determine unordered pairs of real numbers.

of T is the set of all (right) triangles in the plane whose set of interior angles is also $\{\alpha, \beta, \pi/2\}$.

Now notice that the set of interior angles of a right triangle T above can as well be written as $\{\alpha, \pi/2 - \alpha, \pi/2\}$, for suitable α in the range $0 < \alpha \leq \pi/4$. (If you are puzzled about this, note that $\beta = \pi/2 - \alpha$. Also, because we have listed the angles in increasing order, we had $\alpha \leq \beta$ above. But since $\beta = \pi/2 - \alpha$, we find $\alpha \leq \pi/2 - \alpha$, i.e., $2\alpha \leq \pi/2$, or $\alpha \leq \pi/4$.) Since all right triangles in the equivalence class of a fixed triangle T with the set of interior angles $\{\alpha, \pi/2 - \alpha, \pi/2\}$ also have the set of interior angles $\{\alpha, \pi/2 - \alpha, \pi/2\}$, the smallest angle α is the same for elements of $[T]_R$. Hence, we may define a function $f : A/R \to (0, \pi/4]$ that sends the equivalence class of a triangle T with set of interior angles $\{\alpha, \pi/2 - \alpha, \pi/2\}$, $\alpha \leq \pi/2 - \alpha < \pi/2$, to its smallest angle α.

Exercise 9.22.1. Verify that the map f above is a bijection between A/R and the open interval $(0, \pi/4]$, and thus show that A/R has the same cardinality as $\mathbb{R}$. (See Examples 8.15 and 8.20 in Chapter 8.)

The result of this exercise can be interpreted as saying that "up to similarity" there is one right triangle for each real number α in the range $(0, \pi/4]$.

Example 9.23. The triangles in Example 9.22 were restricted to right triangles, since the example arose in the context of considering trigonometric ratios. But more generally, we could consider the set A of *all* triangles in the plane, and study them "up to similarity." Thus, just as in Example 9.22, we define a relation R on A by the rule $(T_1, T_2) \in R$ if T_1 is similar to T_2—the only difference from Example 9.22 is that we are now considering all triangles instead of just right triangles. Verify that R is an equivalence relation on A. How would you describe the equivalence class of a given triangle T?

> **Exercise 9.23.1.** In a manner similar to Example 9.22, construct a bijection between A/R and the subset of $(0, \pi) \times (0, \pi)$ consisting of all (x, y) such that $0 < x \leq y \leq \pi - x - y$. Sketch this region in the xy plane. Show that A/R has the same cardinality as $\mathbb{R}$. (See Exercise 8.34 in Chapter 8.)

The result of the exercise above can be interpreted as saying that "up to similarity" there is one triangle for each point in the region of the xy plane that you would have sketched.

Example 9.24. Let A be the plane, and for any point P, write $|P|$ for the distance of P from the origin. Now define R by the rule $(P, Q) \in R$ if $|P| = |Q|$. Verify that R is an equivalence relation. Verify that the equivalence class of a point P is the circle with center at the origin that passes through P (i.e., of radius $|P|$). There is an obvious bijection between A/R and $[0, \infty)$. Find it!

Example 9.25. Let A be $\mathbb{R}$, and define R by the rule $(x, y) \in R$ if $x - y \in \mathbb{Z}$. (For instance, $\sqrt{5}$ is related to $\sqrt{5} - 1$, to $\sqrt{5} + 3$, etc. Similarly, 0.41 is related to 1.41, to 2.41, to $0.41 - 1 = -0.59$, to -1.59, etc.) Verify that this is an equivalence relation.

> **Exercise 9.25.1.** There is an obvious bijection between A/R and $[0, 1)$. Find it!

There is a natural and more interesting way to view A/R than as just $[0, 1)$—we can view A/R as a circle! To see this, note that under this equivalence relation, 1 is "the same" as 0. (In other words, 1 and 0 are in the same equivalence classes.) Now imagine yourself in $[0, 1)$, which we've seen is in bijection with A/R. Imagine yourself starting at $x = 0$, and moving right with constant velocity. You proceed along $[0, 1)$, and as you move past 0.9, 0.99, 0.999, etc., you are looking forward to reaching $x = 1$, but the point $x = 1$ is absent in $[0, 1)$! However, not to despair. After all, A/R arose from A by collecting together all elements of an equivalence class into one, so, since 1 is "the same" as 0 under this relation as observed above, you simply decree that as you go further and further out past 0.99999, 0.99999999, etc., instead of hitting $x = 1$, you'll simply land back at $x = 0$.

In other words, you have taken the "open necklace" $[0, 1)$, and strung the ends together, and made it into a "closed necklace." Viewed another way, you have found a way of viewing A/R as a circle! See Figure 9.3.

This is not far-fetched at all, and the entire process can be made water-tight using notions of homeomorphisms and quotient topologies, subjects you will study in an introductory course on topology. The critical thing is that it endows A/R with a natural

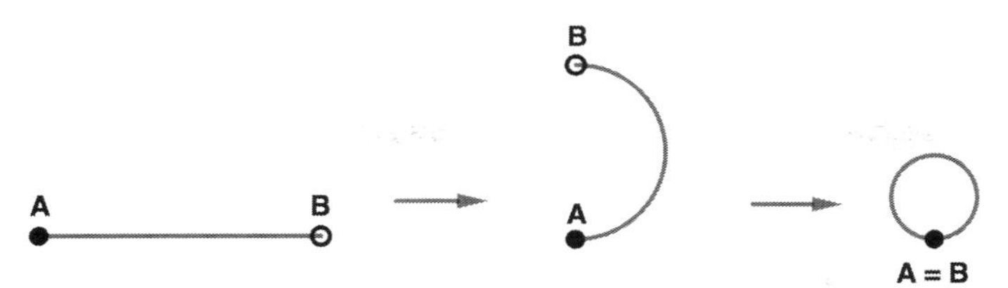

Figure 9.3. The set A/R in Example 9.25: By wrapping $[0, 1)$ onto itself we get a circle.

notion of "nearness" compatible with the equivalence relation. In this relation, points like $x = 0.99$, $x = 0.99999$, etc., are considered as being very close to $x = 0$. Further, with this natural notion, A/R is "the same as" the circle with its natural notion of nearness.

Example 9.26. Example 9.25 can be generalized to $\mathbb{R}^2$ (and to $\mathbb{R}^n$ as well for $n \geq 3$, but we will study just the $n = 2$ case here). Take A to be $\mathbb{R}^2$, and view points via their coordinates (x, y). We define a relation R by the rule that (x_1, y_1) is related to (x_2, y_2) if both $x_1 - x_2$ and $y_1 - y_2$ are in $\mathbb{Z}$. (For instance, the point $(0.1, 0.3)$ is related to $(0.1, 1.3)$, $(0.1, 2.3)$, $(1.1, 0.3)$, $(2, 1, 0.3)$, $(5.1, -11.7)$, $(-17.9, -6.7)$, etc.)

Exercise 9.26.1. There is an obvious bijection between A/R and the set $[0, 1) \times [0, 1)$. Find it!

Just as with Example 9.25, there is a more natural and interesting way of viewing A/R than as just $[0, 1) \times [0, 1)$. Note that $[0, 1) \times [0, 1)$ is a rectangle, with the right and top edges (corresponding to points where $x = 1$ and where $y = 1$ respectively) missing. However, according to the equivalence relation, points of the form $(1, y)$ are "the same" as those of the form $(0, y)$, and similarly, points of the form $(x, 1)$ are "the same" as those of the form $(x, 0)$. Working exactly as in Example 9.25, we have a natural way of first identifying the right edge of the rectangle and the left edge—doing so, we get a cylinder—and then of identifying the top edge and bottom edge of the cylinder—doing so we get a torus or donut!!! See Figure 9.4. Thus, A/R can be viewed as a torus in a natural way. As with Example 9.25, this process endows A/R with a notion of nearness that is compatible with the equivalence relation, and with this notion, A/R is "the same as" the torus with its natural notion of nearness. This process can be made water-tight using the notions of homeomorphisms and quotient topologies.

Example 9.27. In this example we will consider a very well-known object in mathematics, the *Möbius strip*. It is an example of a *non-orientable surface:* if you walk all around on the surface till you get back to the starting point, your head will be pointing in the direction opposite to where it was pointing when you started! Thus, it has no "front" and "back!"

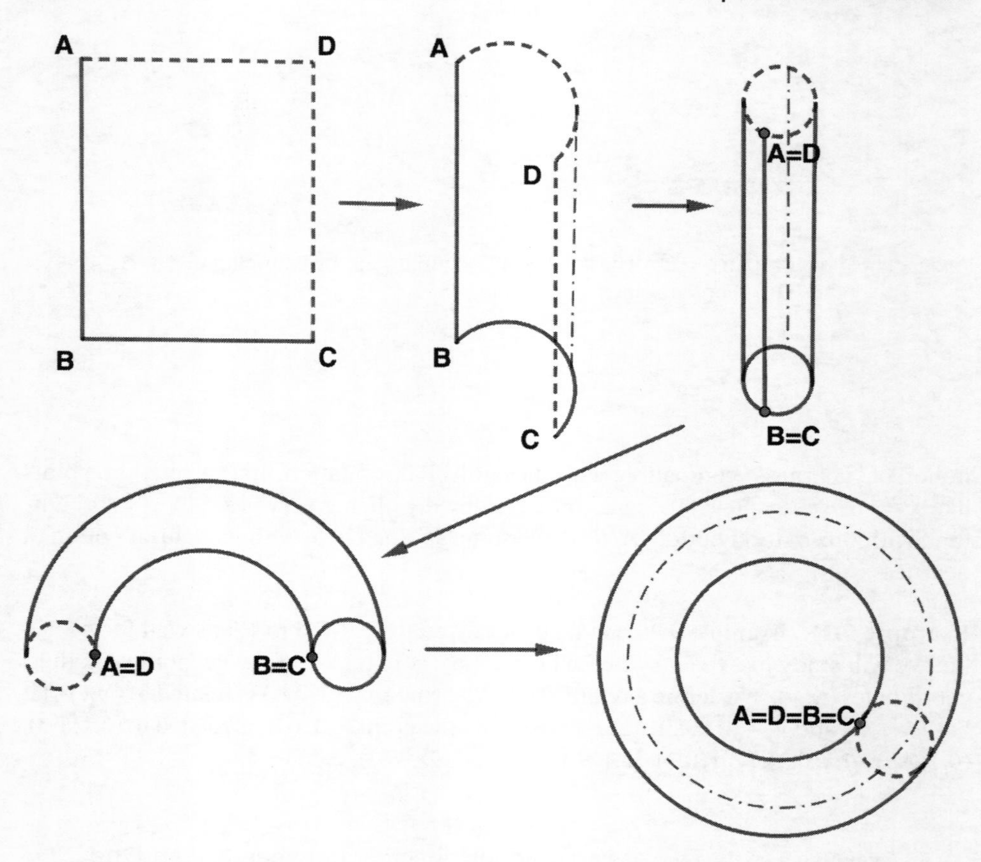

Figure 9.4. The set A/R in Example 9.26: By wrapping the right and left edges of the rectangle together we get a cylinder, and then wrapping the top edge and bottom edge of the cylinder we get a torus.

Let A be the rectangle $PQRS$ as shown in Figure 9.5, with the edges PQ and SR *identified in the reverse direction.* What this means is the following: With the coordinates as shown in the figure, we define a relation $\sim$ on A by saying that

$$(x_1, y_1) \sim (x_2, y_2) \text{ if } \begin{cases} y_1 - y_2 = \pm 5 \text{ and } x_1 + x_2 = 1 \\ \text{OR} \\ x_1 = x_2 \text{ and } y_1 = y_2. \end{cases}$$

We can check that this is an equivalence relation. (We keep in mind that two points (x_1, y_1) and (x_2, y_2) in our rectangle can satisfy the condition $y_1 - y_2 = \pm 5$ only if $y_1 = 0$ and $y_2 = 5$ or vice-versa.) The equivalence class of any point (x_1, y_1) in the rectangle that does not lie on the edges PQ or SR (so $y_1 \neq 0$ or 5) consists of just itself. The equivalence class of a point $(x, 5)$ on the edge PQ consists of two points: itself and $(1 - x, 0)$, and symmetrically, the equivalence class of a point $(x, 0)$ on SR consists of itself and $(1 - x, 5)$. Thus, $A/\sim$ can be thought of as described earlier: the rectangle

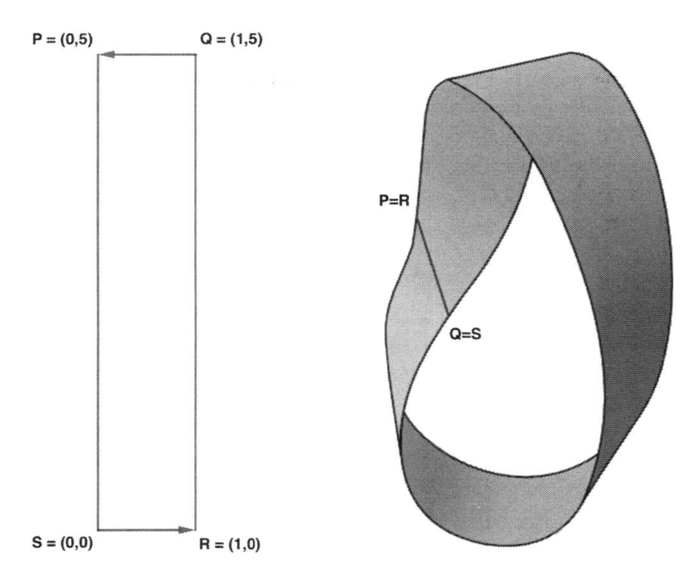

Figure 9.5. The Möbius strip. The edge SR is twisted around and laid on edge PQ so that P and R coincide, as do Q and S. (With thanks to `http://www.i2clipart.com/` for the public domain image.)

PQRS with the edges *PQ* and *SR* identified in the reverse direction, $(x, 0)$ and $(1-x, 5)$ thought of as the same points.

As with previous examples, there is a way to visualize $A/\sim$ as a continuous object in which the identification of the edges happens naturally. Just as we did at the first stage of creating the torus in Example 9.26, we shape the rectangle into something that is almost a cylinder: the difference being that we twist the edge *SR* around and lay it on the edge *PQ* so that *P* and *R* coincide, and *Q* and *S* coincide. The effect is to glue together points of the form $(x, 0)$ and $(1-x, 5)$. Doing so, we get the Möbius strip; see Figure 9.5.

Exercise 9.27.1. Work through the easier example of a rectangle *PQRS* as above but now with the edges *PQ* and *SR* identified in the *same* direction. You should get a cylinder.

Example 9.28. Here is a further generalization of the Möbius strip, another well-known example in mathematics. This is the *Klein bottle*, a *non-orientable surface without boundary.* Analogous to Example 9.27, the term non-orientable now refers to the fact that it has no "inside" and "outside!" The term without boundary refers to the fact that it closes up on itself. We start with the same rectangle *PQRS* as in Example 9.27. We will continue to identify the edges *PQ* and *SR* in the opposite direction, but this time, we will also identify the edges *PS* and *QR* in the same direction. See Figure 9.6.

Exercise 9.28.1. Write down an equivalence relation on the rectangle *PQRS* so that the equivalence classes can be associated with identifying one set of opposite edges in the reverse direction and the other set of opposite edges in the same direction.

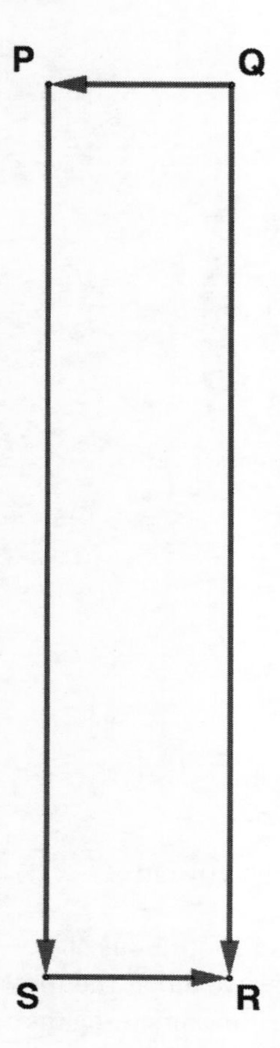

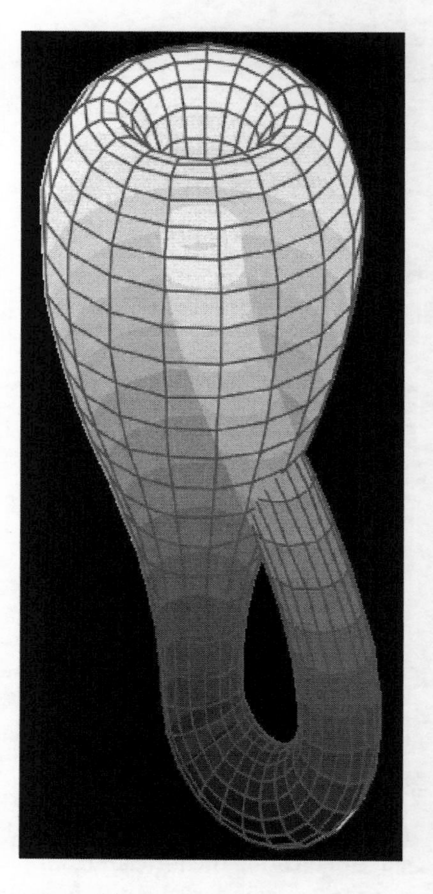

Figure 9.6. The Klein bottle. https://commons.wikimedia.org/wiki/File:Klein_bottle.svg Used under the terms of the GNU Free Documentation License, licensed under Creative Commons License Attribution-Share Alike 3.0 Unported https://creativecommons.org/licenses/by-sa/3.0/deed.en license, Author Tttrung.

If we were to fold $PQRS$ into a cylinder using the identification of PS and QR in the same direction, and then fold the cylinder, much like how we construct the torus in Example 9.26 except we twist it around so that PQ and SR (which are now circles) align themselves in the opposite direction, we get the Klein bottle shown in Figure 9.6.

This folding is harder to visualize, since it cannot be done without self-intersection in three dimensions, but can be done in four dimensions without self-intersection. The sequence of figures in Figure 9.7 shows in greater detail how this folding is accomplished and how the edges align themselves as they are supposed to.

Example 9.29. Let A be the plane with the origin removed, i.e., $\mathbb{R}^2 \setminus (0,0)$. Although we have removed the origin from the set, we will nevertheless use the origin to help us define relations: we will define our relation R by the rule $(P,Q) \in R$ if Q lies on the line that passes through the origin and P. Verify that R is an equivalence relation. Verify

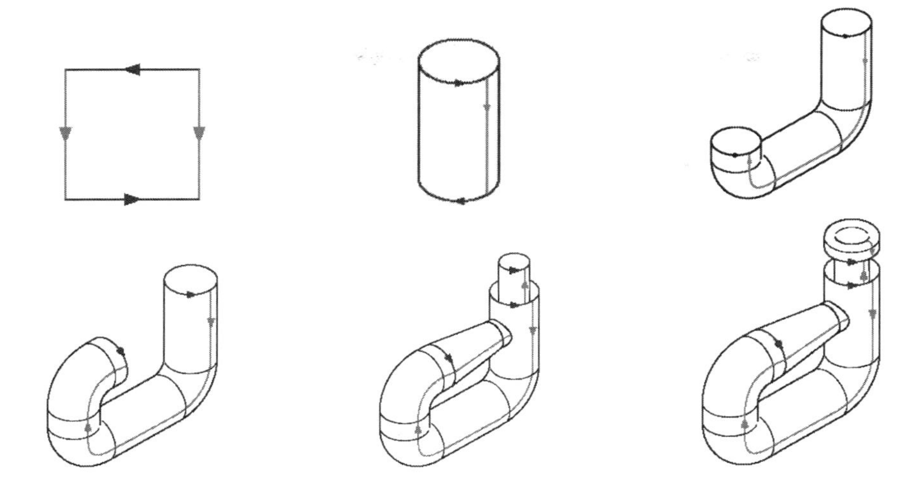

Figure 9.7. Folding a rectangle, with the side pair of opposite edges identified in the same direction and the top pair in the opposite direction, into a Klein bottle. Follow the directional arrows and work through the pictures to convince yourself that the top edges of the rectangle indeed end up identified in the opposite direction. (With thanks to Inductiveload - Own work (Own drawing), Public Domain, `https://commons.wikimedia.org/w/index.php?curid=1679924`.)

that the equivalence class of a point P is the line, with the origin removed, that passes through the origin and P. Why do you think we did not include the origin in our set?

> **Exercise 9.29.1.** There is an obvious bijection between A/R and $[0, \pi)$. Find it!

The result of this exercise can be interpreted, just as in Examples 9.22 and 9.23, as saying that "after identifying points that are on the same line through the origin," there is one point on the plane (with the origin removed) for each real number in the range $[0, \pi)$. This set of points of $\mathbb{R}^2 \setminus (0, 0)$ after the given identification (i.e., the set A/R) is a very important object in mathematics, known as the projective one-dimensional space over $\mathbb{R}$ (denoted $\mathbf{P}^1(\mathbb{R})$ or $\mathbf{RP}^1$). There is a natural way to impose a notion of nearness and continuity on A/R, which makes A/R a "geometric object" in its own right. Since the line that makes an angle of π with the positive x axis is the same as the line that makes an angle of 0, in the bijection between $\mathbf{P}^1(\mathbb{R})$ and $[0, \pi)$ we need to think of points near π as also being near 0. This is exactly as in Example 9.25 above. Thus, in a very natural way, $\mathbf{P}^1(\mathbb{R})$ becomes a circle! (You will study all this in introductory courses on topology.)

Example 9.30. Example 9.29 above can be generalized to arbitrary dimensions: Let A be $\mathbb{R}^{n+1}$ with the origin removed, and once again, define the relation R by the rule $(P, Q) \in R$ if Q is on the line through the origin and P. The equivalence classes are lines through the origin (again, with the origin itself removed), and the resulting set

A/R is known as the projective n-dimensional space over $\mathbb{R}$ (denoted $\mathbf{P}^n(\mathbb{R})$ or $\mathbf{RP^n}$). Once again, there is a natural notion of nearness and continuity that can be imposed on A/R. Under this notion, $\mathbf{P}^n(\mathbb{R})$ can be viewed as the unit disk in n dimensions (i.e., the set of of points $(x_1, \dots, x_n)$ in $\mathbb{R}^n$ with $x_1^2 + \cdots + x_n^2 \leq 1$) *with antipodal points on the boundary identified!* What this last phrase means is that the points on the boundary of the unit disk, i.e., the points with $x_1^2 + \cdots + x_n^2 = 1$, are identified as follows: we think of the points $(x_1, \dots, x_n)$ and $(-x_1, \dots, -x_n)$ on the boundary as the *same*. Exercise 9.30.1 leads you through this visualization of $\mathbf{P}^n(\mathbb{R})$.

You'll notice that for the case $n = 1$, the corresponding 1-dimensional disk is just the interval $[-1, 1]$. Up to a stretch factor and translation we can think of this interval as the segment $[0, 1]$. There are only two points on the boundary of this segment, namely 0 and 1, which are naturally antipodal, and if we identify these two points we are in the situation of Example 9.25. Alternatively, we can think of $[-1, 1]$ as the segment $[0, \pi]$ up to a different stretch factor and translation. Once again, there are only two points on the boundary, namely 0 and π, and these are naturally antipodal. Identifying these two points, we are now in the situation of Example 9.29. As either of these examples show, $\mathbf{P}^1(\mathbb{R})$ is in bijection with a circle.

Although the case $n = 1$ yields a simple object, namely the circle, it turns out that $\mathbf{P}^n(\mathbb{R})$ for $n \geq 2$ has strange twists and "holes" in it. (Again, this is something you will study in topology.)

Exercise 9.30.1. Here is how to build up to the visualization of $\mathbf{P}^n(\mathbb{R})$ described above:

(1) Show that starting from the definition of the equivalence relation R, $\mathbf{P}^n(\mathbb{R})$ can be identified with the unit sphere in $n+1$ dimensions with antipodal points identified, i.e., identified with the set of points $(x_1, \dots, x_{n+1})$ in $\mathbb{R}^{n+1}$ satisfying $x_1^2 + \cdots + x_{n+1}^2 = 1$, with the points $(x_1, \dots, x_{n+1})$ and $(-x_1, \dots, -x_{n+1})$ considered the *same*.

(2) By considering just the "upper hemisphere" of the sphere in part (1) above (with its described identification of antipodal points), show that $\mathbf{P}^n(\mathbb{R})$ can indeed be described as in the discussions above this exercise: as the unit disk in n dimensions with antipodal boundary points identified. (Hint: The "upper hemisphere" may be thought of, in analogy with the ordinary sphere in three dimensions, as the points where $x_{n+1} \geq 0$. View x_{n+1} as a parameter that ranges in $[0, 1]$.)

Example 9.31. We will consider three related examples here. These examples are all situated in the context of the plane, with points now viewed as row vectors. (We can just as easily think of these as column vectors; the choice is arbitrary.) Thus each point P will be viewed as the row vector $\overline{P} = (x, y)$, where x and y are the x and y coordinates respectively of P. (Our set A is therefore $\mathbb{R}^2$.)

Exercise 9.31.1. Define a relation R_x by the rule $(\overline{P}, \overline{Q}) \in R$ if the y-coordinate of the vector $\overline{P} - \overline{Q}$ is zero. Verify that R_x is an equivalence relation. Describe $[P]_{R_x}$ geometrically. Can you construct a very visual bijection between A/R_x and the y-axis? Between A/R_x and the line $x - y = 0$? Between A/R_x and the line $x + y = 0$?

Exercise 9.31.2. Define a relation R_y by the rule $(\overline{P}, \overline{Q}) \in R$ if the x-coordinate of the vector $\overline{P} - \overline{Q}$ is zero. Verify that R_y is an equivalence relation. Describe $[P]_{R_y}$ geometrically. Can you construct a very visual bijection between A/R_y and the x-axis? Between A/R_y and the line $x - y = 0$? Between A/R_y and the line $x + y = 0$?

Exercise 9.31.3. Define a relation R_d by the rule $(\overline{P}, \overline{Q}) \in R_d$ if the x and y-coordinates of the vector $\overline{P} - \overline{Q}$ are equal. Verify that R_d is an equivalence relation. What is the equivalence class of a vector $\overline{P}$? Describe $[P]_{R_d}$ geometrically. Can you construct a very visual bijection between A/R_d and the line $x + y = 0$? Between A/R_d and the y-axis? Between A/R_d and the x-axis?

Example 9.32. This example is for those who have studied vector spaces and subspaces, and directly generalizes the three subexamples in Example 9.31 above. Let V be a vector space (over $\mathbb{R}$) and let W be a subspace. Define a relation R on V by the rule $(\overline{v}, \overline{w}) \in R$ if $\overline{v} - \overline{w} \in W$. Verify that this is an equivalence relation. Why is this a generalization of the subexamples in Example 9.31? (Hint: In Exercise 9.31.1 for instance, what if you took W to be the subspace defined by the x-axis?)

If U is any complementary subspace of W in V, then the set of equivalence classes V/R of R is naturally in bijection with elements of U. To see this, note that by definition of being a complementary subspace, any $\overline{v} \in V$ can be written in one and only one way as $\overline{u} + \overline{w}$ with $\overline{u} \in U$ and $\overline{w} \in W$. We write $\overline{v} \in V$ symbolically as the pair $(\overline{u}, \overline{w})$, with $\overline{u}$ and $\overline{w}$ uniquely determined as above. Now, by definition of the relation R, $\overline{v_1} = (\overline{u_1}, \overline{w_1})$ is related to $\overline{v_2} = (\overline{u_2}, \overline{w_2})$ if and only if the difference $(\overline{u_1} - \overline{u_2}, \overline{w_1} - \overline{w_2})$ is in W, which by the uniqueness of the decomposition happens if and only if $\overline{u_1} - \overline{u_2} = 0$, that is, $\overline{u_1} = \overline{u_2}$. Thus, the equivalence class $[(\overline{u}, \overline{w})]_R$ of any $\overline{v} = (\overline{u}, \overline{w})$ consists of all vectors $(\overline{u}, \overline{w'})$, where $\overline{w'} \in W$ is arbitrary. The set of equivalence classes V/R is therefore in bijection with the set of $\overline{u} \in U$, in other words with U.

In Exercise 9.31.1 for instance, we took three different choices for U: the y-axis (the line $x = 0$), and the two lines $x + y = 0$ and $x - y = 0$.

Example 9.33. We have considered this example at the beginning of this chapter, in the case where $n = 2$: Take $A = \mathbb{Z}$, and for any integer $n \geq 2$, define a relation R_n by $(a, b) \in R_n$ if $a - b$ is divisible by n.

Exercise 9.33.1. Verify that R_n is an equivalence relation.

The following exercise helps give an alternative description of this relation:

Exercise 9.33.2. Show that if a and b are integers, and $n \geq 2$ is a given integer, then $a - b$ is divisible by n if and only if a and b leave the same remainders on dividing by n. (Hint: Write a as $nq + r$, where q is the quotient on dividing by n and r is the remainder, so $0 \leq r < n$. Similarly, write b as $nq' + r'$, where $0 \leq r' < n$. Then $a - b = n(q - q') + (r - r')$. Note that $-n < r - r' < n$.)

Thanks to Exercise 9.33.2 above, we may also define the relation R_n by the rule $(a, b) \in R$ if a and b leave the same remainder on dividing by n. Thus, the various equivalence classes of R_n consist of those integers that leave the same remainder on dividing by n.

Now let us compare the objects in this example with the objects in $\mathbb{Z}/n\mathbb{Z}$ that we considered in Example 1.4 in Chapter 1. There we spoke about collecting all integers that leave a remainder of 0 when divided by n into one set, calling that set $[0]_n$, all integers that leave a remainder of 1 into one set, calling that $[1]_n$, and so on, to get n new "numbers" $[0]_n, [1]_n, \dots, [n-1]_n$. But we have just seen that the various equivalence classes of R_n are precisely the set of integers that leave the same remainder when dividing by n. Thus, under R_n, there are n equivalence classes: those integers that leave a remainder of 0, those that leave a remainder of 1, and so on, up to the set of those integers that leave a remainder of $n-1$. In other words, now that we have the appropriate language, the objects $\mathbb{Z}/n\mathbb{Z}$ that we considered in Chapter 1 are precisely the equivalence classes of the relation R_n that we have introduced here. In Chapter 1, to keep things at an introductory level, we only used the notation $[0]_n, [1]_n, \dots, [n-1]_n$ to denote these objects, but now that we have more comprehensive language, we can denote, for instance, $[1]_5$ by $[6]_5$, or $[-9]_5$, or $[5k+1]_5$ for any integer k, since all integers of the form $5k+1$ leave the same remainder (of 1) when divided by 5.

In Chapter 1, we defined addition and multiplication in $\mathbb{Z}/n\mathbb{Z}$ by working with the specific representatives 0 for $[0]_n$, 1 for $[1]_n$, …, and $n-1$ for $[n-1]_n$. We said that for $r, s \in \{0, 1, \dots, n-1\}$, $[r]_n + [s]_n = [r+s]_n$ and $[r]_n[s]_n = [rs]_n$, with the understanding that if $r+s$ or rs were greater than or equal to n, we first replace them by their remainders on dividing by n. As you will see in Exercise 9.44 ahead, we may define addition and multiplication in $\mathbb{Z}/n\mathbb{Z}$ by the rules $[r]_n + [s]_n = [r+s]_n$ and $[r]_n[s]_n = [rs]_n$ using *any* representatives r and s (that is, not requiring them to be in $\{0, 1, \dots, n-1\}$) and not requiring that we replace $r+s$ and rs by their remainders. In Exercise 9.45 ahead, you will also see that we can use this more general definition of addition and multiplication to establish commutativity, associativity and distributivity.

These objects $\mathbb{Z}/n\mathbb{Z}$ are of vital importance in many areas of mathematics and computer science.

Example 9.34. Now is the time to revisit an old friend—the set of rational numbers $\mathbb{Q}$—and reinterpret it as a set of equivalence classes! How exactly do we put into firm footing the convention that fractions like $\frac{1}{2}$, $\frac{2}{4}$, $\frac{100}{200}$, etc., are all equal? As we know, these fractions are deemed to be equal because after canceling common factors in the numerator and denominator, they all reduce to the same fraction $\frac{1}{2}$. (And of course, this fits our intuitive understanding of ratios as well: two parts out of four is the same as one part out of two, etc.) A different but equivalent way to define the equality of two fractions $\frac{a}{b}$ and $\frac{c}{d}$ that does not invoke cancellation is to say that they are equal precisely if $ad = bc$. We will use this interpretation in what follows.

Take A to be the set $\mathbb{Z} \times (\mathbb{Z} \setminus \{0\})$. Thus, A consists of all ordered pairs (a, b), where a and b are integers, and $b \neq 0$. We will identify the set of fractions with A by thinking of the fraction $\frac{a}{b}$ as the ordered pair (a, b). (Notice that the second slot in the set of ordered pairs in A is not allowed to be zero—this fits precisely with the fact that when considering fractions, we do not allow the denominators to be zero.) On A we define the relation R by the rule (a, b) is related to (c, d) if $ad = bc$.

Exercise 9.34.1. Verify that R is an equivalence relation.

Now, view $\frac{a}{b}$ as the ordered pair (a, b) and observe that R has been designed precisely to mimic our rule in the previous paragraph for equality of fractions!

The set of equivalence classes A/R is then constructed by simply taking one representative for each class: for instance, the various fractions $\frac{1}{2}, \frac{2}{4}, \frac{100}{200}$, etc. are all grouped as one fraction, and while they can be represented by any element in the collection, they are typically represented by the reduced form $\frac{1}{2}$, in which all common factors have been canceled from the numerator and the denominator and the denominator is positive. In fact, every equivalence class can be represented by a uniquely determined fraction $\frac{a}{b}$ with $\gcd(a, b) = 1$ and $b > 0$; see Exercise 10.38, Chapter 10.

See also Exercise 9.43 ahead where we consider the operations of addition and multiplication in $\mathbb{Q}$.

Example 9.35. Let A be the set of polynomials in the variable x with coefficients from $\mathbb{R}$, i.e., the set $\mathbb{R}[x]$. Let f, g be polynomials in $\mathbb{R}[x]$. Define the relation R by the rule $(f, g) \in R$ if $f - g$ has zero constant coefficient.

Exercise 9.35.1. Verify that this is an equivalence relation. Verify that the equivalence class of a polynomial f is the set of all polynomials that have the same constant coefficient as f. Can you find an obvious bijection between A/R and $\mathbb{R}$?

Example 9.36. As with Example 9.35, let A be the set $\mathbb{R}[x]$. For $p, q \in \mathbb{R}[x]$, define the relation S by the rule $(p, q) \in S$ if $p - q$ is a polynomial multiple of $x^2 + 1$. (Thus, p is related to q if there exists a polynomial $h \in \mathbb{R}[x]$ such that $p - q = (x^2 + 1) \cdot h$.)

Exercise 9.36.1. Verify that this is an equivalence relation.

This may not be obvious, but the set of equivalence classes A/S in this example is in natural bijection with the set of complex numbers $\mathbb{C}$! In fact, this bijection can be used to *define* the set of complex numbers! (See Example 9.40 ahead, where we show that there is a well-defined function from A/S to $\mathbb{C}$. See also Exercise 1.22, Chapter 1. Note that $p - q$ is a multiple of $x^2 + 1$ precisely if p and q leave the same remainder on dividing by $x^2 + 1$.)

Example 9.37. This example is closely related to Example 9.29 above. Let A be the set of invertible 2×2 matrices with entries in $\mathbb{R}$. (We saw 2×2 matrices with entries in $\mathbb{C}$ in Example 5.44 in Chapter 5. Here we are restricting the entries to come from $\mathbb{R}$. You may have also seen 2×2 matrices with entries in $\mathbb{R}$ in a linear algebra course. An invertible 2×2 matrix $\begin{pmatrix} a & b \\ c & d \end{pmatrix}$ is one whose "determinant," i.e. the quantity $ad - bc$, is nonzero.) We define a relation on A by the rule $(M, N) \in R$, for two matrices M and N, if $M = \lambda N$ for some nonzero real number λ. (Just as in Exercise 5.62, Chapter 5, λN is the matrix obtained by multiplying every entry of N by λ.)

Exercise 9.37.1. Verify that this is an equivalence relation.

The equivalence class $[M]_R$ is the set of all invertible matrices that are (nonzero) scalar multiples of M. The set of equivalence classes A/R, denoted $\mathbf{PGL}_2(\mathbb{R})$, is a very important object in mathematics. For instance, there is a natural way in which elements of $\mathbf{PGL}_2(\mathbb{R})$ "act" on (intuitively, "push around") the elements of $\mathbf{P}^1(\mathbb{R})$ (an object you have already considered in Example 9.29).

This example can be generalized to higher dimensions. The set $\mathbf{PGL}_n(\mathbb{R})$ is the set of equivalence classes of invertible $n \times n$ invertible matrices with entries in $\mathbb{R}$, under the same equivalence relation of "multiplication by scalars" as in the $n = 2$ case. It acts in a natural way (intuitively, "pushes around" in a natural way) on the elements of $\mathbf{P}^{n-1}(\mathbb{R})$ of Example 9.30.

Functions defined on equivalence classes: well-definedness. We have seen many instances where the set of equivalence classes under some equivalence relation itself becomes a mathematically interesting object. For instance, we have considered the "integers mod n," or $\mathbb{Z}/n\mathbb{Z}$ in Example 9.33, and the objects $\mathbf{P}^n(\mathbb{R})$ in Example 9.30, the objects $\mathbf{PGL}_n(\mathbb{R})$ in Example 9.37 above. Since equivalence classes are mathematical objects, functions from them to other mathematical objects arise naturally, and care needs to be given as to how such functions are defined. This is best seen by an example such as the following:

Example 9.38. Consider the following definition of a function from $\mathbb{Z}/n\mathbb{Z}$ to $\mathbb{Z}/n\mathbb{Z}$:

$$f : \mathbb{Z}/n\mathbb{Z} \to \mathbb{Z}/n\mathbb{Z},$$
$$[a]_n \mapsto [a^2]_n.$$

On the surface, this rule appears simple enough: take the equivalence class of a and send it to the equivalence class of a^2. But wait! There could be a problem here! For, the equivalence class of a is also the same as the equivalence class of any integer that leaves the same remainder on dividing by n as a. (For instance, the equivalence class of $[3]_{12}$ in $\mathbb{Z}/12\mathbb{Z}$ is the same as that of $[15]_{12}$, or that of $[147]_{12}$.) According to the rule, f is supposed to take $[a]_n$ to $[a^2]_n$, but if we were to represent $[a]_n$ as $[a+n]_n$ for instance, then according to the way it is defined, f would take the equivalence class of $a + n$ to the equivalence class of $(a+n)^2$. (In the context of $\mathbb{Z}/12\mathbb{Z}$ above, this map would send the class $[3]_{12}$ to the class $[3^2]_{12} = [9]_{12}$, and the equivalence class of $[15]_{12}$, *which is the same class as* $[3]_{12}$, to $[225]_{12}$.) Now—and this is where there could be a problem—what if the equivalence class of $(a + n)^2$ were not the same as that of a^2? If so, we would have a problem—this definition would take the same element of $\mathbb{Z}/n\mathbb{Z}$, namely the equivalence class once denoted as $[a]_n$ and once denoted $[a+n]_n$, to two different elements: the equivalence class of a^2 if we used the notation $[a]_n$, and the equivalence class of $(a+n)^2$ if we used the notation $[a+n]_n$. This would be clearly absurd: the rule for defining f should not be ambiguous.

We have the following:

Definition 9.39. Let A be a set and R an equivalence relation on A. Let X be an arbitrary set. Let $f : A/R \to X$ be a rule that assigns each element of A/R to some element of X, and assume that f is defined in terms of the representative used to denote an equivalence class. If for some $a, b \in A$ with $[a]_R = [b]_R$ we find the element of X obtained by applying f to the representative a is different from the element obtained by applying f to the representative b, we say that *the function f is not well-defined.* (Although this rule is clearly not a function, it is the cultural practice to say *the function f* is not well-defined;

the correct way to have described the situation is to say *the rule f is not well-defined.*) On the other hand, if whenever $[a]_R = [b]_R$ for elements $a, b \in A$, we find that the element of X obtained by applying f to the representative a is the same as the element obtained by applying f to the representative b, we say that *the function f is well-defined.*

See also Remark 9.41 ahead.

Let us now show that the rule f specified on $\mathbb{Z}/n\mathbb{Z}$ in Example 9.38 is indeed well-defined. Let $[a]_n$ be an arbitrary element of $\mathbb{Z}/n\mathbb{Z}$. Any other representative b of $[a]_n$ has to be an element in the equivalence class of a, thus, $b = a + kn$ for suitable $k \in Z$. (You may wish to review Example 9.33 to see how the relation R_n on $\mathbb{Z}$ is defined.) If we apply the rule f to b, we find $f([b]_n) = [b^2]_n = [(a+kn)^2]_n = [a^2+2akn+k^2n^2]_n$. However, $a^2+2akn+k^2n^2 = a^2+n(2ak+k^2n)$, so a^2 and b^2 are in the same equivalence class as they differ by a multiple of n. It follows that the element in $\mathbb{Z}/n\mathbb{Z}$ obtained by applying f to any two representatives of the same equivalence class are the same, so f is well-defined.

In general, if $f : A/R \to X$ is a rule specified in terms of a representative of each equivalence class, and we need to show that f is well-defined, we need to proceed along a direction similar to what we did in Example 9.38 above. We need to show that the elements in X obtained by applying f to any two representatives of the same equivalence class are the same. Here is another example:

Example 9.40. Consider the set of equivalence classes in Example 9.36. Recall that our set A was $\mathbb{R}[x]$, and for $p, q \in \mathbb{R}[x]$ we defined the relation S by the rule $(p, q) \in S$ if $p-q$ is a multiple of x^2+1. You were asked to verify that S is an equivalence relation.

Now consider the assignment $\phi : A/S \to \mathbb{C}$ given by the rule $\phi([p]_S) = p(i)$. (Thus, ϕ simply plugs in i for x in p. Also, we will use the notation "ϕ" instead of "f" for our assignment because in an example like $\mathbb{R}[x]$, "f" is often used to denote a polynomial!) This assignment therefore is defined in terms of the representative p of an equivalence class, and we wish to verify that ϕ is well-defined, that is, ϕ yields the same complex number no matter which representative is used for an equivalence class.

So suppose p and q are both representatives of the same equivalence class (so they are both in the same equivalence class, and hence related to each other, by Theorem 9.19(4)). Thus, by definition, $p - q$ is of the form $(x^2 + 1) \cdot h$ for some polynomial $h \in \mathbb{R}[x]$. Now ϕ applied to the representative p yields $p(i)$, and when applied to the representative q yields $q(i)$. We wish to show that $p(i) = q(i)$, because if that happens, ϕ will yield the same element in $\mathbb{C}$ independent of which representative is used for an equivalence class. But since $p - q = (x^2 + 1) \cdot h$, we find on substituting i that $p(i) - q(i) = (i^2 + 1) \cdot h(i)$. (Recall from high school algebra that you can substitute i into a product of polynomials like $(x^2 + 1) \cdot h$ by substituting i into each factor and multiplying out the result.) But $i^2 + 1 = 0$. So, $p(i) - q(i) = 0 \cdot h(i) = 0$, which is the same as saying $p(i) = q(i)$.

Remark 9.41. There are situations other than $f : A/R \to X$ where we consider functions defined in terms of representatives of equivalence classes: for instance, we can consider functions $f : A/R \times A/R \to X$. Here, f takes the ordered pair $([x]_R, [y]_R)$ to some element of X defined in terms of the representatives x and y, and once more, one needs to show that this assignment is independent of which representatives we use for the pairs of equivalence classes.

Such functions arise for instance in the definition of addition and multiplication in A/R. Here, the target set X is also A/R, and addition for example can be thought of as a function that takes an ordered pair $([x]_R, [y]_R)$ and gives you some element of A/R. Multiplication can be interpreted similarly. In such instances, one needs to take care that the rules that give the sum and product do not depend on the specific representatives x and y used for the equivalence classes $[x]_R$ and $[y]_R$.

See Exercises 9.43, 9.44, and 9.46 ahead for examples. In Exercise 9.44 we will define sums and products in $\mathbb{Z}/n\mathbb{Z}$ in terms of an arbitrary representative of an equivalence class, and show that this definition yields the same sum and product as the ones we defined in Chapter 1. In Exercise 9.45 we will use this new definition of sums and products to show that $\mathbb{Z}/n\mathbb{Z}$ satisfies properties like commutativity and associativity of addition and multiplication, as also distributivity of multiplication over addition. The results of these exercises thus put the calculations of Chapter 1 on a firm footing.

9.3 Further Exercises

Exercise 9.42. Practice Exercises: In each of the following, a set A and a relation R on A is given. For each, determine which of the three properties—reflexivity, symmetry, and transitivity—R possesses.

Exercise 9.42.1. $A = \mathbb{Z}$, R defined by the rule $(m, n) \in R$ if $m \leq n$.

Exercise 9.42.2. $A = \mathbb{Q}$, R defined by the rule $(q_1, q_2) \in R$ if, after writing q_1 and q_2 in lowest terms, i.e., after canceling out all common factors in the numerator and denominator, the denominators of q_1 and q_2 are equal. (Note: conventionally, a negative fraction is written with the negative sign attached to the numerator. For example, we write $\frac{-5}{6}$, not $\frac{5}{-6}$. So all denominators of fractions will be positive integers.)

Exercise 9.42.3. $A = \mathbb{R}^2$, R defined by the rule $((x, y), (z, u)) \in R$ if $x + y = z + u$.

Exercise 9.42.4. $A = \mathbb{Z}$, R defined by the rule $(m, n) \in R$ if $n = 3^k m$ for some $k \in \mathbb{N}$.

Exercise 9.42.5. $A = \mathbb{Z}$, R defined by the rule $(m, n) \in R$ if $n = 3^k m$ for some $k \in \mathbb{Z}_{\geq 0}$.

Exercise 9.42.6. $A = \mathbb{Z}$, R defined by the rule $(m, n) \in R$ if $n = 3^k m$ for some $k \in \mathbb{Z}$.

Exercise 9.42.7. $A = \mathbb{Z}/6Z$, R defined by the rule $([m]_6, [n]_6) \in R$ if $[n]_6 = 2^k[m]_6$ for some $k \in \mathbb{Z}_{\geq 0}$.

Exercise 9.42.8. $A = \mathbb{Z}/5Z$, R defined by the rule $([m]_5, [n]_5) \in R$ if $[n]_5 = 2^k[m]_5$ for some $k \in \mathbb{Z}_{\geq 0}$.

Exercise 9.42.9. Take A to be the set of all functions $f : \mathbb{R} \to \mathbb{R}$. Define R by the rule $(f(x), g(x)) \in R$ if the set $S = \{x \in \mathbb{R} \mid f(x) \neq g(x)\}$ is contained in the subset $\mathbb{Q}$ of $\mathbb{R}$.

Exercise 9.42.10. (For those who have learned some calculus and are familiar with integration.) Take A to be the set of all open intervals (a, b) $(a < b)$ in $\mathbb{R}$. Let $p(x)$ be a fixed function on $\mathbb{R}$ for which $\int_a^b p(x)\,dx$ exists for any $a, b \in \mathbb{R}$ with $a < b$ — if you don't know what that means, just take $p(x)$ to be a fixed polynomial. Define R by the rule $((a, b), (c, d)) \in R$ if $\int_a^b p(x)\,dx = \int_c^d p(x)\,dx$.
(In the case where $p(x)$ is the constant polynomial 1, can you think of a more straightforward interpretation of this relation that doesn't involve the integral sign?)

Exercise 9.43. As discussed in Example 9.34, the set $\mathbb{Q}$ of rational numbers may be viewed as the set of equivalence classes of the set $A = \mathbb{Z} \times (\mathbb{Z} \setminus \{0\})$, under the relation R defined by the rule (a, b) is related to (c, d) if $ad = bc$. We have of course grown up with a more intuitive understanding of $\mathbb{Q}$ as numbers of the form $\frac{a}{b}$ with a, b integers and $b \neq 0$, with addition defined by $\frac{a}{b} + \frac{c}{d} = \frac{ad+bc}{bd}$ and multiplication defined by $\frac{a}{b} \cdot \frac{c}{d} = \frac{ac}{bd}$. But in our intuitive approach to $\mathbb{Q}$ we have probably never thought carefully about whether these operations of addition and multiplication on $\mathbb{Q}$ are well-defined; that is, if we replace $\frac{a}{b}$ and $\frac{c}{d}$ with equivalent fractions, whether their sum and product change or remain the same. Now that you have some expertise in this matter (see Example 9.38 earlier, as also Remark 9.41), prove that indeed addition and multiplication are well-defined on $\mathbb{Q}$.

(Recalling that $\mathbb{Q} = A/R$, we can describe this exercise in the language of Remark 9.41 as follows: if we let $\psi : A/R \times A/R \to A/R$ be defined by $\psi(\frac{a}{b}, \frac{c}{d}) = \frac{ad+bc}{bd}$ and $\phi : \mathbb{Q} \times \mathbb{Q} \to \mathbb{Q}$ be defined by $\phi(\frac{a}{b}, \frac{c}{d}) = \frac{ac}{bd}$, you need to show that ψ and ϕ are well-defined functions.)

Both ψ and ϕ are examples of *binary operations* on $\mathbb{Q}$. See Definition 12.80, Chapter 12 ahead, and in an expanded form, Definition 13.5, Chapter 13.)

Exercise 9.44. In Chapter 1, we defined addition and multiplication in $\mathbb{Z}/n\mathbb{Z}$ by working with the specific representatives 0 of $[0]_n$, 1 of $[1]_n$, ..., and $n-1$ of $[n-1]_n$. We said that for $r, s \in \{0, 1, \ldots, n-1\}$, $[r]_n + [s]_n = [r+s]_n$ and $[r]_n[s]_n = [rs]_n$, with the understanding that if $r+s$ or rs were greater than or equal to n, we first replace them by their remainders on dividing by n. As explained in Example 9.33 above, the elements of $\mathbb{Z}/n\mathbb{Z}$ are really equivalence classes under the relation R_n on $\mathbb{Z}$ defined by $(a, b) \in R$ if $a - b$ is divisible by n, or equivalently, $(a, b) \in R_n$ if a and b leave the same remainder on dividing by n. Thus, an element $[j]_n$, $0 \leq j < n$, of $\mathbb{Z}/n\mathbb{Z}$ can be written as $[a]_n$ for *any* integer a that leaves a remainder of j on dividing by n. It would be pleasing to define sums and products in $\mathbb{Z}/n\mathbb{Z}$ when the elements are described as $[a]_n$ without the restriction that $0 \leq a < n$. We will do so here, and show that the operations are well-defined and yield the same definitions of sums and products as in Example 1.4, Chapter 1.

(1) Define $[a]_n + [b]_n$ to be $[a+b]_n$. Show that this sum is well-defined.

(2) Show that this definition in part (1) above yields the same answer as the definition of Example 1.4, Chapter 1.

(3) Define $[a]_n \cdot [b]_n$ to be $[a \cdot b]_n$. Show that this sum is well-defined.

(4) Show that this definition in part (3) above yields the same answer as the definition of Example 1.4, Chapter 1.

Exercise 9.45. We are familiar with the following properties of addition and multiplication in $\mathbb{Z}$:

(1) Addition is *commutative,* that is, for all integers p and q, $p+q = q+p$.

(2) Addition is *associative,* that is, for all integers p q, and r, $p+(q+r) = (p+q)+r$.

(3) Multiplication is *commutative,* that is, for all integers p and q, $p \cdot q = q \cdot p$.

(4) Multiplication is *associative,* that is, for all integers p q, and r, $p \cdot (q \cdot r) = (p \cdot q) \cdot r$.

(5) Multiplication *distributes over addition,* that is, for all integers p q, and r, $p \cdot (q+r) = (p \cdot q) + (p \cdot r)$.

Use the formulations of sum and product in $\mathbb{Z}/n\mathbb{Z}$ in Exercise 9.44 above to show that the same properties hold for $\mathbb{Z}/n\mathbb{Z}$ for any integer $n \geq 2$. (Hint: $[a]_n + ([b]_n + [c]_n) = [a]_n + [b+c]_n = [a+(b+c)]_n$, by the way we have formulated addition in Exercise 9.44. Now proceed, invoking properties of the integers.)

Exercise 9.46. Consider the situation in Example 9.36, where A was $\mathbb{R}[x]$ and for $p, q \in \mathbb{R}[x]$, R was defined by the rule p is related to q if $p-q$ is a polynomial multiple of x^2+1.

(1) Define an addition on the set of equivalence classes A/R by the rule $[p]_R + [q]_R = [p+q]_R$. (Alternatively, view this as a function $\psi\colon A/R \times A/R \to A/R$, defined by $\psi([p]_R, [q]_R) = [p+q]_R$.) Show that addition is a well-defined operation on A/R (or in other words, show that ψ is a well-defined function).

(2) Define a multiplication on the set of equivalence classes A/R by the rule $[p]_R \cdot [q]_R = [p \cdot q]_R$. Show that multiplication is a well-defined operation on A/R.

This definition of addition and multiplication makes the set A/R into a "number system" in a manner analogous to how the addition and multiplication operations on the equivalence classes of the relations R_n of Example 9.33 earlier (see Exercise 9.44 above) make $\mathbb{Z}/R_n$ into the by-now familiar number systems $\mathbb{Z}/n\mathbb{Z}$. The set A/R in this example is traditionally written $\mathbb{R}[x]/\langle x^2+1\rangle$.

More is true: the map ϕ of Example 9.40 turns out to be both injective and surjective (see Exercise 9.47 below), and moreover, ϕ "respects the addition and multiplication operations on both sides" (that is, $\phi([p]_R + [q]_R) = \phi([p]_R) + \phi([q]_R)$ and $\phi(([p]_R \cdot [q]_R) = \phi([p]_R) \cdot \phi([q]_R))$. We interpret this as the following: $\mathbb{R}[x]/ < x^2+1 >$ and $\mathbb{C}$ are really the "same object" as far as addition and multiplication are concerned! You will study such concepts in introductory courses on abstract algebra, where objects that are seemingly different turn out to have the same underlying (algebraic) structure. The same ideas occur in other parts of mathematics as well: for instance, two seemingly different "spaces" can turn out to have the same underlying (topological) structure, two seemingly different "manifolds" can turn out to have the same underlying (geometric) structure, and so on.

Exercise 9.47. Show that the map $\phi : A/R \to \mathbb{C}$ of Example 9.40 is both injective and surjective. (Hint: for the injectivity, suppose that $\phi([p]_R) = \phi([q]_R)$. Thus, $p(i) = q(i)$. We wish to show that $[p]_R = [q]_R$. Dividing $p - q$ by $x^2 + 1$ using long division, we may write $p-q = h\cdot(x^2+1)+r$, where h is the quotient polynomial, and r is the remainder polynomial—-what is critical here is that r is of degree at most 1. (You may wish to review how long division works.) What happens if you put $x = i$ on both sides of $p - q = h \cdot (x^2 + 1) + r$?)

This exercise shows that ϕ provides a bijection between A/R of this example (written traditionally as $\mathbb{R}[x]/\langle x^2+1\rangle$) and $\mathbb{C}$. It is easy to see—knowing how substitution of $x = i$ works from your high school studies—that ϕ respects addition and multiplication on both sides (see Exercise 9.46 above for what this means). For example, we find $\phi([p]_R + [q]_R) = \phi([p + q]_R)$ by the definition of addition in $\mathbb{R}[x]/\langle x^2 + 1\rangle$. This equals $(p + q)(i)$, which in turn equals $p(i) + q(i)$, because substituting i into a sum of two polynomials is the same as substituting into each and adding the results. But $p(i) + q(i)$ is just $\phi([p]_R) + \phi([q]_R)$. (We similarly find that ϕ respects multiplication on both sides.) The upshot of all this is what we described at the end of Exercise 9.46: $\mathbb{R}[x]/ < x^2 + 1 >$ and $\mathbb{C}$ are really the "same object" as far as addition and multiplication are concerned!

10

Unique Prime Factorization in the Integers

In this chapter we will revisit a result we are familiar with from our school days, and which we have used already in exercises earlier in the book (e.g. in Exericse 1.17, Chapter 1, or in Exercise 4.24, Chapter 4). This is the theorem of unique prime factorization of the integers, a profound theorem actually. In fact, this theorem does not hold in every number system (for instance, for numbers of the form $m + n\sqrt{-5}$, where m and n are arbitrary integers). The failure of this theorem to hold in other number systems in turn leads to a whole body of mathematics called algebraic number theory. We will prove this theorem carefully here, and then examine some consequences.

First, recall from Example 5.38, Chapter 5, the definition of primes: they are those positive integers (≥ 2) whose only divisors are 1 and themselves. We now state the theorem:

Theorem 10.1. (*Unique Prime Factorization Theorem.*) *Every integer greater than* 1 *can be factored into a product of primes. The primes that appear in the factorization are uniquely determined, except for the order in which they appear in the factorization.*

What we mean by "uniqueness of the primes except for the order in which they appear" is the following: if we consider the factorization $108 = 2^2 \cdot 3^3$, then no matter how someone else arrives at a factorization of 108 into a product of primes, there must be two 2s and three 3s and no other primes in the factorization. At worst, this other person may write the factorization in a different order as, say $108 = 2 \cdot 3 \cdot 2 \cdot 3 \cdot 3$, or say $108 = 3 \cdot 3 \cdot 2 \cdot 3 \cdot 2$.

As it turns out, the existence of the factorization is quite easy to prove, but the uniqueness (up to order, that is) takes a bit more work. But the reward for the work is that we will see new ideas that we would not have seen earlier, ideas that have far reaching generalizations to other numbers than just the integers.

Let us prove the existence part of Theorem 10.1 before taking up these new ideas:

Proof of existence of unique prime factorization: Let n be an integer greater than 1; write $P(n)$ for the statement "The integer n can be factored into a product of primes." We will prove that $P(n)$ is true for all $n > 1$ by induction. The smallest integer we need to consider is $n = 2$, and $P(2)$ is clearly true: 2 is a prime number, so "$2 = 2$" is the factorization of 2 into primes! Now assume that the statement has been proven for all integers k, with $2 \leq k \leq n$. We will show that $P(n+1)$ must also be true. If $n+1$ is already a prime, there is nothing to prove: just as with $P(2)$, "$n+1 = n+1$" is the factorization of $n+1$ into primes! If $n+1$ is not prime, then there exist integers a and b, with $1 < a, b < n+1$, such that $n+1 = ab$. Notice that both a and b belong to $\{2, \ldots, n\}$. By the induction hypothesis, both $P(a)$ and $P(b)$ are true. Hence, a and b admit factorizations into primes: say $a = p_1 p_2 \cdots p_k$ and $b = q_1 q_2 \cdots q_l$. Multiplying these together we find $ab = n+1 = p_1 p_2 \cdots p_k \cdot q_1 q_2 \cdots q_l$, a factorization of $n+1$ into primes! By induction, $P(n)$ is true for all $n > 1$. □

10.1 Notion of Divisibility

With the existence of prime factorization behind us, we need to turn our attention to the uniqueness (up to order) of the factorization, and for this, we will start by considering more carefully the notion of divisibility in the integers. Before we formally define what it means for one integer to divide another, let us review the process of "division" that we did in school: if we were given positive integers a and b, we "divided" a by b by removing as many multiples of b from a as we could, till we got a remainder that is less than b, possibly zero. In other words, we found a "quotient" q (the number of multiples of b we removed from a) and a remainder r satisfying $0 \leq r < b$, such that $a = bq + r$. There are two things we (likely) glossed over in school: are the integers q and r uniquely determined by a and b? (That is, by using a different technique for division, could someone else not have found a different pair (q', r') with $0 \leq r' < b$, such that $a = bq' + r'$?) Also, although none of us really had any difficulty performing the division in school for any positive integers a and b once we mastered the technique, is it possible that there exist integers a and b (obviously larger than any a and b that we may have tried in our school days!) such that there do not exist a quotient q and remainder r satisfying $a = bq + r$? We will show here formally that we need not worry about either matter!

(If you did Exercise 1.15.1 in Chapter 1, you will recognize that the first part of the proof below provides a solution to that exercise!)

Theorem 10.2. *Let a and b be positive integers. There exist unique integers q and r, with $0 \leq r < b$, such that $a = bq + r$.*

Proof. We will prove the uniqueness first.

If $a = bq + r$ and also $a = bq' + r'$, then we find $(q' - q)b = r - r'$. Suppose that $q \neq q'$. Note that $|q' - q| \geq 1$ in this case, so $|(q' - q)b| = |q' - q||b| \geq |b|$. On the other hand, since $0 \leq r, r' < b$, the difference $r' - r$ lies somewhere in the interval $[-(b-1), b-1]$. We can see this as follows: if $r = 0$, then $r' - r = r'$ lies in the interval $[0, b-1]$. If $r = 1$, then $r' - r = r' - 1$ lies in the interval $[-1, b-2]$. Proceeding this way, we see that if $r = b-1$, then $r' - r = r' - (b-1)$ lies in the range $[-(b-1), 0]$. Putting all these cases together, we find indeed that $r' - r$ must lie somewhere in $[-(b-1), b-1]$. Or put differently, $|r' - r| \leq (b-1)$. Thus, the left side of $(q' - q)b = r - r'$ has absolute value at least b while the right side has absolute value at most $b - 1$: a contradiction! Hence, q must equal q'.

Now that $q = q'$, the left side of $(q' - q)b = r - r'$ is zero, so we conclude that $r - r' = 0$, that is, $r = r'$. Thus, q and r are indeed uniquely determined, as was to be proved.

As for the existence, consider the set $S = \{a - by \mid y \in \mathbb{Z}_{\geq 0}, a - by \geq 0\}$. Thus, S is the set of all nonnegative integers that are expressible as $a - by$ for some nonnegative integer y. Note that S is nonempty, as a $(= a - b \cdot 0)$ is in S. By the Well Ordering principle (see Definition 8.12, Chapter 8), S has a least element, call it r. Thus, $r = a - bq$ for some integer q. We claim that $r < b$. (Of course, $0 \leq r$ already holds, as r is an element of S, all of whose elements are nonnegative.) Assume to the contrary that $r \geq b$. Then $a - b(q+1) = a - bq - q = r - b \geq 0$. Thus, $a - b(q+1)$ is also in S as $a - b(q+1)$ is of the form $a - by$ and is nonnegative. But $a - b(q+1) = r - b$, and $r - b < r$. This contradicts the fact that r was the least element of S. Hence, our assumption that $r \geq b$ is flawed, and indeed $0 \leq r < b$. Since r was just $a - bq$, we find $a = bq + r$, with $0 \leq r < b$. This shows the existence of q and r. □

Recall that if a and b are positive integers, we say b divides a precisely when the remainder r above is zero. We will now extend this definition to the case where a and b are no longer required to be positive (except that, for reasons we will see immediately, we do not allow b to be zero).

Definition 10.3. Let a and b be integers, with $b \neq 0$. We say b divides a (written $b|a$) if there exists an integer q such that $bq = a$. If $b|a$, we also say b is a *factor* or *divisor* of a, and a is a *multiple* of b.

Remark 10.4. Notice that we did not make any restriction here on the sign of the integers, even though we are accustomed to divisibility of only positive integers. Indeed, we will find it useful to consider divisibility for negative integers (as well as for 0). Thus, according to this definition, -2 divides 6 since $(-2) \cdot (-3) = 6$, and -2 divides -6 since $(-2) \cdot 3 = -6$, and so on.

Note that a is allowed to be 0. In fact, all nonzero integers divide 0, since, for any nonzero integer b there exists the integer 0 (our "q" in the definition) with the property that $b \cdot 0 = 0$.

On the other hand, we do not allow b to be 0, that is, we do not consider the notion of 0 dividing any integer. This is because for any possible q, $0 \cdot q = 0$, so first of all $0 \cdot q = a$ is possible only if $a = 0$ (that is, *potentially,* 0 can only divide 0). But then, when $a = 0$, there is no uniqueness to q (our "quotient"): *for any integer q, $0 \cdot q = 0$,* and this is awkward! So we simply decree that 0 cannot divide 0 because there is no unique quotient, and of course, as we just observed, it is impossible for 0 to divide any nonzero integer either. Put together, we say *that division by 0 is impossible.*

Here is some easy results, that are actually quite powerful, especially the one in part (4):

Lemma 10.5. *We have the following:*

(1) *Let d and a be nonzero integers. If $d|a$, then $d|-a$, $-d|a$ and $-d|-a$.*

(2) *Let a and b be nonzero integers. If $a|b$ and $b|a$, then $a = \pm b$.*

(3) *Let a, b, and c be integers, with a and b nonzero. If* $a|b$ *and* $b|c$*, then* $a|c$.

(4) *Let k be a nonzero integer. If* $k|x$ *and* $k|y$*, for some integers x and y, then* $k|(ax+by)$ *for all integers a and b.*

Proof. We will consider each statement above:

(1) If $d|a$, then $a = dq$ for some integer q. Multiplying by -1, we may write this as $-a = d(-q)$, and $-q$ is an integer, so by the definition of divisibility, $d|-a$. The other assertions in this part are proved similarly.

(2) Since $a|b$, we may write $b = aq$ for some integer q, by the definition of divisibility. Similarly, since $b|a$, we may write $a = br$ for some integer r. Putting these two together, we find $b = aq = (br)q = b(rq)$. Canceling the nonzero integer b from both sides (cancellation is a process which we can do in the integers; in later courses you will learn that you cannot do this in general number systems!), we find $1 = rq$. Now, the only pairs of integers r and q that multiply out to 1 are $(r,q) = (1,1)$ and $(r,q) = (-1,-1)$. In the first case, we have $a = b \cdot 1 = b$, and in the second, we have $a = b \cdot (-1) = -b$, as desired.

(3) Since $a|b$, we can write $b = aq$ for some integer q, and since $b|c$, we can write $c = bs$ for some integer s. Putting these two together, we find $c = bs = (aq)s = a(qs)$. Since qs is an integer, we find c is of the form a times an integer. Thus, by the definition of divisibility, $a|c$.

(4) Since $k|x$, we may write $x = kq$ for some integer q. Similarly, we may write $y = kr$ for integer r. Then, $ax + by = a(kq) + b(kr) = k(aq + br)$. Since a, q, b, and r are all integers, $aq + br$ is an integer, so we find $ax + by$ is of the form k times an integer. Thus, by the definition of divisibility, $k|(ax + by)$.

□

The power of the result in part (4) above is that a and b are allowed to be *arbitrary* integers. Thus, for instance, taking $a = b = 1$, we find $k|(x + y)$, taking $a = 1$ and $b = -1$, we find $k|(x - y)$, taking $a = 2$ and $b = 3$, we find $k|(2x + 3y)$, etc.

We note something about divisors of a *nonzero* integer a: Suppose d is a divisor of a. Then by definition, there exists an integer q such that $dq = a$. Now, it is possible that some of these integers d, q, and a are negative (for instance, it is possible that d and a are negative while q is positive, or that d and q negative while a is positive, or that q and a are negative while d is positive). Taking absolute values, however, we find $|d||q| = |a|$. Now, q cannot be zero, since we said $a \neq 0$, so $|q| > 0$. Since $|q|$ is now a positive integer, we find $|d| \leq |a|$. It follows from this that any divisor of a must lie somewhere in the set $\{-|a|, -|a|+1, \ldots, -1, 1, \ldots, |a|-1, |a|\}$. (Note that we have excluded 0 from this set, since we know that 0 cannot divide a or any other integer for that matter.) Note, too, that $|a|$ is already a divisor of a (for, $|a| = a$ if $a > 0$, and $|a| = -a$ if $a < 0$, and both a and $-a$ divide a as is readily seen by taking $q = 1$ and $q = -1$ respectively). It follows that the largest divisor of a is $|a|$.

10.2 Greatest Common Divisor, Relative Primeness

We now define the greatest common divisor, a notion that you are familiar with from high school, but one that will be critical to what follows:

Definition 10.6. Let a and b be integers, with at least one of them nonzero. The *greatest common divisor* of a and b is the *largest* of the *common divisors* of a and b, that is, it is the largest integer d such that d divides both a and b. It is denoted $\gcd(a, b)$.

Let us note something about $\gcd(a, b)$: it is defined as long as at least one of a and b is a nonzero integer. The two integers a and b are allowed to be equal: when that happens, $\gcd(a, a)$ simply becomes the largest divisor of a, and we have seen above that this is just $|a|$. Thus, $\gcd(a, a) = |a|$. Now, if say a is nonzero but b were 0, then since every integer divides 0, the common divisors of a and b are now simply the divisors of a, and the largest among these is $|a|$ as we noted above. Hence, $\gcd(a, 0) = |a|$. The greatest common divisor is not defined if both a and b are both zero, since as we have already noted, every integer divides 0, so there can be no largest divisor of 0. Finally, if both a and b are nonzero, then, since any d that divides a is bounded above by $|a|$, we find $d \leq |a|$ and (similarly) $d \leq |b|$, so putting these together, $d \leq \min(|a|, |b|)$. Of course, in all cases where $\gcd(a, b)$ is defined, since 1 divides all integers, 1 is already a common divisor of a and b, so $\gcd(a, b) \geq 1$.

Example 10.7. The concept of the greatest common divisor is familiar to us from high school, for instance, $\gcd(9, 12) = 3$. Here are some examples that include negative integers or 0, that you may not have seen in school:

(1) $\gcd(-9, 12) = \gcd(9, -12) = \gcd(-9, -12) = 3$.

(2) $\gcd(9, 0) = \gcd(-9, 0) = 9$.

(3) $\gcd(9, 1) = \gcd(-9, 1) = \gcd(9, -1) = \gcd(-9, -1) = 1$.

Let us collect together some properties of the greatest common divisor, most of them exemplified above:

Proposition 10.8. *Let a and b be integers, and assume $a \neq 0$. Then*

(1) *$\gcd(a, b) = \gcd(\pm a, \pm b)$.*

(2) *$\gcd(a, 0) = |a|$.*

(3) *$\gcd(a, 1) = 1$.*

(4) *$\gcd(a, b) = \gcd(b, a - bq)$ for any integer q.*

Proof. To prove part (1), note first from Lemma 10.5(1) that the set of divisors of a and the set of divisors of $-a$ are the same, as are the set of divisors of b and the set of divisors of $-b$. It follows that the set of common divisors of a and b is the same as the set of common divisors of a and $-b$, and is the same as the set of common divisors of $-a$ and b, and is the same as the set of common divisors of $-a$ and $-b$. Since these sets are all equal, the greatest element of each of these sets are equal. Hence, $\gcd(a,b) = \gcd(a,-b) = \gcd(-a,b) = \gcd(-a,-b)$.

We have already proved part (2) in the paragraph above Example 10.7 above. We will leave part (3) for you to prove. (Hint: What are the divisors of 1?)

As for part (4), we will show that the set of common divisors of a and b is the same as the set of common divisors of b and $a - bq$. As in the proof of part (1), it will follow that the greatest element of these two sets will be equal, that is, $\gcd(a,b)$ will equal $\gcd(b, a-bq)$. Accordingly, let k divide both a and b. Then, by part (4) of Lemma 10.5, k must divide $1 \cdot a + (-q) \cdot b$, so k must divide $a - bq$. Thus, k is a common divisor of b and $a - bq$, showing that the set of common divisors of a and b is a subset of the set of common divisors of b and $a - bq$. To show the reverse containment, let k be a common divisor of b and $a - bq$. Then by part (4) of Lemma 10.5 again, k must divide $q \cdot b + 1 \cdot (a - bq)$, that is, k must divide a. Thus, k is a common divisor of a and b, showing that the set of common divisors of b and $a - bq$ is a subset of the set of common divisors of a and b. The two sets are therefore equal. □

Exercise 10.9. Here is a quick exercise: If a and b are two integers, not both zero, show that $-\gcd(a,b)$ is the smallest of the set of common divisors of a and b.

Definition 10.10. Two nonzero integers a and b are said to be *relatively prime* if $\gcd(a,b) = 1$.

For instance, 9 and 16 are relatively prime, as the divisors of 9 are $\pm 1, \pm 3$, and ± 9 and those of 16 are $\pm 1, \pm 2, \pm 4$, and ± 16.

Note that for any nonzero a $\gcd(a, \pm 1) = 1$, so a and 1 are relatively prime.

(We could have extended Definition10.10 to the case where one of a or b is zero, but, if say $b = 0$, then since $\gcd(a, 0) = |a|$, we'll find that $\gcd(a, 0) = 1$ precisely when $a = \pm 1$, thus, this extension of the definition would not have been particularly useful.)

We will now consider the greatest common divisor from a different perspective: we will give a characterization of it that allows us to describe the greatest common divisor of two integers without knowing what their divisors are! First, some terminology:

Definition 10.11. If a and b are integers, an *integer linear combination* of a and b is any integer that is expressible as $xa + yb$ for some two integers x and y. Typically, while working solely in the context of integers, we drop the word "integer" in front, and simply say "linear combination" of a and b.

For instance, 3 is a linear combination of 9 and 12: it is expressible as $(-1)\cdot 9 + 1 \cdot 12$ (as also $3 \cdot 9 + (-2) \cdot 12$). Similarly, 1 is a linear combination of 9 and 16: it is expressible as $(-7) \cdot 9 + 4 \cdot 16$ (as also $9 \cdot 9 + (-5) \cdot 16$).

Here is an exercise you should do:

Exercise 10.12. Let a and b be integers, and let S be the set of all linear combinations of a and b, T the set of all linear combinations of a and $-b$, U the set of all linear combinations of $-a$ and b, and V the set of all linear combinations of $-a$ and $-b$. Show that $S = T = U = V$. Conclude that the set of linear combinations of a and b equals the set of linear combinations of $|a|$ and $|b|$. (Hint: Any linear combination $xa+yb$ of a and b can be written as $xa + (-y)(-b)$, which is a linear combination of a and $-b$. Similarly, any linear combination $xa + y(-b)$ of a and $-b$ can be written as $xa + (-y)(b)$, which is a linear combination of a and b.)

Here is the promised characterization of the greatest common divisor:

Theorem 10.13. *The greatest common divisor of two integers (at least one of which is nonzero) is the smallest positive linear combination of the two integers.*

Wow! Did you see that coming? For those who have only seen the greatest common divisor defined in terms of actual factors of the two integers, this characterization must appear very baffling. What on earth does being a linear combination have to do with greatest common divisors? Mathematics is full of very surprising results like this: results that are not at all obvious, but, that provide alternative and very useful perspectives (as this result will also do).

One hint of the connection between the greatest common divisor, or at least, between common divisors, and linear combinations, lies in a result we have already seen: part (4) in Lemma 10.5. Indeed, that result, in our new language, says that every common divisor of a and b divides every linear combination of a and b! Indeed, we will use this result along the way in the proof of our theorem.

Proof of Theorem 10.13. Let a and b be the two integers, at least one of which is nonzero. We may assume $a \neq 0$. First note that by Proposition 10.8(1), $\gcd(a, b) = \gcd(|a|, |b|)$. Moreover, the set of linear combinations of a and b equals the set of linear combinations of $|a|$ and $|b|$, by Exercise 10.12. In particular, the smallest positive linear combination of a and b will also be the smallest positive linear combination of $|a|$ and $|b|$. Hence, it is harmless while considering the greatest common divisor to replace a by $|a|$ and b by $|b|$, and therefore assume that a, which is already assumed to be nonzero, is actually positive (and b is nonnegative).

Next, note that there is something that is swept under the rug in the statement of Theorem 10.13: how do we even know that there is indeed a smallest positive linear combination of a and b? One of the first things we need to do is to establish the existence of a smallest positive linear combination. What is the problem here? After all, if we collect all positive integers that are expressible in the form $xa + yb$ for suitable integers x and y into a set S, then, since S would be a subset of the positive integers, the Well Ordering principle (see Definition 8.12, Chapter 8) would guarantee that S must have a least element... or would it? Not yet! The Well Ordering principle applies to *nonempty* subsets of $\mathbb{N}$. We do not know yet that S is nonempty, that is, we do not know yet that there exist positive integers that are expressible as $xa + yb$ for suitable integers x and y.

Fortunately, it is trivial to show that S is nonempty. We have assumed that a is positive (see the argument two paragraphs above). Now a is already a linear combination of a and b, since $a = 1 \cdot a + 0 \cdot b$. Hence $a \in S$, and S is nonempty. At this point, we can apply the Well Ordering principle to S with a clear conscience, and be certain that there is a least positive linear combination of a and b.

Write d for this least positive linear combination. We need to show that $d = \gcd(a, b)$. Thus, we need to show that d is the largest of the common divisors of a and b. There are two tasks here: we need to show that first of all d is a common divisor of a and b, and next, that d is the largest of all common divisors.

To show that d is a common divisor, let us show that $d|a$. Dividing d into a using the usual long-division process for positive integers we learn in school, we may write $d = aq + r$, where q is the quotient and r is the remainder. (Alternatively, we may view the existence of q and r as simply given by Theorem 10.2.) Thus, r lies in the range $0 \leq r < d$. We wish to show that $r = 0$, that is, that d divides a evenly without a remainder. Since d is a linear combination of a and b, we may write $d = xa + yb$ for suitable integers x and y. Plugging this into the relation $d = aq + r$, we find $xa + yb = aq + r$ or $(x - q)a + yb = r$. Since $x - q$ is also an integer, this shows that r is also a linear combination of a and b. If r were not zero, then the relation $0 \leq r < d$ would now show that $1 \leq r < d$, and in particular, that r is positive. But d was selected as the *smallest* positive linear combination of a and b, while we have found another positive linear combination, namely r, that is *less than* d. This is clearly a contradiction, so r must equal 0, that is, $d|a$.

If $b = 0$, then d already divides b. Otherwise, we may assume (as noted above) that b is positive, and then, by an identical argument as for a, we find $d|b$ as well. Hence d is a common divisor of a and b.

To show that d is the largest of all common divisors, let e be any other common divisor of a and b. Then, $|e|$ divides both a and b (this follows from Lemma 10.5(1), since $|e| = e$ if $e \geq 0$ and $|e| = -e$ if $e < 0$). By Lemma 10.5(4), we find that $|e|$ divides every linear combination of a and b, so $|e|$ divides d. Write $d = |e|k$ for suitable k, and notice that since d and $|e|$ are both positive, k is positive. Hence $|e| \leq d$, and since $e \leq |e|$, we find $e \leq d$. Since e was an arbitrary common divisor of a and b, we find that indeed d is the largest of the common divisors, that is, $d = \gcd(a, b)$, as desired. □

The beauty of this result, as already mentioned, is that it gives a characterization of the greatest common divisor of a and b that does not involve the factors of a and b. Although it appears to be an abstract characterization, it proves to be enormously useful! Here is an immediate corollary:

Corollary 10.14. *Two nonzero integers are relatively prime if and only if* 1 *is a linear combination of them.*

Proof. If the two nonzero integers a and b are relatively prime, then $\gcd(a, b) = 1$ by definition. By Theorem 10.13, 1 is a linear combination of a and b (and in fact, the smallest positive linear combination). On the other hand, suppose 1 is a linear combination of a and b. Since 1 is the smallest positive integer, 1 then *has to be* the smallest positive linear combination of a and b. By Theorem 10.13, 1 is the greatest common divisor of a and b. □

Remark 10.15. It is worth emphasizing that Theorem 10.13 describes $\gcd(a, b)$ as the *smallest* positive linear combination of a and b, and not simply as a positive linear combination of a and b. In particular, if you can write a positive integer d as a linear combination of a and b, you cannot yet conclude that d is their gcd: you need to further show that no positive integer smaller than d can be a linear combination of a and b. For instance 2 is a linear combination of 9 and 16, since $2 = 2 \cdot 9 + (-1) \cdot 16$, but you cannot conclude from this that $\gcd(9, 16) = 2$. (In fact, as we have seen 1 is also a linear combination of 9 and 16: $1 = (-7) \cdot 9 + 4 \cdot 16$, so by Corollary 10.14 above, $\gcd(9, 16) = 1$.)

Remark 10.16. The proof of Theorem 10.13 does not provide an algorithm to determine $d = \gcd(a, b)$ as a linear combination of a and b. There is a fast algorithm known as

Euclid's algorithm for writing d as a linear combination, that is, to determine integers x and y such that $d = xa + yb$. This is developed in Exercise 10.37 ahead.

Remark 10.17. We remarked just before we began the proof of Theorem 10.13 that every common divisor of a and b divides every linear combination of a and b, and we used this fact in the proof to show that the least positive linear combination of a and b equals $\gcd(a, b)$. It follows therefore that every common divisor of a and b divides $\gcd(a, b)$.

Earlier (Example 5.38, Chapter 5), we had defined primes as those positive integers (≥ 2) whose only divisors are 1 and themselves. Since we have expanded the notion of divisibility to allow for division by negative integers as well, we need to specify in the definition of prime that the only *positive* divisors are 1 and themselves. Note that if d is an integer that divides p, then $-d$ also divides p, as we have seen already (Lemma 10.5), so to say that the only positive divisors of p are 1 and p is equivalent to saying that the only divisors of p are ± 1 and $\pm p$. Nevertheless, we will choose to write the definition in terms of just the positive divisors, and we will record it here:

Definition 10.18. A *prime* integer (often just referred to as a prime number, or even more simply, as a prime) is an integer $p > 1$ whose only positive divisors are 1 and p. A *composite* integer is an integer $n > 1$ that is not prime.

Remark 10.19. It is worth remarking why 1 is not taken to be a prime (even though 1 satisfies the requirement that its only positive divisors be 1 and itself!). This is done for convenience, so as to make the statement of the unique factorization theorem (Theorem 10.1) come out "clean." If 1 were considered prime, then, for example, 10 would have several factorizations, differing from one another in more than just order: $10 = 2 \cdot 5$, $10 = 2 \cdot 5 \cdot 1$, $10 = 2 \cdot 5 \cdot 1 \cdot 1$, etc. At the same time, there is a "triviality" to these extra factorizations beyond the standard $10 = 2 \cdot 5$: they come for free simply because multiplying an integer by 1 any number of times does not change the integer. These extra factors of 1 thus do not provide any new information on how 10 decomposes. Having to consider these extra factorizations each time and then to say each time that we will ignore those factors of 1 is awkward. So, we define our way out of this awkwardness by declaring 1 to not be a prime!

Remark 10.20. We may define *any* integer $p \neq 0, \pm 1$ to be a prime if its only factors are ± 1 and $\pm p$. For example, by this definition, $-2, -3, -5$, etc. are also primes. We will not adopt this wider definition in this book for the sake of simplicity. Also note that if $n > 1$ is composite, then necessarily, n has a positive divisor x with $1 < x < n$, and therefore, there exists another positive integer y, with $1 < y < n$ such that $n = xy$.

We use Corollary 10.14 now to prove a very useful result: a different characterization of prime numbers. (We have already used one direction of this result informally in Exercise 1.17 of Chapter 1.)

Proposition 10.21. *An integer (greater than* 1*) is prime if and only if whenever it divides the product of two integers, it must divide one or the other.*

Proof. Let $p > 1$ be the integer. First assume that p is prime. We need to show that whenever p divides the product ab of two integers a and b, then either $p|a$ or $p|b$. Given that $p|ab$, if $p|a$, we are done. So assume that p does not divide a. Note that if p does not divide a, then $-p$ also does not divide a (this follows from the contrapositive form of Lemma 10.5(1)). Now since the factors of p are ± 1 and $\pm p$, and since p and $-p$ do not divide a, while of course ± 1 divide a, we conclude that the only common divisors of p and a are ± 1. Hence, $\gcd(p, a)$, which is the largest of their common divisors, is 1. By Corollary 10.14, there exist integers x and y such that $1 = xp + ya$. Multiplying both sides by b and grouping as shown, we find $b = (bx)p + y(ab)$. Now p clearly divides p, and we are given that p divides ab. By part (4) of Lemma 10.5, p must divide their linear combination $(bx)p + y(ab)$, or in other words, p must divide b. This proves one direction of the result!

As for the other direction, assume that whenever p divides a product of integers ab, either p must divide a or p must divide b. We need to show that p is prime. Assume to the contrary. Then p is composite, that is, p can be written as the product mn of two positive integers, with $m < p$ and $n < p$. The obvious fact $p|p$ now reads $p|mn$, and by the hypothesis, p must divide m or p must divide n. But we arrive at a contradiction: as $m < p$ and both p and m are positive, $p|m$ is impossible. Similarly, $p|n$ is also impossible. We conclude that p must be prime. □

Exercise 10.22. Extract the appropriate arguments out of the proof of one direction of Proposition 10.21 above to prove the following: if d is a nonzero integer that divides the product ab of two integers a and b, and if $\gcd(d, a) = 1$, then $d|b$.

Observe that while proving Proposition 10.21, what we showed first was that if a prime p does not divide the integer a, then p and a are relatively prime. After that, we essentially proved the statement of this exercise!

See Exercise 10.31.3 ahead.

We will need the result of the following corollary in the proof of the uniqueness part of the prime factorization theorem (Theorem 10.1) below:

Corollary 10.23. *If a prime divides the product of k integers ($k \geq 2$), then it must divide one of these k integers.*

Proof. Let p be a prime. Write $S(k)$ ($k \geq 2$) for the statement "if p divides $a_1 \cdot a_2 \cdot \cdots \cdot a_k$, where the a_i are arbitrary integers, then $p|a_i$ for some i, $1 \leq i \leq k$." We will establish the truth of $S(k)$ by induction.

$S(2)$ is clearly true: this is the contention of Proposition 10.21. Now assume that $S(2), \ldots, S(k)$ have been proven true for some $k \geq 2$. We will show that $S(k+1)$ must then be true. $S(k+1)$ is the statement "if p divides $a_1 \cdot a_2 \cdot \cdots \cdot a_{k+1}$, where the a_i are arbitrary integers, then $p|a_i$ for some i, $1 \leq i \leq k+1$." Write the product of the $k+1$ integers as $(a_1 \cdot a_2 \cdot \cdots \cdot a_k) \cdot a_{k+1}$. Viewing the product in the parentheses as a single integer, the truth of $S(2)$ shows that either p divides the integer $(a_1 \cdot a_2 \cdot \cdots \cdot a_k)$ or p divides a_{k+1}. If $p|a_{k+1}$ we are done, if not, p must divide $(a_1 \cdot a_2 \cdot \cdots \cdot a_k)$. But this is a product of k integers, so since $S(k)$ is true, p must divide some a_i, for $1 \leq i \leq k$. Putting these two results together, we find that p must divide some a_i, $1 \leq i \leq k+1$. Hence $S(k+1)$ is true, so by induction we are done. □

10.3 Proof of Unique Prime Factorization Theorem

We are finally ready to complete the proof of Theorem 10.1:

Proof of the uniqueness of prime factorization, Theorem 10.1. For each integer $n \geq 2$, let $P(n)$ be the statement "the integer n factors into a product of primes that is unique except for the order in which the primes appear in the factorization." We have already proved the existence of the factorization, now we'll prove the uniqueness by induction on n.

When $n = 2$ (the lowest possible value of n), we have one factorization of 2 as a product of primes: "$2 = 2$." Since the only divisors of 2 are ± 1 and ± 2, this has to be the only possible way of writing 2 as a product of primes. (To spell this out: if we wrote $2 = q_1^{a_1} \cdots q_s^{a_s}$, where the q_i are primes and the a_i are positive integers, then these primes all have to divide 2, and hence, there can only be one such q_i, namely q_1, and that q_1 has to be 2. Moreover, a_1 then has to be 1, as every higher power of 2 exceeds 2.) Hence $P(2)$ is true.

Notice that more generally, this argument applies to any prime number p. Since p only has factors ± 1 and $\pm p$, the only way of writing p as a product of primes is "$p = p$."

Now assume that $P(2), \cdots, P(n)$ have all been proved to be true for some $n \geq 2$, and we'll show that $P(n+1)$ must also be true in that case. If $n+1$ is prime, then, as discussed already above, "$n+1 = n+1$" is the only possible way of writing $n+1$ as a product of primes. So suppose that $n+1$ is not prime, and it factors into the product of primes $p_1^{a_1} \cdot \cdots \cdot p_s^{a_s}$ as also into the product of primes $q_1^{b_1} \cdot \cdots \cdot q_t^{b_t}$. To claim that the primes in these two products are the same except for the order in which they appear is to claim the following: $s = t$, and after renumbering if necessary, $p_1 = q_1, p_2 = q_2, \ldots, p_s = q_s, a_1 = b_1, a_2 = b_2, \ldots, a_s = b_s$. Accordingly, this what we will prove.

Consider the prime p_1. Clearly $n+1$ is of the form p_1 times an integer, so p_1 divides $n+1$. It hence divides the product $q_1^{b_1} \cdot \cdots \cdot q_t^{b_t}$ which is just $n+1$ written a different way. By Corollary 10.23, p_1 must divide some prime $q_1, \ldots, q_t$. Renumbering, if necessary, we may assume that p_1 divides q_1. But the only positive divisors of q_1 are 1 and q_1. Since p_1 divides q_1 and $p_1 > 1$, we conclude that $p_1 = q_1$.

Now note that $n+1$ is not prime by assumption, so p_1 (and therefore q_1, as $p_1 = q_1$) are strictly less than $n+1$. Hence, $\frac{n+1}{p_1}$ (and therefore $\frac{n+1}{q_1}$, as $\frac{n+1}{p_1} = \frac{n+1}{q_1}$) are both integers greater than or equal to 2. Consider the equation

$$n+1 = p_1^{a_1} \cdot \cdots \cdot p_s^{a_s} = q_1^{b_1} \cdot \cdots \cdot q_t^{b_t}.$$

We set $q_1 = p_1$ and divide by p_1 to find

$$\frac{n+1}{p_1} = p_1^{a_1-1} p_2^{a_2} \cdot \cdots \cdot p_s^{a_s} = p_1^{b_1-1} q_2^{b_2} \cdot \cdots \cdot q_t^{b_t}.$$

As we have already noted, $\frac{n+1}{p_1}$ is an integer greater than or equal to 2, and of course less than $n+1$. By induction, $P\left(\frac{n+1}{p_1}\right)$ is true. It follows that $s = t$, and after renumbering if necessary, $p_2 = q_2, \ldots, p_s = q_s, a_1 - 1 = b_1 - 1$ (so $a_1 = b_1$), $a_2 = b_2, \ldots, a_s = b_s$. This proves the uniqueness. □

10.4 Some Consequences of the Unique Prime Factorization Theorem

We present several results in this section that are consequences of Theorem 10.1. We start by determining the set of divisors of an integer:

Theorem 10.24. *Let n be an integer greater than 1, and let $n = p_1^{a_1} \cdot p_2^{a_2} \cdot \cdots \cdot p_k^{a_k}$ be its factorization into primes, where the p_i are distinct primes and the exponents a_i are positive integers. If m is a divisor of n that is greater than 1, then the prime factorization of m is of the form $m = p_{l_1}^{e_1} \cdot \cdots \cdot p_{l_u}^{e_u}$, where $u \le k$, the primes $p_{l_1}, \dots, p_{l_u}$ form a subset of the primes $p_1, \dots, p_k$, and $1 \le e_i \le a_{l_i}$, for $i = 1, \dots, u$. Hence, the set of positive divisors of n consists of all integers of the form $m = p_1^{b_1} \cdot p_2^{b_2} \cdot \cdots \cdot p_l^{b_k}$, where $0 \le b_i \le a_i$.*

Proof. Let m (> 1) be a positive divisor of n. If $m = n$, then of course $m = p_1^{a_1} \cdot p_2^{a_2} \cdot \cdots \cdot p_k^{a_k}$, and this is clearly of the desired form. Now assume that $1 < m < n$, and write $n = mx$ for some integer x. Let m have the prime factorization $q_1^{e_1} \cdot \cdots \cdot q_u^{e_u}$, for suitable (distinct) primes q_i and exponents e_i, $1 \le i \le u$. Since $m < n$, x must be greater than 1, so x also has a prime factorization: assume that the prime factorization of x reads $x = r_1^{f_1} \cdot \cdots \cdot r_v^{f_v}$, for suitable (distinct) primes r_i and exponents f_i, $1 \le i \le v$. Putting together the prime factorizations of n, m and x, we find

$$\underbrace{p_1^{a_1} \cdot p_2^{a_2} \cdot \cdots \cdot p_k^{a_k}}_{n} = \underbrace{q_1^{e_1} \cdot \cdots \cdot q_u^{e_u}}_{m} \cdot \underbrace{r_1^{f_1} \cdot \cdots \cdot r_v^{f_v}}_{x}. \tag{10.1}$$

Note that it is quite possible that some of the prime factors r_i of x could be the same as some of the prime factors q_j of m. But that will not concern us, except that we will note that if say, after re-indexing, $q_i = r_i$, then the total exponent of the prime q_i on the right side of equation (10.1) above will be $e_i + f_i$, which is greater than e_i. Now let us invoke the unique prime factorization theorem. The same integer n has one prime factorization on the left side of equation (10.1), and another prime factorization on the right side of equation (10.1). The uniqueness part of the theorem says that these two prime factorizations must be the same, except possibly for the order in which the primes are listed. Since the right side contains the prime q_1 with exponent e_1 (or possibly even higher than e_1 if some prime factor of x were also equal to q_1 as we noted above), the left side must also contain at least e_1 copies of q_1. So, one of the primes on the left, call it p_{l_1}, must equal q_1, and since p_{l_1} occurs a_{l_1} on the left, we find $a_{l_1} \ge e_1$, or what is the same, $e_1 \le a_{l_1}$. Proceeding, for $i = 1, \dots, u$, we find q_i equals some p_{l_i}, and $e_i \le a_{l_i}$. Thus, $m = p_{l_1}^{e_1} \cdot \cdots \cdot p_{l_u}^{e_u}$. It is clear that $u \le k$. Hence m is of the desired form. This proves the first statement of the theorem.

For the second statement of the theorem, we note that if a divisor $m > 1$ has the factorization $m = p_{l_1}^{e_1} \cdot \cdots \cdot p_{l_u}^{e_u}$ as described above, and if $u = k$, then (except for a possible rearrangement of the primes) it is already in the form $p_1^{b_1} \cdot p_2^{b_2} \cdot \cdots \cdot p_l^{b_k}$, with $0 \le b_i \le a_i$ (in fact, $1 \le b_i \le a_i$). If $u < k$, then some primes p_i do not appear in the list $p_{l_1}, \dots, p_{l_u}$. In that case, simply write the remaining primes to the exponent 0 (note that $p^0 = 1$) and multiply them into the product $p_{l_1}^{e_1} \cdot \cdots \cdot p_{l_u}^{e_u}$ to find that indeed, after rearrangement if necessary, m is of the form $p_1^{b_1} \cdot p_2^{b_2} \cdot \cdots \cdot p_l^{b_k}$, where now $0 \le b_i \le a_i$. And, if $m = 1$, then taking all $b_i = 0$, we find m is once again of the form

$p_1^{b_1} \cdot p_2^{b_2} \cdot \cdots \cdot p_l^{b_k}$. This shows that every positive divisor of n is in the set consisting of integers of the form $p_1^{b_1} \cdot p_2^{b_2} \cdot \cdots \cdot p_l^{b_k}$, where $0 \le b_i \le a_i$. For the reverse inclusion, note that given any integer of the form $p_1^{b_1} \cdot p_2^{b_2} \cdot \cdots \cdot p_l^{b_k}$, where $0 \le b_i \le a_i$, we may write $n = \left(p_1^{b_1} \cdot p_2^{b_2} \cdot \cdots \cdot p_l^{b_k}\right) \cdot \left(p_1^{a_1-b_1} \cdot p_2^{a_2-b_2} \cdot \cdots \cdot p_l^{a_k-b_k}\right)$. Since $a_i \ge b_i$, the difference $a_i - b_i$ is nonnegative for all i, so $\left(p_1^{a_1-b_1} \cdot p_2^{a_2-b_2} \cdot \cdots \cdot p_l^{a_k-b_k}\right)$ is an integer, and therefore, $p_1^{b_1} \cdot p_2^{b_2} \cdot \cdots \cdot p_l^{b_k}$ is a divisor of n. Thus, the second statement of the theorem is also true. □

Exercise 10.25. Suppose $n = ab$, where n, a, and b are positive integers. If x is a divisor of a and y is a divisor of b, show that xy is a divisor of n. Conversely, show that every divisor of n is of the form xy, where x is a divisor of a and y is a divisor of b.

See also Exercise 10.39 ahead.

Next, we present Euclid's proof that there are infinitely many primes, a veritable gem of mathematical reasoning!

Theorem 10.26. *The set of primes is infinite.*

Proof. Assume to the contrary that there are only finitely many primes, say there are k of them. (We know that 2 is a prime, so certainly $k \ge 1$.) We label them $p_1, \ldots, p_k$. Consider the integer $N = (p_1 \cdot \cdots \cdot p_k) + 1$. This is an integer that is greater than 1 (since each $p_i \ge 2$), so it admits factorization into primes. Consider any prime factor q of N. By assumption, there are only finitely many primes $p_1, \ldots, p_k$, so q must be one of these. Say $q = p_1$ (after re-indexing if necessary). Then $N = q(p_2 \cdot \cdots \cdot p_k) + 1$. Of course, $1 < q$. Thus, q cannot divide N since it leaves a remainder of 1 when divided into N. But q was given to be a divisor of N. We therefore arrive at a contradiction, so the set of primes must be infinite!

(Note that we only used the *existence* of prime factorization, but did not have to invoke the *uniqueness* of the factorization for this result!) □

Theorems 10.24 and 10.26 are both instances where the unique prime factorization theorem leads to some nontrivial results about integers! The following are further examples of interesting results that arise from this theorem:

Example 10.27. *Perfect squares:* For instance, if n is an integer greater than 1, we can recognize from its prime factorization whether n is a perfect square (i.e., $n = m^2$ for some integer m) by the following: n is a perfect square if and only if every prime p_i that appears in its prime factorization $n = p_1^{a_1} \cdot p_2^{a_2} \cdot \cdots \cdot p_k^{a_k}$ ($a_i \ge 1$) appears with *even* exponent, i.e., each a_i is even.

To see this, note that if each each exponent is even, say $a_i = 2b_i$ for a suitable positive integer b_i, then taking $m = p_1^{b_1} \cdot p_2^{b_2} \cdot \cdots \cdot p_k^{b_k}$, we find $m^2 = (p_1^{b_1} \cdot p_2^{b_2} \cdot \cdots \cdot p_k^{b_k})^2$, which of course equals $p_1^{2b_1} \cdot p_2^{2b_2} \cdot \cdots \cdot p_k^{2b_k}$. Since $2b_i$ is just a_i, we find $m^2 = n$, so n is a perfect square. Conversely, if n is a perfect square, say $n = m^2$, then if m has the prime factorization $p_1^{b_1} \cdot p_2^{b_2} \cdot \cdots \cdot p_k^{b_k}$ for suitable $k \ge 1$ and $b_i \ge 1$, we find $n = m^2 = p_1^{2b_1} \cdot p_2^{2b_2} \cdot \cdots \cdot p_k^{2b_k}$. This is a product of primes, and by the uniqueness part of the unique prime factorization theorem, this must be the prime factorization of n. It is clear then that each prime factor of n appears with even exponent.

Clearly, similar characterizations can be given for perfect cubes, perfect fourth powers, and so on.

Example 10.28. *$\sqrt{2}$ is not a rational number:* This is another gem of reasoning! Assume to the contrary that $\sqrt{2}$ is rational. Thus, we may write $\sqrt{2} = a/b$, where a and b are positive integers, and $\gcd(a, b) = 1$ (i.e., we may assume we have cancelled all common factors in the denominator and the numerator). We square and cross multiply to find $2b^2 = a^2$. If $a = 1$ we find that 2 times a positive integer is 1, but positive multiples of 2 are at least as big as 2, so this cannot happen. Hence $a > 1$. If $b = 1$, we find that 2 is the square of an integer, but this cannot happen because the perfect squares are either 1 or are at least as big as 4, so $b > 1$. Now consider the prime factorizations on both sides of the relation $2b^2 = a^2$. By Example 10.27 above, the exponent of 2 in both a^2 and b^2 must be even. (Here, if it turns that b is odd, then since $2^0 = 1$, we may view 2 as appearing to power 0 in b and therefore with power 0 in b^2 as well—and of course 0 is even. Note that a cannot be odd—why?) We now have a contradiction: Consider the total exponent of 2 in the prime factorizations of both sides. On the left, we have one 2 from the multiplier 2 in front of $2b^2$, and an even number of 2s, say $2k$ of them, from b^2. Hence, the total exponent of 2 on the left is $2k + 1$, an odd integer. On the right, we have an even number of 2s coming from a^2. By the uniqueness part of the prime factorization, 2 must appear to the same power on both sides, but we have just seen that 2 occurs to an odd power on the left and an even power on the right! Hence, our assumption that $\sqrt{2}$ is rational is flawed, and indeed, $\sqrt{2}$ must be irrational!

Example 10.29. *Divisors of p^a:* Here is a result we can derive from Theorem 10.24 (which in turn is derived from the unique prime factorization theorem): Let $n \geq 2$ be of the form p^a for some prime p and some $a \geq 1$ (i.e., a *prime power*). Then the number of positive divisors of n is $a + 1$.

To prove this, recall that by Theorem 10.24, the positive divisors of p^a are precisely the integers of the form p^b with $0 \leq b \leq a$. There are $a + 1$ choices for b, so there are $a + 1$ divisors!

> **Exercise 10.29.1.** Generalizing the result above, show that if $n = p_1^{a_1} \cdot p_2^{a_2} \cdot \cdots \cdot p_k^{a_k}$ is the factorization of a positive integer n, then the number of positive divisors of n is $(a_1 + 1) \cdot (a_2 + 1) \cdots (a_k + 1)$.

Example 10.30. *Number of positive integers relatively prime to p^a:* Here is another result that falls out of Theorem 10.24: Let $n = p^a$ as in Example 10.29 above. The number of positive integers m in the range $1 \leq m \leq p^a$ such that m and n are relatively prime is $p^a - p^{a-1}$.

To prove this, recall that m is relatively prime to n if $\gcd(m, n) = 1$. We use our basic definition of the greatest common divisor: it is the largest of all the common divisors of m and n. Now the divisors of n are precisely the various powers $1, p, p^2, \ldots, p^a$, by Theorem 10.24. So, $\gcd(m, n)$ is the highest of these $a + 1$ powers $1, p, p^2, \ldots, p^a$ that divides m.

If $\gcd(m, n) = 1$, then obviously, p is already not a common divisor, and hence, p does not divide m. Conversely, if p does not divide m, then no p^i can divide m for

$i \geq 1$ (as p divides p^i and p would therefore divide m by Lemma 10.5(3)). It follows that the largest common divisor from the list of $a + 1$ powers of p above is 1, that is, $\gcd(m, n) = 1$.

Thus, the set of positive integers m in the range $1 \leq m \leq p^a$ such that m and n are relatively prime is precisely the set of integers m in this range that are not divisible by p. To count these, let us instead look at the complementary set of *multiples* of p in the range 1 through $n = p^a$: they are $p, 2 \cdot p, 3 \cdot p, \ldots, p \cdot p = p^2, (p+1) \cdot p$, and so on. How many are there in the range 1 through p^a? A multiple of p is of the form $p \cdot l$ for some integer l, and we want $p \cdot l \leq p^a$. This happens if and only if $l \leq p^{a-1}$. Thus there are p^{a-1} multiples of p in the range 1 through $n = p^a$.

It follows that there are $p^a - p^{a-1}$ integers in the range 1 through $n = p^a$ that are *not* multiples of p. From our discussions above, these are precisely the integers in our range that are relatively prime to n. This establishes the result.

10.5 Further Exercises

Exercise 10.31. Practice Exercises:

These exercises are centered around the definition of divisibility and the application of Lemma 10.5, particularly of part (4).

Exercise 10.31.1. Let a, b, c, and d be nonzero integers with $c \neq 0$ and $d \neq 0$. Show that if $c|a$ and $d|b$, then $cd|ab$.

Exercise 10.31.2. Let a, b, and c be nonzero integers with $c \neq 0$. Show that if $c|a$ and $c|b$, then $c^2|a^2 + b^2$. (Hint: Use the result of Exercise 10.31.1 above, along with Lemma 10.5(4).)

Exercise 10.31.3. Let a, b, and d be integers, with $d \neq 0$. It is a very common mistake in the beginning study of divisibility to say that if $d|ab$, then $d|a$ or $d|b$. This is not true: for example, $6|(4 \cdot 9)$, yet 6 does not divide 4, and 6 does not divide 9. Give at least five examples of integer triples a, b, and d, with $d \neq 0$, such that $d|ab$, yet, $d \nmid a$ and $d \nmid b$.

Remark: In this context, see Exercise 10.22 earlier in the text. What that exercise shows is that in the situation of the current exercise, if further d and a are relatively prime, then we can conclude that $d|b$. (And similarly of course, if d and b are relatively prime, we can conclude that $d|a$.) But hopefully the examples you generated would have convinced you that without such further assumptions, we cannot conclude from $d|ab$ that $d|a$ or $d|b$.

Exercise 10.31.4. Show that any two consecutive integers must be relatively prime. (Hint: If d divides both the integers, it must divide a suitable linear combination of the two.)

Exercise 10.31.5. Show that any two consecutive odd integers must be relatively prime.

Exercise 10.31.6. Suppose a and b are integers, and suppose $a = bq + r$ for some integers q and r. Show that an integer d divides both a and b if and only if d divides both b and r.

Exercise 10.31.7. In the context of Exercise 10.31.6, assume further that $b \neq 0$. Use the result of that same Exercise 10.31.6 to show that $\gcd(a, b) = \gcd(b, r)$. (Note: the assumption $b \neq 0$ was only brought in so that we can define the quantities $\gcd(a, b)$ and $\gcd(b, r)$.)

Exercise 10.31.8. Suppose p and q are distinct primes. Show that $\frac{1}{p} + \frac{1}{q}$ is not an integer. (Hint: Assume that $\frac{1}{p} + \frac{1}{q} = n$ for some integer n. What happens if you clear all fractions by suitable cross multiplications?)

Exercise 10.32. Let a and b be integers with at least one of them nonzero, and let $h = a/\gcd(a, b)$ and $k = b/\gcd(a, b)$. Show that $\gcd(h, k) = 1$.

Exercise 10.33. Show that if a and b are nonzero integers with $\gcd(a, b) = 1$, and if c is an arbitrary integer, then $a|c$ and $b|c$ together imply $ab|c$. Give an example to show that this result is false if $\gcd(a, b) \neq 1$. (Hint: Use the fact that $\gcd(a, b) = 1$ to write $1 = xa + yb$ for suitable integers x and y, and then multiply both sides by c.)

Exercise 10.34. Suppose that a and b are nonzero integers that are relatively prime. Show that $\gcd(a+b, a-b)$ is either 1 or 2. (Hint: Show that no odd prime can divide both $a + b$ and $a - b$ by considering suitable linear combinations of the two. Using the same technique, show that 2^k cannot divide both $a + b$ and $a - b$ if $k \geq 2$.)

Exercise 10.35. Show that 3, 5, and 7 are the only three consecutive odd integers such that each is prime. (Hint: if (a, b, c) is another triple with this property, consider these integers modulo 3.)

Exercise 10.36. Use the result of Exercise 7.21 in Chapter 7 to show that if $2^n - 1$ is a prime, then n must be prime.

Exercise 10.37. Theorem 10.13 shows us that if a and b are integers with at least one of them being nonzero, then $\gcd(a, b)$ is a linear combination (in fact, the least positive linear combination) of a and b. But the proof of the theorem does not indicate how to find the gcd as a linear combination, except by exhaustive search, since even if we try out a large number of linear combinations of a and b, we will have no guarantee that we have hit the smallest positive one! (This is particularly so when a and b are large.) In the following exercises, we will lay out the basis for a very efficient algorithm for determining $d = \gcd(a, b)$ in the form $xa + yb$ for suitable integers x and y, known as the *Euclidean algorithm.*

(1) Show that we may "reduce to the situation $a > b > 0$," that is, show that it is sufficient if we solved this problem in the case $a > b > 0$ (Hint: Proposition 10.8 and the discussions preceding that proposition. Note, for example, that if $a < 0$ or $b < 0$, and if we are able to write $d = \gcd(|a|, |b|)$ as the linear combination $d = x|a|+y|b|$, then we can rewrite d as $d = (\pm x)a+(\pm y)b$ as appropriate. Note too that $\gcd(b, a) = \gcd(a, b)$, so we may assume $a \geq b \geq 0$ at this stage. Now dispose of the special cases where $a = b$, or $b = 0$, by explicitly displaying the greatest common divisor in these cases as a linear combination of a and b.)

(2) We assume that $a > b > 0$ as in part (1). By Theorem 10.2, we may write $a = bq + r$ for uniquely determined integers q and r with $0 \leq r < b$. Show

that $\gcd(a, b) = \gcd(b, r)$. (Hint: You would have shown this already in the practice exercises!)

(3) Continuing with the situation in part (2) above, suppose you are told that $\gcd(b, r) = x'b + y'r$ for suitable integers x' and y'. Use this to show that we can write $\gcd(a, b)$ as $xa + yb$ for suitable integers x and y.

We now illustrate how this simple result plays out in practice. We write $a = bq_0 + r_0$, $0 \le r_0 < b$ (as in part (2) above, except we add subscripts to q and r). We have $d = \gcd(a, b) = \gcd(b, r_0)$. We next write $b = q_1 r_0 + r_1$, with $0 \le r_1 < r_0$. We now have $d = \gcd(b, r_0) = \gcd(r_0, r_1)$. We similarly write $r_0 = q_2 r_1 + r_2$, $0 \le r_2 < r_1$, and so on, till we get a reminder of 0. (Note that indeed we will get a reminder of 0 at some stage: We have $r_0 > r_1 > r_2 > \ldots \ge 0$, and if at the i-th stage we find that r_i is still greater than zero, we can perform one more division to find that $r_i > r_{i+1} \ge 0$. This process cannot go one forever as the r_i are getting smaller and smaller. Therefore, at some stage, say the $(k+1)$-th stage as indexed below, we will find $r_{k+1} = 0$.)

We thus have:

$$\begin{aligned}
a &= bq_0 + r_0 & d &= \gcd(a, b), \\
b &= r_0 q_1 + r_1 & d &= \gcd(b, r_0), \\
r_0 &= r_1 q_2 + r_2 & d &= \gcd(r_0, r_1), \\
&\vdots = \vdots \\
r_{k-3} &= r_{k-2} q_{k-1} + r_{k-1} & d &= \gcd(r_{k-3}, r_{k-2}), \\
r_{k-2} &= r_{k-1} q_k + r_k & d &= \gcd(r_{k-2}, r_{k-1}), \\
r_{k-1} &= r_k q_{k+1} + 0 & d &= \gcd(r_{k-1}, r_k) = r_k.
\end{aligned}$$

Since $d = r_k$, we find from the last but one equation that

$$d = r_{k-2} - r_{k-1} q_k,$$

so we have now written d as a linear combination of r_{k-2} and r_{k-1}. Now substituting for r_{k-1} from two equations before the last, we find

$$d = r_{k-2} - (r_{k-3} - r_{k-2} q_{k-1})\, q_k.$$

After cleaning up, this reads

$$d = r_{k-3}(-q_k) + r_{k-2}\,(1 + q_{k-1} q_k),$$

so we have written d as a linear combination of r_{k-3} and r_{k-2}. Now substituting for r_{k-2} from *three* equations before the last, and continuing in the same vein, we will ultimately write d as a linear combination of a and b.

(4) Use the scheme described in part (3) above to write $1 = \gcd(128, 729)$ as a linear combination of 128 and 729.

Exercise 10.38. Consider the equivalence relation on fractions in Example 9.34, Chapter 9. Prove that every equivalence class is represented by a uniquely determined fraction $\frac{a}{b}$ with $\gcd(a,b) = 1$ and $b > 0$. (Hint: First show that every equivalence class contains a fraction of the form described. The result of Exercise 10.32 may be useful. So every class can indeed be represented by such a fraction. Next, to show uniqueness, assume that two fractions $\frac{a}{b}$ and $\frac{c}{d}$, satisfying $\gcd(a,b) = 1$ and $b > 0$ and $\gcd(c,d) = 1$ and $d > 0$, are in the same equivalence class. Invoke Exercise 10.22 to show that $a = c$ and $b = d$.)

Exercise 10.39. This is a continuation of Exercise 10.25. Let $n = ab$ as in that exercise. We have seen there that every pair of divisors x of a and y of b leads to a divisor xy of n, and conversely, every divisor of n is expressible as the product of a divisor of a and a divisor of b. Now assume further that $\gcd(a,b) = 1$. Show that the function

$$\text{Positive divisors of } a \times \text{Positive divisors of } b \to \text{Positive divisors of } n$$

that sends a pair (x,y), with x a positive divisor of a and y a positive divisor of b to xy, is both injective and surjective.

Exercise 10.40. Let n be a positive integer. Show that the integers $(n+1)!+2$, $(n+1)!+3, \ldots, (n+1)!+(n+1)$ are all composite. (Hint: Look at some obvious divisors of $(n+1)!$)

Since n can be any positive integer, this result shows that we can create arbitrary long sequences of consecutive integers that are all composite!

Exercise 10.41. Let m and n be positive integers with prime factorizations $p_1^{m_1} p_2^{m_2} \cdots p_k^{m_k}$ and $p_1^{n_1} p_2^{n_2} \cdots p_k^{n_k}$ respectively. Here, we have written the factorizations of m and n using a common set of primes as follows: if for example some prime p_i divides m but does not divide n, we write n as having the prime factor p_i^0. So, in this scheme, the various exponents m_i and n_i are allowed to be zero as well as positive. (Note that m_i and n_i cannot both be zero, as that would mean that p_i does not appear either as a prime factor of m or of n.) Now use Theorem 10.24 to show that $\gcd(m,n)$ equals the product $p_1^{\min(m_1,n_1)} p_2^{\min(m_2,n_2)} \cdots p_k^{\min(m_k,n_k)}$.

Exercise 10.42. If m and n are positive integers, we define the *least common multiple* of m and n, denoted $\operatorname{lcm}(m,n)$, to be, as the name suggests, the smallest positive integer that is both a multiple of m and a multiple of n.

(1) Show that $\operatorname{lcm}(m,n)$ exists.

(2) Let m and n have the prime factorizations $p_1^{m_1} p_2^{m_2} \cdots p_k^{m_k}$ and $p_1^{n_1} p_2^{n_2} \cdots p_k^{n_k}$ respectively, where the factorizations are written using a common set of primes and the m_i are allowed to be zero (see Exercise 10.41 above for an explanation). Show that $\operatorname{lcm}(m,n)$ is the product $p_1^{\max(m_1,n_1)} p_2^{\max(m_2,n_2)} \cdots p_k^{\max(m_k,n_k)}$.

Exercise 10.43. Let n be an integer ≥ 2 such that some prime factor of n appears to an odd exponent in the prime factorization of n. (For instance, $500 = 2^2 \cdot 5^3$, and 5 occurs to an odd power.) Show that $\sqrt{n}$ is not rational.

Exercise 10.44. It follows from Theorem 10.2 that if p is an odd prime, then $p = 4n + 1$ or $4n + 3$ for suitable n. Thus, odd primes are either congruent to 1 mod 4 (we describe this as being *of the form* $4n+1$) or to 3 mod 4 (we describe this as being *of the form* $4n + 3$). Borrow ideas from the proof of Theorem 10.26 to show that there are infinitely many primes of the form $4n + 3$. (Hint: It would be helpful for you to review the multiplication table for $\mathbb{Z}/4\mathbb{Z}$ before doing this exercise. Assume there are only finitely many primes of the form $4n + 3$ and consider the integer obtained by adding suitable small integers to their product.)

Note that there are infinitely many primes of the form $4n + 1$ as well, although that is a bit harder to prove at this stage.

Exercise 10.45. Show that odd primes (greater than 3) are either congruent to 1 mod 6 (we describe this as being *of the form* $6n + 1$) or to 5 mod 6 (we describe this as being *of the form* $6n + 5$). Show that there are infinitely many primes of the form $6n+5$. (Hint: Mimic the ideas in Exercise 10.44 above. Just as in that exercise with primes of the form $4n + 1$, there are infinitely many primes of the form $6n + 1$ as well, although that is a bit harder to prove at this stage.)

Exercise 10.46. Let $n \geq 2$ be an integer with prime factorization $n = p_1^{a_1} \cdot p_2^{a_2} \cdot \cdots \cdot p_k^{a_k}$. Let $\sigma(n)$ denote the sum of the positive divisors of n. (For instance, $\sigma(4) = 1 + 2 + 4 = 7$ and $\sigma(6) = 1 + 2 + 3 + 6 = 12$. We will derive a formula for $\sigma(n)$ in terms of its prime factorization as follows:

(1) First assume $k = 1$, that is $n = p^a$, a power of a single prime p. Show that $\sigma(n)$ in this case is given by $\dfrac{p^{a+1} - 1}{p - 1}$. (Hint: Use Example 10.29 and Example 7.3, Chapter 7.)

(2) Using induction on k (with the $k = 1$ step proven in part (1) above) show that for $k \geq 1$, $\sigma(n) = \prod_{i=1}^{k} \dfrac{p_i^{a_i+1} - 1}{p_i - 1}$. (Hint: Write n as $(p_1^{a_1} \cdot p_2^{a_2} \cdot \cdots \cdot p_{k-1}^{a_{k-1}}) \cdot p_k^{a_k}$ and use Exercise 10.39.)

Exercise 10.47. How many integers from 1 to 10^6 are neither perfect squares nor perfect cubes? (Hint: Count the number of integers that are either perfect squares or perfect cubes. A warning here, the word *or* is used as an inclusive *or*; see Example 3.21 in Chapter 3. You need to account for the integers that are both perfect squares and perfect cubes. This is the simplest case of the Inclusion-Exclusion Principle we considered in Exercise 7.27 of Chapter 7.)

Exercise 10.48. Recall the number systems $\mathbb{Z}/n\mathbb{Z}$ from Chapter 1. Let us assume n is a prime, and let us call it p instead. We will show that if $[a]_p \neq [0]_p$ in $\mathbb{Z}/p\mathbb{Z}$, then there exists a *unique* $[b]_p$ such that $[a]_p[b]_p = [1]_p$. This element $[b]_p$ is known as the *multiplicative inverse of* $[a]_p$.

(1) Fix any representative $\tilde{a}$ of the equivalence class $[a]_p$ (see Example 9.33 in Chapter 9; for instance, we may choose an $\tilde{a}$ with $0 \leq \tilde{a} < p$ although we are really free to choose any other representative). Show that $\gcd(\tilde{a}, p) = 1$ for any such representative.

(2) Apply Theorem 10.13 to write 1 as an integer linear combination of $\tilde{a}$ and p. Use Exercise 9.44, Chapter 9, to read this linear combination in $\mathbb{Z}/p\mathbb{Z}$ and find the required $[b]_p$ such that $[a]_p[b]_p = [1]_p$.

(3) Suppose there exists some element $[b']_p$ such that $[a]_p[b']_p = [1]_p$. Show that $[b]_p = [b']_p$, that is, the element $[b]_p$ such that $[a]_p[b]_p = [1]_p$ is *unique*. (Hint: Bring terms to one side and use Proposition 10.21.)

(4) For an illustration of this method in part (2) above, find the multiplicative inverse of $[7]_{29}$. Try to write the inverse as $[b]_{29}$, with $0 \leq b < 29$.

Exercise 10.49. We will consider a generalization of Exercise 10.48 in this exercise: Let $n \geq 2$ be an integer, not assumed to be a prime. When n is a prime, we saw in Exercise 10.48 above that $[a]_n$ has a multiplicative inverse in $\mathbb{Z}/n\mathbb{Z}$ as long as $[a]_n \neq [0]_n$. We will determine the elements in $\mathbb{Z}/n\mathbb{Z}$ that have multiplicative inverses, for all $n \geq 2$, without the assumption of n being prime. (Elements that have multiplicative inverses are also known as *invertible* elements.)

(1) Let $[a]_n$ be any nonzero element of $\mathbb{Z}/n\mathbb{Z}$. Show that if $\tilde{a}$ and $\tilde{\tilde{a}}$ are two representatives of $[a]_n$, then $\gcd(\tilde{a}, n) = 1$ if and only if $\gcd(\tilde{\tilde{a}}, n) = 1$.

(2) Fix a representative $\tilde{a}$ of $[a]_n$. Show that $[a]_n$ is invertible if and only if $\gcd(\tilde{a}, n) = 1$.

Exercise 10.50. Let $n \geq 2$ be an integer, and let $a \in Z$ be relatively prime to n. Show that there exists some positive integer t such that $[a]_n^t = [1]_n$ in $\mathbb{Z}/n\mathbb{Z}$. (Hint: Apply the ideas in Exercise 2.11, Chapter 2, to the powers of a. You may find Exercise 10.22 useful.)

11

Sequences, Series, Continuity, Limits

If you have had a calculus course before, you would likely have seen the notions of sequences and series, continuity of functions, and limits of functions, if only informally. We will study these notions in slightly greater depth and with slightly greater rigor here than is typical in a first (computationally oriented) calculus course. As we will see, the last two notions of continuity of functions and limits of functions, which are closely related, are easiest to understand via the notion of convergent sequences (although there are alternative approaches as well). Accordingly, we will begin the chapter with sequences.

Sequences form an integral part of the foundations of calculus: they allow the microscopic study of functions (from $\mathbb{R}$ to $\mathbb{R}$, but more generally, also from $\mathbb{C}$ to $\mathbb{C}$). They enable us to approximate the continuous (and uncountable) aspect of functions defined on $\mathbb{R}$ by discrete behaviors and thus allow us to *slow down* such functions to a human scale. But more is true: sequences do not merely serve as approximations of functions from $\mathbb{R}$ to $\mathbb{R}$, they actually enable us to *define* attributes of functions in terms of attributes of sequences obtained from them. A good understanding of sequences is therefore vital to the further study of mathematics. Especially vital is the notion of convergence of sequences: we will focus on a rigorous definition of this notion, and on rigorous proofs that various sequences converge.

In the later portion of this chapter, we will study how sequences can be used to pin down the notion of continuity of real-valued functions on subsets of the real numbers, more precisely, functions $f : I \to \mathbb{R}$, where I is an open interval in $\mathbb{R}$. The more intuitive definition of continuity is expressed via sequences, although there is a different definition that is easier to work with when showing that specific functions are continuous. We will consider both definitions, and then follow this by showing that the two definitions of continuity are equivalent!

We will end by considering another notion that is easiest understood via sequences and is closely related to the notion of continuity: the notion of the limit of a function.

Here too, the more intuitive definition is via sequences, although there is an alternative definition that is easier to work with.

11.1 Sequences

We will consider only *real* sequences in this chapter. This is the definition:

Definition 11.1. A sequence is a function $s : \mathbb{N} \to \mathbb{R}$.

Remark 11.2. Given such a sequence s as in the definition above, we say s is *indexed by the positive integers.* Often, one writes s_n for the real number $s(n)$, $n = 1, 2, \dots$, and refers to the s_n as the *terms* of the sequence. The sequence is then listed as $(s_1, s_2, \dots)$. Often, we write (s_n) for the sequence.

Remark 11.3. It is often convenient to consider sequences indexed by the nonnegative integers, that is, functions $s : \mathbb{Z}_{\geq 0} \to \mathbb{R}$. The terms of such a sequence are now s_0, s_1, etc. As a matter of convention, we will tacitly assume that our sequences are indexed by the positive integers, and if there are exceptions, we will explicitly point them out. In particular, we will define concepts such as convergence and boundedness for sequences indexed by $n = 1, 2, 3 \dots$. The modification of these definitions for sequences that are indexed by $n = 0, 1, 2, \dots$ should be clear.

Here are various examples of sequences:

Example 11.4. Consider the sequence $s\colon \mathbb{N} \to \mathbb{R}$ given by $s_n = n$ for all $n \in \mathbb{N}$. It can be listed as $(1, 2, 3, \dots)$. We recognize this sequence as being obtained by taking discrete snapshots of the function $y = x$, defined on $\mathbb{R}$, at the integer values $x = 1, 2, 3, \dots$.

Example 11.5. Consider the sequence $s\colon \mathbb{N} \to \mathbb{R}$ given by $s_n = n^2$ for all $n \in \mathbb{N}$. It can be listed as $(1, 4, 9, \dots)$. As in Example 11.4, we recognize this sequence as being obtained by taking discrete snapshots of the function $y = x^2$, defined on $\mathbb{R}$, at the integer values $x = 1, 2, 3, \dots$.

Example 11.6. Consider the sequence $s\colon \mathbb{N} \to \mathbb{R}$ given by $s_n = \left(1 + \frac{1}{n}\right)^n$ for all $n \in \mathbb{N}$. It can be listed as $(2, 2.25, 2.37037037\dots, \dots)$. One interpretation is that this sequence is being obtained by taking discrete snapshots of the function $y = \left(1 + \frac{1}{x}\right)^x$, defined for say positive x, at the integer values $x = 1, 2, \dots$. You may however remember this sequence in connection with e, the base of the natural logarithm.

Example 11.7. Here is another sequence connected with e: Consider the sequence $s\colon \mathbb{N} \to \mathbb{R}$ given by $s_n = 1 + \frac{1}{1!} + \frac{1}{2!} + \cdots + \frac{1}{n!}$. It can be listed as $(1, 2, 2.5, 2.66\dots, \dots)$. It is harder to view this sequence as derived from discrete snapshots of any function we know, but sequences are very general, and do not have to arise as snapshots of some function from $\mathbb{R}$ to $\mathbb{R}$.

Example 11.8. Here is yet another sequence connected with e! Consider the sequence $s\colon \mathbb{Z}_{\geq 0} \to \mathbb{R}$ given by the listing $(2, 2.7, 2.71, 2.712, \dots)$. We may be able to recognize the pattern here: These are just the digits of e, up to the n-th place after the decimal, $n = 0, 1, \dots$. Note that it is convenient to index this sequence with the nonnegative integers, that is, to number the terms of this sequence from 0 onward: $s_0 = 2$ represents e to zero places after the decimal! As with Example 11.7 above, it is hard to view this sequence as derived from discrete snapshots from any function we know, but it is worth repeating: sequences are very general and do not have to arise as snapshots of some function from $\mathbb{R}$ to $\mathbb{R}$.

11.2 Convergence

Although sequences are allowed to be very general functions from $\mathbb{N}$ (or $\mathbb{Z}_{\geq 0}$) to $\mathbb{R}$, the most useful sequences from the point of view of calculus are those that "converge." We will study some examples below to get an idea of what convergence means: intuitively, we say a sequence converges if its terms, *on the whole,* come closer and closer to some fixed real number. We will consider some examples to see what this means, and then formally define convergence after these examples.

It is helpful to picture the terms of a sequence as points P_n, $n = 1, 2, \ldots$, on the plane, with P_n having coordinates (n, s_n). Thus, the terms themselves appear as the y-coordinates of the points P_i. The convergence of a sequence (or its failure to converge, as the case may be) can often be seen immediately using this technique for visualization.

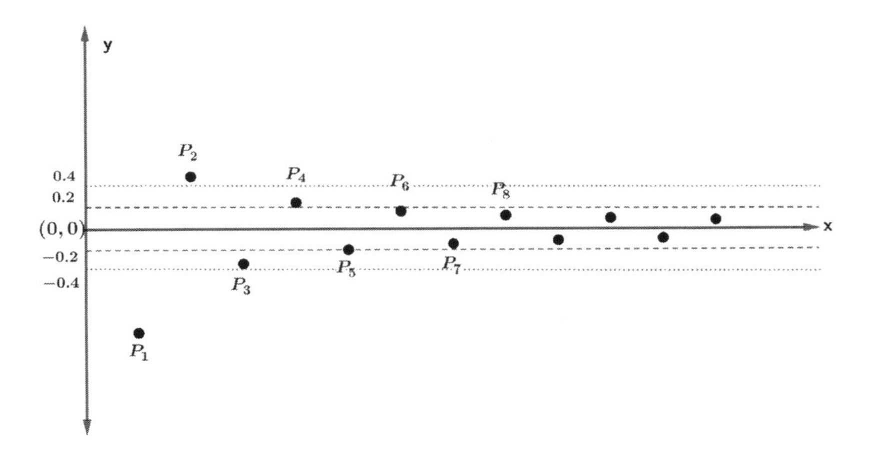

Figure 11.1. The sequence $s_n = \dfrac{(-1)^n}{n}$, pictured as points on the plane. Here, P_n has coordinates (n, s_n), so the n-th term is the y-coordinate of the n-th point shown. It is visually clear that the terms converge to 0. Moreover, we can see from the positions of the P_i that $|s_n| < 0.4$ for all $n \geq 3$, and $|s_n| < 0.2$ for all $n \geq 6$.

Example 11.9. Consider the sequence $s \colon \mathbb{N} \to \mathbb{R}$ defined by $s_n = \dfrac{(-1)^n}{n}$. See Figure 11.1, where we have plotted the terms of the sequence as described above: point P_n has coordinates (n, s_n). We know just from our experience with division that as n grows larger and larger, $\dfrac{(-1)^n}{n}$ grows smaller and smaller. Eventually, as n grows to be very large, s_n becomes very small, and then *stays small.* In fact, we can make s_n as small as we want, and make it stay small, by taking n large enough. We describe this informally by saying that $\frac{(-1)^n}{n}$ *stays arbitrarily close to zero* as n increases.

But we can go a step further. We can quantify that s_n stays arbitrarily close to zero as follows: if we want s_n to be, say, within 10^{-3} of zero in absolute value (that is, if we want $|0 - s_n|(= |s_n|) < 10^{-3}$, or what is the same thing, $-10^{-3} < s_n < 10^{-3}$), we simply consider terms of the sequence from $n = 1001$ on: we will find $|s_n| = \frac{1}{n} < 10^{-3}$ if $n \geq 1001$. Similarly, if we want s_n be within, say 10^{-8} of zero in absolute value, we consider terms of the sequence from $n = 10^8 + 1$ on: we will find $|s_n| = \frac{1}{n} < 10^{-8}$ for $n \geq 100000001$.

Figure 11.1 shows this phenomenon for larger values that will fit into its frame: it shows that $|s_n| < 0.4$ for all $n \geq 3$, and $|s_n| < 0.2$ for all $n \geq 6$.

We describe this property of staying arbitrarily close to zero as n increases by saying that (s_n) converges to zero.

Example 11.10. If instead we had considered the sequence defined by $s_n = 1 + \frac{(-1)^n}{n}$, we would have found that (s_n) stays arbitrarily close to 1. The same calculations would apply: if we want s_n to be, say, within 10^{-3} of *one* in absolute value (that is, $|1 - s_n| < 10^{-3}$, or $-10^{-3} < 1 - s_n < 10^{-3}$), we simply consider terms of the sequence from $n = 1001$ on: from $n = 1001$ on, s_n will satisfy $|1 - s_n| = \frac{1}{n} < 10^{-3}$. Similarly, if we want s_n be within, say 10^{-8} of 1 in absolute value, we consider terms of the sequence from $n = 10^8 + 1$ on: from $n = 100000001$ on, we will find $|1 - s_n| = \frac{1}{n} < 10^{-8}$.

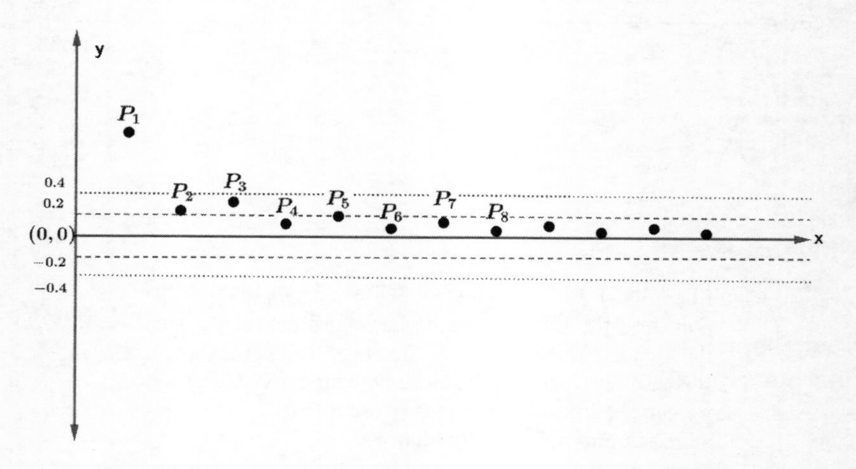

Figure 11.2. The sequence of Example 11.11 pictured as points on the plane. It is visually clear that the terms *on the whole* converge to 0, even though P_{2n+1} is higher than P_{2n} for $n = 1, 2, \ldots$. Just as in Figure 11.1, we can see from the positions of the P_i that $|s_n| < 0.4$ for all $n \geq 2$, and $|s_n| < 0.2$ for all $n \geq 6$, even though we had this slight up and down behavior of the points.

Example 11.11. If instead of the sequence in Example 11.9, suppose we had considered the sequence defined by

$$s_n = \begin{cases} \dfrac{1}{n} \text{ if } n \text{ is odd,} \\ \\ \dfrac{1}{2n} \text{ if } n \text{ is even.} \end{cases}$$

See Figure 11.2. The first few terms of this sequence are 1, 1/4, 1/3, 1/8, 1/5, 1/12, 1/7, 1/16, ... Here, the terms are not getting smaller in succession: it is true that $|1/4| < |1|$, but $|1/3|$ is *greater* than $|1/4|$. Similarly, it is true that $|1/8| < |1/3|$, but $|1/5|$ is *greater* than $|1/8|$. This pattern continues throughout the sequence: in one pair of terms where the first term is in odd position, (s_{2k-1}, s_{2k}), $k = 1, 2, \ldots$, the second term is smaller than the first in absolute value, but in the very next pair of terms (s_{2k}, s_{2k+1}), the second term is larger than the first in absolute value. Yet, if we consider the sequence *on the whole,* at a macroscopic level so to speak, the terms are indeed getting smaller and smaller. This can be quantified as follows: we find

$$|s_n| = \begin{cases} \left|\dfrac{1}{n}\right| = \dfrac{1}{n} \text{ if } n \text{ is odd,} \\ \\ \left|\dfrac{1}{2n}\right| = \dfrac{1}{2n} < \dfrac{1}{n} \text{ if } n \text{ is even.} \end{cases}$$

Thus, we find $|s_n| \leq \dfrac{1}{n}$ for all n. It is now clear, just like in Example 11.9, that the terms of this sequence, *on the whole,* are getting smaller. Exactly like in that example, we can quantify this as follows: if we want our terms, in absolute value, to be smaller than 10^{-k} for any $k \geq 1$ (in Example 11.9 we looked at $k = 3$ and $k = 8$), we simply consider the sequence from $n = 10^k + 1$ on: s_n will satisfy $|s_n| < 10^{-k}$ for $k \geq 10^k + 1$ (this despite the fact that successive pairs of terms will decrease, then increase, then decrease, etc. in absolute value).

We will turn the calculations of Examples 11.9, 11.10, and 11.11 into our definition of convergence:

Definition 11.12. A sequence (s_n), $n = 1, 2, \ldots$ (or $n = 0, 1, 2, \ldots$) is said to *converge* to the real number l if, given any $\epsilon > 0$, there exists a positive integer N_ϵ with the property that for all $n \geq N_\epsilon$, $|l - s_n| < \epsilon$, or what is the same thing, $l - \epsilon < s_n < l + \epsilon$. If (s_n) converges to l, we write $\lim_{n\to\infty} s_n = l$, and we also say (s_n) *has limit* L or (s_n) *converges to the limit* L. A sequence (s_n) that converges to some real number l is also called a *convergent* sequence. If no real number l exists such that (s_n) converges to l, we say that the sequence (s_n) *diverges* (or is a *divergent* sequence).

Remark 11.12.1. We use the notation "N_ϵ" to emphasize that the integer N_ϵ depends on ϵ: a different ϵ may need a different value of N_ϵ.

Remark 11.12.2. Note, if a sequence (s_n), $n = 1, 2, \ldots$, converges to the real number l as per this definition above, and if it also converges to the real number m, then necessarily, m must equal l. We will prove this in Proposition 11.26 ahead.

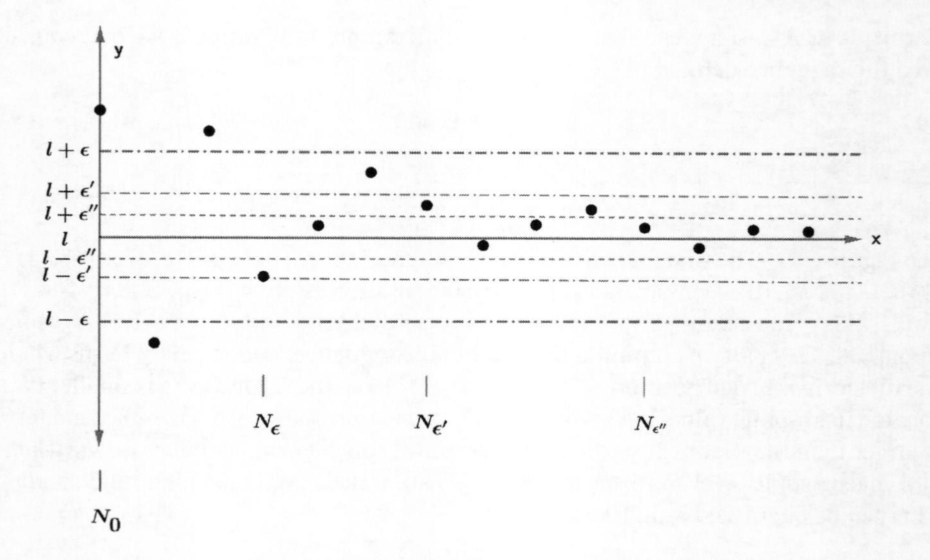

Figure 11.3. Definition of sequence convergence. A sequence (s_n) is pictured on the plane as points (n, s_n); our focus is on points to the right of some "N_0"-th term. Notice that for $n \geq N_\epsilon$ the points of the sequence lie within the lines $y = l \pm \epsilon$, for $n \geq N_{\epsilon'}$ the points lie within the lines $y = l \pm \epsilon'$, and for $n \geq N_{\epsilon''}$ the points lie within the lines $y = l \pm \epsilon''$.

Remark 11.13. In Examples 11.9 and 11.10 above, we considered $\epsilon = 10^{-3}$, for which N_ϵ was 1001, and $\epsilon = 10^{-8}$, for which N_ϵ was $10^8 + 1$. In Example 11.11 we considered, in one shot, $\epsilon = 10^{-k}$ for $k = 1, 2, \ldots$, for which the corresponding N_ϵ was $10^k + 1$.

Remark 11.14. A pictorial way of thinking about convergence to a real number l is that given any $\epsilon > 0$, the points (n, s_n) must lie within the lines $y = l - \epsilon$ and $y = l + \epsilon$ for all $n \geq N_\epsilon$. See Figure 11.3. If ϵ were tightened, say if the new ϵ—call it ϵ'—is taken as the old ϵ divided by 2, then going further out if need be on the x-axis to $N_{\epsilon'}$, the points (n, s_n) must now lie within the lines $y = l - \epsilon'$ and $y = l + \epsilon'$ for all $n \geq N_{\epsilon'}$. If ϵ were further tightened to a new $\epsilon'' = \epsilon/4$, then by going out even farther if need be on the x-axis to $N_{\epsilon''}$, the points (n, s_n) must now lie within the lines $y = l - \epsilon''$ and $y = l + \epsilon''$ for all $n \geq N_{\epsilon''}$, and so on.

Here are some examples of convergence:

Example 11.15. The constant sequence defined by $s_n = a$ for a real number a: this sequence clearly converges to a. For a formal proof, note that for any $\epsilon > 0$, we can take $N_\epsilon = 1$: clearly, for all $n \geq 1$, $|s_n - a| < \epsilon$. So, by definition, $\lim_{n\to\infty} s_n = a$. (Note that N_ϵ is independent of ϵ in this example. This will not be true in general.)

Example 11.16. Consider (s_n), defined by $s_n = 10^{-n}$. It is abundantly clear that this sequence converges to zero: the terms are $0.1, 0.01, 0.001, 0.0001, 0.00001, \ldots$. But here is the formal proof: Given any $\epsilon > 0$, we wish to find an integer N_ϵ such that $|s_n| < \epsilon$ for all $n \geq N_\epsilon$. Since $s_n = 10^{-n}$, it is clear that the sequence decreases, that is, $s_1 >$

$s_2 > s_3 > \cdots$. Thus, if we find that N_ϵ such that $10^{-N_\epsilon} < \epsilon$, then for all $n \geq N_\epsilon$, we will find $s_n = 10^{-n} \leq s_{N_\epsilon} = 10^{-N_\epsilon} < \epsilon$, as desired. But finding N_ϵ such that $10^{-N_\epsilon} < \epsilon$ is equivalent to finding N_ϵ such that $1/\epsilon < 10^{N_\epsilon}$, and such an N_ϵ can of course be found: the powers of 10 increase without a bound, so simply consider the first power of 10 that is bigger than $1/\epsilon$! Hence, given any ϵ we can indeed find an N_ϵ such that $|s_n| < \epsilon$ for all $n \geq N_\epsilon$. By definition, (s_n) converges to zero. (Note here that N_ϵ is very much dependent on ϵ.)

Example 11.17. Consider (s_n) defined by $s_n = 1+\frac{1}{\sqrt{n}}$, $n \in \mathbb{N}$. This sequence converges to 1. To prove this formally, given any $\epsilon > 0$, we need to find N_ϵ such that $|1 - s_n| < \epsilon$ for all $n \geq N_\epsilon$. Since $|1 - s_n| = \frac{1}{\sqrt{n}}$, we need to find N_ϵ such that $\frac{1}{\sqrt{n}} < \epsilon$ for all $n \geq N_\epsilon$. Clearly, $\frac{1}{\sqrt{n}} < \epsilon$ is equivalent to $\sqrt{n} > \frac{1}{\epsilon}$, which is equivalent to $n > \frac{1}{\epsilon^2}$. It follows therefore that if we take N_ϵ to be the least integer greater than $\frac{1}{\epsilon^2}$, then for all $n \geq N_\epsilon$, we will have $|1 - s_n| < \epsilon$. This proves that our sequence converges to 1.

> *Remark* 11.17.1. It is intuitively obvious from our experience with real numbers that indeed, given the quantity $\frac{1}{\epsilon^2}$, there exists an integer greater than it, and therefore, a least integer greater than it. We will study the structure of the real numbers more carefully in Chapter 12 ahead, and we will see that this intuitively clear fact is actually a consequence of a very deep property of the real numbers known as the Least Upper Bound property (Definition 12.12 in that chapter). The fact that we can find an integer greater than a given positive real number is known as the Archimedean property of the real numbers, and Theorem 12.19 in that chapter shows how it is a consequence of the least upper bound property.

Example 11.18. Consider (s_n) defined by $s_n = 1 + \frac{1}{n^2+n}$, $n \in \mathbb{N}$. This sequence also converges to 1. To prove this formally, we would argue as in Example 11.17 above: given any $\epsilon > 0$, we need to find N_ϵ such that $|1 - s_n| = \frac{1}{n^2+n} < \epsilon$ for all $n \geq N_\epsilon$. From this we get the requirement, just as in Example 11.17, that $n^2 + n > \frac{1}{\epsilon}$. Now how do we find such an n? Our expression on the left side is more complicated than simply $\sqrt{n}$, it now involves a sum of terms containing n. But we can reduce our need to an expression containing just one n as follows: Since $n^2 > n$ for $n \geq 1$, we find $n^2 + n > n + n = 2n$. It is sufficient if we have $2n > \frac{1}{\epsilon}$, since we will then automatically have $n^2 + n > 2n > \frac{1}{\epsilon}$. (Mathematicians can be devious!) But it is easy to guarantee the condition $2n > \frac{1}{\epsilon}$, we simply have to take $n > \frac{1}{2\epsilon}$. It follows therefore that if we take N_ϵ to be the least integer greater than $\frac{1}{2\epsilon}$, then for all $n \geq N_\epsilon$, we will have $|1 - s_n| < \epsilon$. Our sequence hence converges to 1.

Example 11.19. Here is another instance where we prove convergence, as in Example 11.18 above, by replacing a complicated expression with a simpler expression using "estimation," that is, by showing that one expression is bigger or smaller than the other. Consider the sequence (s_n) defined by $s_n = \frac{\cos(n)}{n}$, $n \in \mathbb{N}$. This sequence converges to zero. To find N_ϵ such that $|s_n| < \epsilon$ for a given $\epsilon > 0$, note that $|s_n| = \left|\frac{\cos(n)}{n}\right| = \frac{|\cos(n)|}{n} \leq \frac{1}{n}$, since $|\cos(n)| \leq 1$ for all $n \geq 1$. Thus, it is sufficient to guarantee that $\frac{1}{n} < \epsilon$, for then, automatically, $|s_n| = \frac{|\cos(n)|}{n} \leq \frac{1}{n} < \epsilon$. But this is just Example 11.9 above: to guarantee that $\frac{1}{n} < \epsilon$, we just need to take $n > \frac{1}{\epsilon}$, so we need to take N_ϵ to be the least integer greater than $\frac{1}{\epsilon}$.

It would be a good idea for you to practice showing that certain sequences converge, using these techniques above. See Exercise 11.43 at the end of the chapter.

Remark 11.20. While working on these examples above, or while studying Definition 11.12 for the convergence of a sequence, you may have noticed that when establishing convergence, all the emphasis is on the "tail" of a sequence! More precisely, for any $\epsilon > 0$, we only want the terms s_n *from the* N_ϵ-th position on to satisfy $|l - s_n| < \epsilon$ (where l is the limit): the first $N_\epsilon - 1$ terms are irrelevant, at least for this epsilon. It follows from this that if two sequences (s_n) and (t_n) satisfy $s_n = t_n$ for all $n \geq M$, where M is some positive integer, then the sequence (s_n) converges if and only if (t_n) converges. For, suppose (s_n) converges. Then, given $\epsilon > 0$, there exists N_ϵ such that $|l - s_n| < \epsilon$ for all $n \geq N_\epsilon$. If $N_\epsilon \geq M$, then $s_n = t_n$ for all $n \geq N_\epsilon$, so we find $|l - t_n| < \epsilon$ for all $n \geq N_\epsilon$. If $N_\epsilon < M$, replace N_ϵ by M: we find $|l - s_n| < \epsilon$ for all $n \geq M$ as well, and since $s_n = t_n$ for all $n \geq M$, we find $|l - t_n| < \epsilon$ for all $n \geq M$. It follows that (t_n) converges. The situation is symmetric in (s_n) and (t_n), so the same proof (with obvious modifications) shows that if (t_n) converges, then (s_n) converges.

We now establish some results about convergent sequences. First, a definition:

Definition 11.21. A sequence (s_n), $n = 1, 2, \ldots$, is said to *bounded* if there exists a positive real number L such that $|s_n| \leq L$ for all $n \geq 1$. If no such L exists, then we say (s_n) is *unbounded*.

Thus, if the terms of a bounded sequence are plotted as the points (n, s_n) in the plane, then these points stay between the lines $y = L$ and $y = -L$.

Remark 11.22. Here is an equivalent definition for a sequence to be bounded:

> **Definition 11.22.1.** A sequence (s_n), $n = 1, 2, \ldots$, is said to be *bounded* if there exist real numbers A and B such that $A \leq s_n \leq B$ for all $n \geq 1$. If no such (A, B) pair can be found, we say the sequence is unbounded.

The equivalence of the two is easy to see in one direction: if, as in Definition 11.21 above, there exists a positive real number L such that $|s_n| \leq L$ for all $n \geq 1$, then we can take $A = -L$ and $B = L$ to find $A \leq s_n \leq B$. Now suppose that we are given real numbers A and B such that $A \leq s_n \leq B$ for all $n \geq 1$. Let's set $L = \max(|A|, |B|)$. From $L \geq |A|$ we find $-L \leq -|A|$. Now, if A is nonnegative, $|A| = A$, so $-|A| = -A < A$. If A is negative, $|A| = -A$, so $-|A| = A$. Thus, in both cases, $-|A| \leq A$, and $-L \leq -|A|$ leads to $-L \leq -|A| \leq A$. By similar considerations, $B \leq |B|$, so $B \leq |B| \leq L$. It follows that $-L \leq A \leq s_n \leq B \leq L$, or what is the same, $|s_n| \leq L$. Thus, the sequence is bounded according to Definition 11.21 above.

We now show:

Proposition 11.23. *Every convergent sequence of real numbers is bounded.*

Proof. Let (s_n) be a convergent sequence and let $\lim_{n\to\infty} s_n = S$ for some real number S. Let us choose $\epsilon = 1$ in the definition of convergence (Definition 11.12). Then, we can find an integer N such that $|S - s_n| < 1$ for all $n \geq N$. In other words, for all $n \geq N$, s_n lies between $S-1$ and $S+1$. Let $A = \max\{s_1, s_2, \dots, s_{N-1}\}$ and let $B = \min\{s_1, s_2, \dots, s_{N-1}\}$. Thus, the first $n-1$ terms are bounded above by A and all remaining terms are bounded above by $S+1$, so taken together, all terms are bounded above by the greater of A and $S+1$. Write C for $\max\{A, S+1\}$, so all terms are bounded above by C. By a similar argument, all terms are bounded below by $D = \min\{B, S-1\}$. Hence all terms lie between C and D. By Remark 11.22, (s_n) is bounded by $L = \max(|C|, |D|)$. □

Remark 11.24. Note that by contrast, a divergent sequence can either be bounded or unbounded. In the other direction, a bounded sequence can be either convergent or divergent. See Exercise 11.48. On the other hand, an unbounded sequence is necessarily divergent, since this is the contrapositive of the statement in Proposition 11.23.

Remark 11.25. While sequences can diverge simply because the terms fail to converge to a fixed real number, quite often the divergence can occur due to a very specific behavior: the terms of the sequence become larger and larger on the whole, *without bound*, as we move towards the tail end of the sequence. Take the sequence (s_n), where $s_n = n$. The terms of this sequence simply grow in size, without any bound. Clearly (s_n) cannot converge to any real number.

If a sequence (s_n) has the property that given any integer L, there exists an integer N_L such that for all $n \geq N_L$, we find $s_n > L$, we say that $\lim_{n\mapsto\infty} s_n = \infty$. The expression $\lim_{n\mapsto\infty} s_n = -\infty$ has an analogous interpretation with $s_n > L$ replaced by $s_n < L$. Proposition 11.23 shows that a sequence (s_n) for which $\lim_{n\mapsto\infty} s_n = \pm\infty$ cannot be convergent.

Here is a quick exercise to help you assimilate this remark:

Exercise 11.25.1. Let (s_n) and (t_n) be defined by

$$s_n = \begin{cases} n & \text{if } n \text{ is odd,} \\ \sqrt{n} & \text{if } n \text{ is even,} \end{cases} \qquad t_n = \begin{cases} -n & \text{if } n \text{ is odd,} \\ -\sqrt{n} & \text{if } n \text{ is even.} \end{cases}$$

Show that $\lim_{n\mapsto\infty} s_n = \infty$ and $\lim_{n\mapsto\infty} t_n = -\infty$.

Note that if a sequence (s_n) is unbounded, it does not automatically mean that $\lim_{n\mapsto\infty} s_n = \pm\infty$. There can be wild fluctuations in an unbounded sequence. Consider the sequence (s_n) defined by

$$s_n = \begin{cases} n & \text{if } n \text{ is odd,} \\ 0 & \text{if } n \text{ is even.} \end{cases}$$

The sequence is unbounded because of the odd terms. On the other hand, if we take $L = 1$, we cannot find an integer N_1 such that $s_n \geq 1$ for all $n \geq N_1$, because every other term of the sequence is zero. Thus (s_n) diverges (it is unbounded), but we cannot say $\lim_{n\mapsto\infty} s_n = \infty$.

Now that we have had some experience with convergence of sequences, let us prove a technical result alluded to already in Remark 11.12.2: if a sequence converges, then it can only converge to one real number.

Proposition 11.26. *The limit of a convergent sequence is unique.*

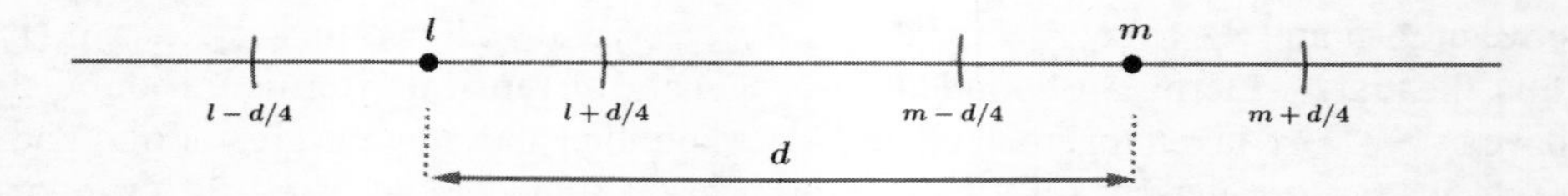

Figure 11.4. For $n > N$, the terms of the sequence are in the interval $(l - d/4, l + d/4)$ as well as in the interval $(m - d/4, m + d/4)$. This is clearly impossible: the two intervals do not overlap.

Proof. Suppose that a sequence (s_n) converges to a limit l, and also to a limit m. Suppose that $l \neq m$. We will arrive at a contradiction. First, since $l \neq m$, $|l - m| = d$, for some real $d > 0$. The idea behind the proof is the following: if we take ϵ to be any positive real number less than or equal to $d/2$, say $d/4$ to feel safe, then all terms of the sequence beyond some "N"-th term must congregate within $d/4$ on either side of l, and also within $d/4$ on either side of m. But this is impossible, since the separation between l and m is d. See Figure 11.4.

So, to write this out formally, take $\epsilon = d/4$ as above. Then, by definition of convergence to l, there exists an integer N_ϵ such that for all $n \geq N_\epsilon$, $|l - s_n| < d/4$. Similarly, by definition of convergence to m, there exists an integer M_ϵ such that for all $n \geq M_\epsilon$, $|m - s_n| < d/4$. Take $N = \max(N_\epsilon, M_\epsilon)$. Then clearly, for all $n \geq N$, $|l - s_n| < d/4$ and $|m - s_n| < d/4$. Recall how absolute values work— $|x + y| \leq |x| + |y|$. The conditions $|l - s_n| < d/4$ and $|m - s_n| < d/4$ now yield: $|l - m| = |(l - s_n) + (s_n - m)| \leq |l - s_n| + |s_n - m| < d/4 + d/4 = d/2$. Reading across the various equalities and inequalities in this last sentence, we discover that $|l - m| < d/2$. But this is absurd, as we know $|l - m| = d$, and $d/2 < d$ as d is positive. This contradiction shows that $l = m$. □

11.3 Continuity of Functions

Recall from Example 5.40 in Chapter 5 that an open interval, denoted (a, b), is a set of the form $\{x \in \mathbb{R} \mid a < x < b\}$. Here, either $a < b$ are real numbers, or $a = -\infty$ and b is some real number, or a some real number and $b = +\infty$, or else, $a = -\infty$ and $b = +\infty$.

Let $f : (a, b) \to \mathbb{R}$ be a function defined on some open interval (a, b). What does it mean for f to be continuous at some point $c \in (a, b)$? Intuitively, on the graph of $y = f(x)$, we do not want a jump at c. Put another way, we want it where if we take a sequence of x-values in (a, b) marching towards (i.e., converging to) c, the corresponding sequence of y-values should march towards $f(c)$. See Figure 11.5. For instance, suppose for convenience of explanation that c is more than 1 unit away from both a and b. We could take the sequence of x-values $c + 1, c + 1/10, c + 1/100, \ldots$, which clearly converges to c, and we would then require that the sequence $f(c + 1), f(c + 1/10), f(c + 1/100), \ldots$ converges to $f(c)$. But this is the not the only way that x-values could march towards c. For instance, we could take the sequence of x-values $c+1, c-1/10, c+1/100, c-1/1000, \ldots$, and for this sequence as well, we would require that the corresponding sequence $f(c + 1), f(c - 1/10), f(c + 1/100), f(c - 1/1000), \ldots$ should converge to $f(c)$. Or we could take some other, less orderly sequence of x-values

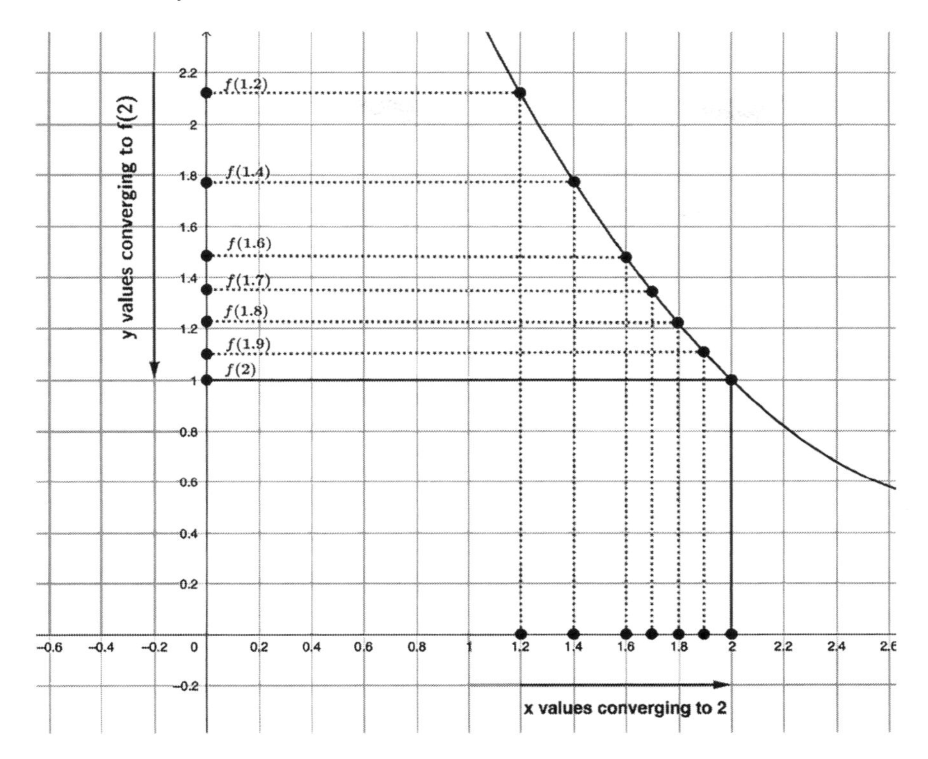

Figure 11.5. The graph of a function f. As the sequence of x-values converges to 2, the sequence of y-values converges to $f(2)$. For the function to be continuous at 2, for *every* such sequence (x_n) that converges to 2, the corresponding sequence $(f(x_n))$ must converge to $f(2)$.

converging to c, and for that sequence as well we would require the corresponding sequence of y-values to converge to $f(c)$. The key here is that *no matter which sequence of x-values that converges to c that we choose,* we want the corresponding y-values to converge to $f(c)$. This is indeed a valid definition of continuity, and we state it here:

Definition 11.27. *Intuitive Definition:* Let $f : (a, b) \to \mathbb{R}$ be a function defined on an open interval (a, b). Given $c \in (a, b)$, we say f is *continuous at* c if for every sequence (x_n), $n = 1, 2, \ldots$, $x_i \in (a, b)$, that converges to c, the corresponding sequence $(f(x_n))$, $n = 1, 2, \ldots$, converges to $f(c)$.

It is crucial in Definition 11.27 that *every* sequence (x_n) that converges to c be considered. Consider for example the following function:

$$f(x) = \begin{cases} x & \text{if } x \le 0, \\ 1 + x & \text{if } x > 0. \end{cases}$$

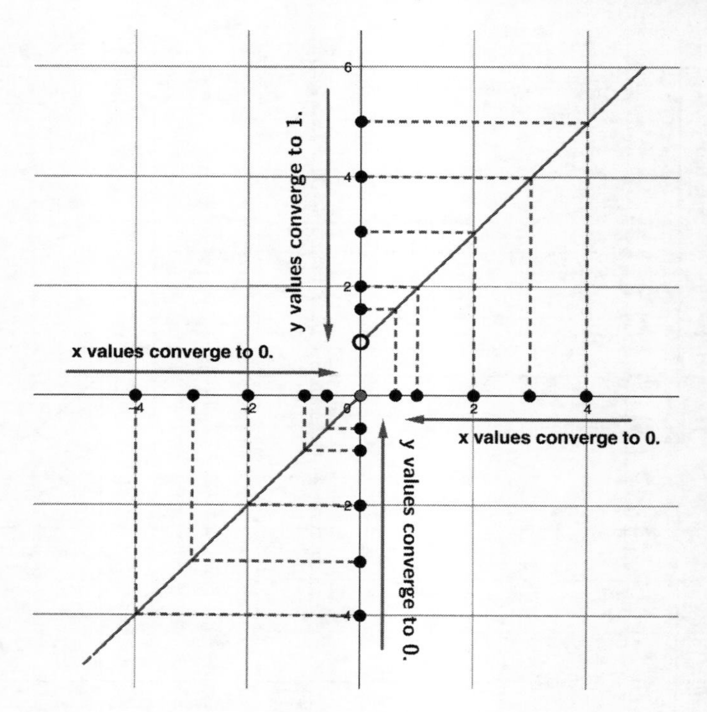

Figure 11.6. The graph of a function f discontinuous at 0. For the sequence of x-values shown that converges to 0 from the left, the corresponding y-values converge to 0, but for the sequence of x-values that converge to 0 from the right, the corresponding y-values converge to 1.

See Figure 11.6. If we consider a sequence (x_n) that converges to 0 from the left, as shown, the corresponding y-values converge to 0 $(= f(0))$. But if we consider a sequence (x_n) that converges to 0 from the right, as also shown, the corresponding y-values converge to 1 $(\neq f(0))$. Such a function cannot be continuous at 0 by Definition 11.27, and indeed, this is borne out in Figure 11.6.

Definition 11.27 is a fine definition of continuity of a function and is very intuitive. The only problem with this definition is that it is hard to work with: to formally prove that a function f is continuous at some $c \in (a, b)$, one needs to test *all* sequences (x_n) that converge to c and show that the corresponding sequence of y-values converges to $f(c)$. It would be more convenient to have an equivalent formulation that replaces testing all possible sequences converging to c with testing the behavior of a *single set* of x-values. Definition 11.28 below is indeed an equivalent definition; it has the advantage that it avoids having to test out all possible sequences, although it is considerably less intuitive than Definition 11.27 above. Definition 11.28 also allows the notion of continuity of a function to be generalized in a meaningful way to functions between other interesting sets besides subsets of $\mathbb{R}$ (and you will see some examples of this in your future mathematics courses), and is the standard way that continuity is now defined.

Definition 11.28. Let $f : (a, b) \to \mathbb{R}$ be a function defined on an open interval (a, b). Given $c \in (a, b)$, we say f is *continuous at* c if given any $\epsilon > 0$, there exists a $\delta_\epsilon > 0$ such that whenever $|x - c| < \delta_\epsilon$, $|f(x) - f(c)| < \epsilon$.

Remark 11.28.1. Just as with the notation "N_ϵ" in the definition of sequence convergence (see Remark 11.12.1), we use the notation "δ_ϵ" to indicate the dependence of δ_ϵ on ϵ.

Before we see that the two definitions are indeed equivalent, let us see some examples of how Definition 11.28 can be used:

Example 11.29. Let us show that the function $f : \mathbb{R} \to \mathbb{R}$ given by $f(x) = 5x - 1$ is continuous at 3 (or, as mathematicians often say, "at $x = 3$"). Here, the function is defined on the open interval $(-\infty, +\infty)$. Given $\epsilon > 0$, we wish to find δ_ϵ such that whenever $|x-3| < \delta_\epsilon$ we will have $|f(x) - f(3)| < \epsilon$. Our strategy for this is as follows: We will expand $|f(x)-f(3)|$ by using the definition of f, and will (miraculously!) factor out an "$|x - 3|$" from this. It will turn out that $|f(x) - f(3)|$ equals $|x - 3|$ times a non-zero constant, call it "C." So, wanting $|f(x) - f(3)| < \epsilon$ becomes equivalent to wanting $|x-3| \cdot C < \epsilon$. This in turn is equivalent to wanting $|x-3| < \epsilon/C$. So working backwards, if we take δ_ϵ to ϵ/C, then we will find that $|x-3| < \delta_\epsilon$ implies $|f(x)-f(3)| < \epsilon$: exactly what we want!

Now it is just a matter of writing things out! We have $|f(x) - f(3)| = |5x - 1 - (5 \cdot 3 - 4)| = |5x - 1 - 14| = |5x - 15| = 5|x - 3|$. Here 5 is the constant C we alluded to in the previous paragraph. So, just as described there, $5|x - 3| < \epsilon$ is equivalent to $|x - 3| < \epsilon/5$, so taking $\delta_\epsilon = \epsilon/5$ will ensure that whenever $|x - 3| < \delta_\epsilon$ we will have $|f(x) - f(3)| < \epsilon$. Hence $f(x)$ is continuous at $x = 3$.

Study this example and notice how easy (if at first unintuitive!) it is to show continuity. For practice, try to show continuity at a different x value, say at $x = -7$. Now try a different linear function, for instance $f(x) = 3x + 11$, and show continuity at various x-values for practice. Then do the following:

Exercise 11.29.1. Let a, b, c be arbitrary real numbers with $a \neq 0$. Show that the function $f : \mathbb{R} \to \mathbb{R}$ given by $f(x) = ax + b$ is continuous at c.

The next few examples need a bit more work to show continuity. One part of the strategy is still the same: to expand $|f(x)-f(c)|$ and somehow factor an $|x-c|$ from the resultant expression. The complication is that we won't get $|f(x)-f(c)| = |x-c| \cdot C$ as we got in Example 11.29 above. Instead, we will get $|f(x)-f(c)| = |x-c| \cdot g(x)$, where $g(x)$ is some other expression involving x, not just a constant. In such a case, we need to invoke *estimation*. We will argue that if we restrict x to begin with to a suitable interval around c, say $(c - \alpha, c + \alpha)$, then $g(x)$ is bounded above by some non-zero constant C, so in this interval, $|f(x) - f(c)| = |x - c| \cdot g(x) < |x - c| \cdot C$. At this point, we have a constant C multiplying $|x - c|$, and it is sufficient to make $|x - c| \cdot C < \epsilon$, for then, we will find $|f(x) - f(c)| < |x - c| \cdot C < \epsilon$. We then argue as before that we can take δ_ϵ to be ϵ/C if $\epsilon/C \leq \alpha$, or just α otherwise.

Example 11.30. Let us show that the function defined on all real numbers except $x = -3/2$ given by $f(x) = \frac{1}{2x+3}$ is continuous at 0. The domain of this function is

$(-\infty, -3/2) \cup (-3/2, \infty)$. Viewing this, in particular as also a function from $(-3/2, \infty)$ to $\mathbb{R}$ (since this is the open interval that contains $x = 0$), we wish to show that given any $\epsilon > 0$, there exists $\delta_\epsilon > 0$ such that whenever $|x - 0| = |x| < \delta_\epsilon$, $|f(x) - f(0)| < \epsilon$. (Note that the δ_ϵ we produce must be less than $3/2$, because otherwise, there would be a point, namely $-3/2$, in the interval $|x| < \delta_\epsilon$ where $f(x)$ is not defined.)

We have

$$|f(x) - f(0)| = \left|\frac{1}{2x+3} - \frac{1}{3}\right| = \left|\frac{-2x}{3(2x+3)}\right| = \left(\frac{2}{3|2x+3|}\right)|x|.$$

Thus, the expression $g(x)$ in the discussion just above this example is $\frac{2}{3|2x+3|}$. As described there, we will restrict x to a suitable interval around 0 and use this to bound $g(x)$ by a suitable non-zero constant C. We will try to bound just the term $|2x + 3|$ in the denominator by some non-zero constant C' (the remaining terms in $g(x)$ are themselves constants and we can adjust for those later).

To find this constant, note that $\frac{1}{|2x+3|} < C'$ is equivalent to $|2x + 3| > \frac{1}{C'}$. Hence, to put an upper bound on $\frac{1}{|2x+3|}$ we need to somehow put a lower bound on $|2x + 3|$. For this, first restrict x to lie in the range $-1 < x < 1$ (notice that this interval contains $x = 0$, the point at which we want to show continuity, and also, does not contain $x = -3/2$, the point at which f is not defined). Then multiplying by 2 (which is positive, it won't change the sign of the inequalities) and then adding 3, we find $2 \cdot -1 + 3 < 2x + 3 < 2 \cdot 1 + 3$, that is, $1 < 2x + 3 < 5$. In particular, $2x + 3$ is positive when $-1 < x < 1$, and $|2x + 3| = 2x + 3 > 1$. Hence, $\frac{1}{|2x+3|} < \frac{1}{1} = 1$. We thus take $C' = 1$, and then, $\frac{2}{3|2x+3|} < \frac{2 \cdot 1}{3}$. Therefore the constant C in the discussion just above this example is $\frac{2 \cdot 1}{3} = \frac{2}{3}$. To summarize, we find that after restricting x to lie in $(-1, 1)$, we have $|f(x) - f(0)| = |\frac{1}{2x+3} - \frac{1}{3}| = \frac{2}{3|2x+3|}|x| < \frac{2}{3}|x|$, and it is sufficient to make $\frac{2}{3}|x| < \epsilon$. Of course, this in turn is equivalent to taking $|x| < \frac{3\epsilon}{2}$.

Finally, we take $\delta_\epsilon = \min(1, \frac{3\epsilon}{2})$—we have to do this because we first need to restrict x to the range $|x| < 1$. (Notice that since $\delta_\epsilon \leq 1$, we will automatically have $\delta_\epsilon \leq 3/2$, a condition that we considered in the first paragraph of this example.) Tracing through the arguments, we find, just as we found in Example 11.29 above, that if we choose δ_ϵ as above, then $|x| < \delta_\epsilon$ will imply that $|f(x) - f(0)| < \epsilon$. The function is hence continuous at 0.

As in Example 11.29, study this example and practice these arguments by showing continuity at different values of x, for instance $x = 1$, $x = -1$, etc. Next, try to show continuity at various x-values of reciprocals of other linear functions, for instance, $f(x) = \frac{1}{5x-2}$. Learn how to restrict x to make the estimation. (For instance, when showing continuity of $f(x) = \frac{1}{2x+3}$ at $x = -1$, we cannot restrict x to be the range $-2 < x < 0$—a range of radius 1 about $x = -1$—because this interval contains $-3/2$, a point of discontinuity of f. So, you must choose a smaller range to restrict x to: for example, $-5/4 < x < -3/4$—a range of radius only $1/4$ about $x = -1$.) Finally, do the following:

Exercise 11.30.1. Let a, b, c be arbitrary real numbers with $a \neq 0$ and $c \neq -b/a$. Show that the function $f : \mathbb{R} \setminus \{-b/a\} \to \mathbb{R}$ given by $f(x) = \frac{1}{ax+b}$ is continuous at c.

Example 11.31. We looked at the continuity of linear functions in Example 11.29 above; here we will look at the general quadratic function. We will show that the function defined on all real numbers given by $f(x) = ax^2 + bx + c$, where a, b, c are fixed real numbers and $a \neq 0$, is continuous at 1.

So, as usual, given $\epsilon > 0$, we wish to show that there exists $\delta_\epsilon > 0$ such that whenever $|x-1| < \delta_\epsilon$, $|f(x)-f(1)| < \epsilon$. Now, $|f(x)-f(1)| = |ax^2+bx+c-a-b-c|$. We need to extract an "$|x-1|$" somehow out of this. We write $|ax^2+bx+c-a-b-c| = |a(x^2-1)+b(x-1)| = |x-1||a(x+1)+b|$. As is standard practice by now, to deal with the term $|a(x+1)+b|$, we will try to find a positive constant k such that $|a(x+1)+b| < k$. If we find such a k, we would have $|f(x)-f(1)| = |x-1||a(x+1)+b| < |x-1|k$, and it would then be sufficient to make $|x-1|k < \epsilon$, which in turn can be guaranteed by taking $|x-1| < \epsilon/k$.

So, to get this upper bound on $|a(x+1)+b|$, first note that $|a(x+1)+b| \leq |a(x+1)| + |b| = |a||x+1| + |b|$. Now, if we first agree to restrict x to be in the range $0 < x < 2$ (an interval of radius 1 about $x = 1$), then we have $1 < x+1 < 3$. In particular, since $x+1$ is positive in this range, we find $|x+1| = x+1 < 3$. Hence, when $0 < x < 2$, $|a(x+1)+b| \leq |a||(x+1)| + |b| < 3|a| + |b|$.

The rest of the proof is as described in the first paragraph. Taking $0 < x < 2$, we find $|f(x)-f(1)| = |x-1||a(x+1)+b| < |x-1|(3|a|+|b|)$. Note that $3|a|+|b| \neq 0$ as $|a| \neq 0$ (because $a \neq 0$). Therefore, if we further take $|x-1| < \frac{\epsilon}{3|a|+|b|}$, then $|f(x)-f(1)| < \epsilon$. Hence, it is sufficient to take $\delta_\epsilon = \min(1, \frac{\epsilon}{3|a|+|b|})$ so that both restrictions $0 < x < 2$ and $|x-1| < \frac{\epsilon}{3|a|+|b|}$ are satisfied.

Exercise 11.31.1. Show that the function defined on all real numbers given by $f(x) = ax^2+bx+c$, where a, b, c are fixed real numbers and $a \neq 0$, is continuous at d, where $d \in \mathbb{R}$ is arbitrary.

Example 11.32. We will show that the function defined for $x \geq -3/2$ by $f(x) = \sqrt{2x+3}$ is continuous at 3.

As usual, given $\epsilon > 0$, we wish to show that there exists $\delta_\epsilon > 0$ such that whenever $|x-3| < \delta_\epsilon$, $|f(x)-f(3)| < \epsilon$. We have $|f(x)-f(3)| = |\sqrt{2x+3}-\sqrt{9}| = |\sqrt{2x+3}-3|$. We wish to somehow extract an "$|x-3|$" from this. But first, we also need to deal with the square root. We use the standard trick of multiplying and dividing by the conjugate, namely $\sqrt{2x+3}+3$. Using the relation $(a-b)(a+b) = a^2-b^2$, we find: $|\sqrt{2x+3}-3| = \left|\frac{2x+3-9}{\sqrt{2x+3}+3}\right| = \frac{2|x-3|}{\sqrt{2x+3}+3}$. (Here, we have ignored the absolute value sign on the denominator because $\sqrt{2x+3}+3$ is positive for all $x \geq -3/2$, by the very definition of the square root function.) By now you know how to proceed: we find a constant C' such that $\frac{1}{\sqrt{2x+3}+3} < C'$ by suitably restricting x. So, for instance, if we first restrict x to lie in $2 < x < 4$, then $\sqrt{2\cdot 2+3}+3 < \sqrt{2x+3}+3 < \sqrt{2\cdot 4+3}+3$. In particular, $\sqrt{2x+3}+3 > \sqrt{7}+3$, so $\frac{1}{\sqrt{2x+3}+3} < \frac{1}{\sqrt{7}+3}$. The rest of the proof is like the proofs you have seen in the examples above.

Exercise 11.32.1. Let a, b, and c be real numbers, with $a > 0$ and $c > -b/a$. Show that the function defined for $x \geq -b/a$ given by $f(x) = \sqrt{ax+b}$ is continuous at c.

We now show that our two definitions of continuity of functions are equivalent:

Proposition 11.33. *Definitions* 11.27 *and* 11.28 *are equivalent, that is, a function* $f : (a, b) \to \mathbb{R}$ *is continuous at* c *for* $c \in (a, b)$ *according to Definition* 11.27 *if and only if it is continuous at* c *according to Definition* 11.28.

Proof. Assume first that the function f is continuous at c according to Definition 11.28, that is, given any $\epsilon > 0$, there exists $\delta_\epsilon > 0$ such that whenever $|x - c| < \delta_\epsilon$, $|f(x) - f(c)| < \epsilon$. Let (x_n), $n = 1, 2, \ldots, x_n \in (a, b)$, be any sequence that converges to c. We wish to show that the sequence $(f(x_n))$ converges to $f(c)$. By the definition of convergence of sequences, this means that given any $\epsilon > 0$, we need to show that there exists an integer N_ϵ such that for all $n \geq N_\epsilon$, $|f(x_n) - f(c)| < \epsilon$. Now, using this same ϵ, Definition 11.28 of continuity of f at $x = c$ shows that there exists $\delta_\epsilon > 0$ such that whenever $|x - c| < \delta_\epsilon$, $|f(x) - f(c)| < \epsilon$. Now, given this $\delta_\epsilon > 0$, the definition of convergence of the sequence (x_n) to c shows that there exists N_{δ_ϵ} such that for all $n \geq N_{\delta_\epsilon}$, $|x_n - c| < \delta_\epsilon$. Putting these together, we find that for all $n \geq N_{\delta_\epsilon}$, $|x_n - c| < \delta_\epsilon$, and because $|x_n - c| < \delta_\epsilon$, $|f(x_n) - f(c)| < \epsilon$. Therefore, taking "$N_\epsilon$" to be N_{δ_ϵ}, we find that for all $n \geq N_\epsilon$, $|f(x_n) - f(c)| < \epsilon$. We have thus shown that f satisfies Definition 11.27 for continuity at $x = c$.

Now assume that f is continuous at c according to Definition 11.27, that is, for any sequence (x_n), $n = 1, 2, \ldots, x_n \in (a, b)$, that converges to c, the corresponding sequence $(f(x_n))$ converges to $f(c)$. We wish to show that f also satisfies Definition 11.28 for continuity at $x = c$, that is, given any $\epsilon > 0$, there exists $\delta_\epsilon > 0$ such that whenever $|x - c| < \delta_\epsilon$, $|f(x) - f(c)| < \epsilon$. Assume to the contrary that there is an $\epsilon > 0$ for which no such δ_ϵ can be found. We will come up with a contradiction as follows: write d_1 for $\min(c - a, b - c)$, with the understanding that if $a = -\infty$ and $b = \infty$, we will set $d_1 = 1$. (So d is the distance from c to the closest "boundary point" of the open interval (a, b), unless $(a, b) = \mathbb{R}$, in which case we simply set $d_1 = 1$.) Then, there exists a point x_1 with $|x_1 - c| < d_1/2$ for which $|f(x_1) - f(c)| \geq \epsilon$. For, if not, we could take δ_ϵ to be $d_1/2$ and find that for all points satisfying $|x - c| < \delta_\epsilon = d_1/2$, $|f(x) - f(c)| < \epsilon$, contrary to what we have assumed. Write d_2 for $|x_1 - c|$. Note that $d_2 < d_1/2$ by design. Then, once again, there exists a point x_2 with $|x_2 - c| < d_2/2$ for which $|f(x_2) - f(c)| \geq \epsilon$ (else, just as before, we could take δ_ϵ to be $d_2/2$ and find that for all points satisfying $|x - c| < \delta_\epsilon = d_2/2$, $|f(x) - f(c)| < \epsilon$, contrary to what we have assumed). Now let $d_3 = |x_2 - c|$. Note that $d_3 < d_2/2 < d_1/4$ by design. Yet once again, there exists a point x_3 with $|x_3 - c| < d_3/2$ for which $|f(x_3) - f(c)| \geq \epsilon$, for exactly the same reasons as for x_1 and x_2. Proceeding thus, we create a sequence $x_1, x_2, x_3, x_4, \ldots,$ with $|x_n - c| < d_n/2 < \cdots < d_1/2^n$. This sequence clearly converges to c (see Exercise 11.47), yet the corresponding sequence $(f(x_n))$ does not converge to $f(c)$ because each $f(x_n)$ is at least ϵ away from $f(c)$. This contradicts Definition 11.27 for continuity at $x = c$. Therefore, our assumption that for this $\epsilon > 0$, no $\delta_\epsilon > 0$ can be found for which $|x - c| < \delta_\epsilon$ implies $|f(x) - f(c)| < \epsilon$ is flawed. Put differently, f satisfies Definition 11.28 for continuity at c. □

11.4 Limits of Functions

In this section we consider another concept that is easiest to understand via sequences: the notion of the *limit* of a function. This notion is very closely connected with the notion of continuity, and in fact, is already embedded in Definition 11.27 of continuity. In that definition, we said that a function $f : (a,b) \to \mathbb{R}$ is continuous at c if for all sequences (x_n) with $x_i \in (a,b)$ that converge to c, the corresponding sequence $f(x_n)$ converges to $f(c)$. Now, there are many situations where for all sequences (x_n) with $x_i \in (a,b)$ that converge to c, the corresponding sequence $(f(x_n))$ converges to *some value* (that is independent of the convergent sequence of (x_n) you pick). This value may or may not equal $f(c)$, but the very fact that there is a value to which all these $(f(x_n))$ sequences converge is itself of interest. Thus, it is convenient to separate out the two properties that are inherent in Defintion 11.27: first, that for every sequence (x_n), $x_i \in (a,b)$ that converges to c, the corresponding sequence $(f(x_n))$ converges to a fixed value independent of which convergent sequence (x_n) you start from, and second the fact that this fixed value is actually $f(c)$. The first property is what the concept of limit is all about.

Interestingly, since after separating the two properties above into two stages, we do not care about the value of $f(c)$ in the first stage, the concept of limit can be considered in a slightly more general setting as well, *where the function is not even defined at* $x = c$. (In fact, it is in this setting that limits are particularly useful: for instance, in calculus, the derivative is defined as the limit of a function at a point where the function is not defined!) We have the following:

Definition 11.34. *Intuitive Definition:* Let (a,b) be an open interval, and let $c \in (a,b)$ be some point. Let S be either the set (a,b) or else the set $(a,b) \setminus \{c\}$. Let $f : S \to \mathbb{R}$ be a function. We say that the *limit of f as x approaches c* equals l, for some real number l, written $\lim_{x\to c} f(x) = l$, if for all sequences (x_n), $n = 1, 2, \dots$, $x_i \in (a,b)\setminus\{c\}$, that converge to c, the corresponding sequence $(f(x_n))$, $n = 1, 2, \dots$, converges to l. If no such real number l exists, we say that the limit of f as x approaches c does not exist.

Example 11.35. Let us consider the following examples first; our approach here will be informal, based on the corresponding graphs:

(1) Consider the function f restricted to the set $S = (-\infty, 0) \cup (0, \infty)$ by $f(x) = x^2 + 1$. The graph of this function is a parabola opening upwards, with its vertex at $(0,1)$, except that there is a hole where the vertex should be. See Figure 11.7. The hole occurs because this function is not defined at $x = 0$, as 0 is not in S. Yet, as x approaches 0 either along the positive x-axis or along the negative x-axis or along a combination of the two, the values of $f(x)$ show a definite pattern: they approach closer and closer to 1. (In Figure 11.7, a sequence of x-values approaching 0 from the left is shown.) Put differently, for any sequence (x_n), with the $x_i \in S$, that converges to 0, the corresponding sequence of y-coordinates $(f(x_n))$ converges to 1. Thus, the limit of $f(x)$ as x approaches 0 is 1, and we write this as $\lim_{x\to 0} f(x) = 1$. Notice that we only considered sequences (x_n) where the x_i are in S, that is, we do not allow x_i to equal 0. Notice too that though the function f is not even defined at $x = 0$, the limit of f as x approaches 0 exists.

(2) Now consider the function g defined on $\mathbb{R}$ by the rule

$$g(x) = \begin{cases} x^2 + 1 & \text{if } x \neq 0, \\ 0 & \text{if } x = 0. \end{cases} \tag{11.1}$$

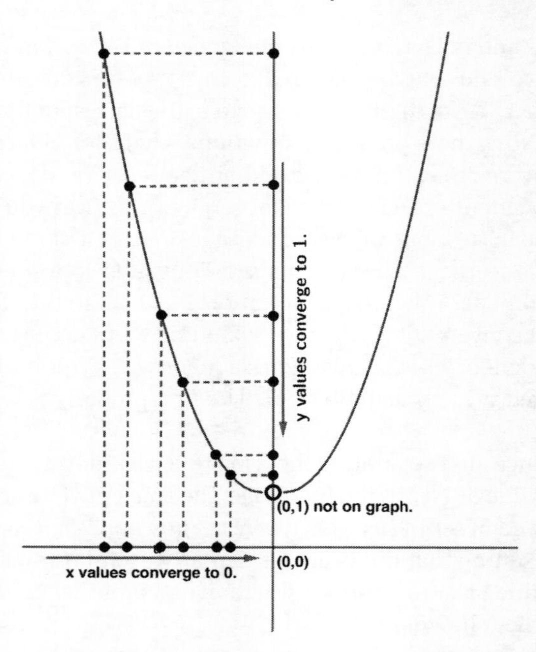

Figure 11.7. The graph of $f(x) = x^2 + 1$ defined on $(\infty, 0) \cup (0, \infty)$. A sequence of x-values approaching 0 from the left is shown; the corresponding y-values approach 1, even though $(0, 1)$ is not on the graph.

The graph of this function is a parabola opening upwards, with its vertex at $(0, 1)$, except that there is a hole where the vertex should be; *instead, the point* $(0, 0)$ *is also on the graph!* The same considerations as in part (1) apply: For any sequence (x_n), with the $x_i \in \mathbb{R} \setminus \{0\}$ (note!), the corresponding sequence of y-coordinates $(g(x_n))$ converges to 1. Thus the limit of $g(x)$ as x approaches 0 is 1, and we write this as $\lim_{x \to 0} g(x) = 1$.

Note two things about this example: First, even though $g(x)$ is defined at $x = 0$, while considering the behavior of g as x approaches 0, we only consider sequences (x_n) where $x_i \in \mathbb{R} \setminus \{0\}$, that is, we do not allow x_i to equal 0. Next, unlike the example in part (1), the function g is certainly defined at $x = 0$. However, the value of g at $x = 0$ has no bearing on the limit, since while deciding what the limit is, we only consider sequences (x_n) where the x_i are nonzero. As it turns out, $g(0) \neq \lim_{x \to 0} g(x)$, and this corresponds to our notion that g is "broken" at $x = 0$ (to be more precise, there is a jump in the graph of g at $x = 0$).

(3) Finally, consider the function h defined on $\mathbb{R}$ by $h(x) = x^2 + 1$. As with parts (1) and (2), if we consider any sequence (x_n), with the $x_i \in \mathbb{R} \setminus \{0\}$, the corresponding sequence of y-coordinates $(h(x_n))$ converges to 1. Thus the limit of $h(x)$ as x approaches 0 is 1, and we write this as $\lim_{x \to 0} h(x) = 1$.

The key difference between this example and the one in part (2) is that $h(x)$ is not only defined at $x = 0$, but also, $h(0)$ equals $\lim_{x \to 0} h(x)$. This corresponds to our notion that $h(x)$ is "unbroken" (or continuous) at $x = 0$.

(4) In a different direction, consider the function $s(x)$ defined on $\mathbb{R} \setminus \{0\}$ by $s(x) = \sin(1/x)$. As x approaches 0 (say along the positive x axis), $1/x$ becomes larger and larger, and runs through intervals of the form $[2n\pi, 2(n+1)\pi]$, for larger and larger integers n. Each time $1/x$ runs through such an interval, $\sin(1/x)$ oscillates through one cycle, and so as x approaches 0, $\sin(1/x)$ oscillates between -1 and 1. This suggests that there should be sequences (x_n) that converge to zero for which the corresponding sequence $(\sin(1/x_n))$ fails to converge. Here is one such sequence: take $x_n = 1/(n\pi + \pi/2)$. Then $\sin(1/x_n)$ takes on values 1 and -1 alternatively, and therefore fails to converge. Thus, the limit of $\sin(1/x)$ as x approaches 0 does not exist.

Now how do we formally prove the assertions in the various examples contained in Example 11.35? Recall that in our study of continuity, we started with Definition 11.27, which described continuity in terms of the behavior of sequences, but switched to Definition 11.28, which was easier to apply. (These two definitions are equivalent, as we showed in Proposition 11.33.) Since the concept of the limit of a function has been extracted from the sequence definition of continuity (by simply weakening the condition that the sequence "$(f(x_n))$" converges to "$f(c)$" by replacing "$f(c)$" with "l"), it should come as no surprise that there is another definition of limit as well:

Definition 11.36. Let (a, b) be an open interval, and let $c \in (a, b)$ be some point. Let S be either the set (a, b) or else the set $(a, b)\setminus\{c\}$. Let $f : S \to \mathbb{R}$ be a function. We say that the *limit of f as x approaches c* equals l, for some real number l, written $\lim_{x\to c} f(x) = l$, if given any $\epsilon > 0$, there exists a $\delta_\epsilon > 0$ such that whenever $0 < |x - c| < \delta_\epsilon$, we have $|f(x) - l| < \epsilon$. If no such real number l exists, we say that the limit of f as x approaches c does not exist.

> *Remark* 11.36.1. Notice that unlike in the corresponding definition for continuity (Definition 11.28), we insist that $0 < |x - c|$. In other words, we insist that $x \neq c$. This is because, when considering limits, we purposely ignore what is happening to the function at $x = c$.

The similarity to Definition 11.28 should not be surprising at all, since, once again, the concept of the limit has been extracted from the sequence definition of continuity. Not surprising as well, Definitions 11.34 and 11.36 are equivalent:

Proposition 11.37. *Definitions* 11.34 *and* 11.36 *are equivalent, that is, with S, a, b, and c as in either Definition* 11.34 *or* 11.36, $\lim_{x\to c} f(x) = l$, *according to Definition* 11.34, *if and only if* $\lim_{x\to c} f(x) = l$, *according to Definition* 11.36.

Proof. Except for some minor changes of language, the proof of Proposition 11.37 is identical to the proof of Proposition 11.33. It is a good exercise for you to go through the proof of Proposition 11.33 and make the changes required so that that proof will work for the current proposition; it will help cement your understanding of the ideas behind the proof. Accordingly, we will formalize it as Exercise 11.62 ahead. □

Remark 11.38. The limit of a function, if it exists at all, is unique. We will leave this for you to show:

Exercise 11.38.1. With S, a, b, and c and f as either in Definition 11.34 or 11.36, prove that if $\lim_{x\to c} f(x) = l$ and $\lim_{x\to c} f(x) = m$, then $l = m$. (Hint: Work with Definition 11.34 and apply Proposition 11.26.)

Now let us turn our attention to showing that the limits in the four examples contained in Example 11.35 are as described, using Definition 11.36.

Example 11.39. *Example* 11.35 *revisited:* We will first consider part (1) of Example 11.35. To show that $\lim_{x\to 0} f(x) = 1$, we need to show that given any $\epsilon > 0$, there exists $\delta_\epsilon > 0$ such that whenever $0 < |x - 0| < \delta_\epsilon$, $|f(x) - 1| < \epsilon$. Let us analyze what this means: $|f(x) - 1| = |x^2 + 1 - 1| = x^2$ (remember $x^2 \geq 0$, so we may ignore the absolute value sign). Thus, to say that $|f(x) - 1| < \epsilon$ is to say that $x^2 < \epsilon$, and this happens precisely when $-\sqrt{\epsilon} < x < \sqrt{\epsilon}$, i.e., $|x| < \sqrt{\epsilon}$. Thus, we should take $\delta_\epsilon = \sqrt{\epsilon}$, and, ignoring the value $x = 0$, we will find that $0 < |x - 0| < \delta_\epsilon$ implies $|f(x) - 1| < \epsilon$. So indeed, $\lim_{x\to 0} f(x) = 1$.

The proof that $\lim_{x\to 0} g(x) = 1$ and $\lim_{x\to 0} h(x) = 1$ in parts (2) and (3) (respectively) are identical. This is because the value of $f(0)$ is irrelevant when considering the limit as x approaches 0.

Now let us consider part (4). We wish to show, using Definition 11.36, that the limit of $s(x)$ as x approaches 0 does not exist. So, let l be an arbitrary real number. We will show that $\lim_{x\to 0} s(x) \neq l$. Take $\epsilon = 0.5$. Then, the interval $[l - 0.5, l + 0.5]$ has width 1, and therefore, at least one of -1 or 1 cannot be in this interval. Let us assume that 1 is not in $[l - 0.5, l + 0.5]$, so $|l - 1| > 0.5$. Then, no matter what $\delta_\epsilon > 0$ we pick, we can find an n large enough such that $(2n + 1/2)\pi > 1/\delta_\epsilon$. Taking $x = 1/(2n + 1/2)\pi$, we find $0 < |x| < \delta_\epsilon$, yet $|s(x) - l| = |1 - l| > \epsilon$. Thus, no $\delta_\epsilon > 0$ will work for this ϵ, so l cannot be the limit of $s(x)$ as x approaches 0. (If 1 is in the interval $[l - 0.5, l + 0.5]$ we work with -1: given any $\delta_\epsilon > 0$, we find an n large enough so that $(2n + 3/2)\pi > 1/\delta_\epsilon$, and argue similarly.)

11.5 Relation between limits and continuity

Recall from the discussion just before Definition 11.34 that we introduced the concept of the limit of a function by separating the definition of continuity into two parts: first, that for every sequence (x_n), $x_i \in (a, b)$, the corresponding sequence $(f(x_n))$ converges to a fixed value independent of which convergent sequence (x_n) you start from, and second the fact that this fixed value is actually $f(c)$. The first of these two properties led to the notion of a limit. It is not surprising then that we have the following:

Proposition 11.40. *A function* $f : (a, b) \to \mathbb{R}$ *is continuous at* $c \in (a, b)$ *if and only if* $\lim_{x\mapsto c} f(x)$ *exists, and equals* $f(c)$.

Proof. It is clearest if we work with the $\epsilon - \delta$ definitions, namely Definitions 11.28 and 11.36.

So first suppose f is continuous at c. Given any $\epsilon > 0$, by Definition 11.28, there exists a $\delta_\epsilon > 0$ such that $|x - c| < \delta_\epsilon$ implies $|f(x) - f(c)| < \epsilon$. In particular, this means that if $0 < |x - c| < \delta_\epsilon$ then $|f(x) - f(c)| < \epsilon$. By Definition 11.36, this means that $\lim_{x\mapsto c} f(x)$ exists to begin with, and further, $\lim_{x\mapsto c} f(x) = f(c)$.

In the reverse direction, assume that $\lim_{x \mapsto c} f(x)$ exists and equals $f(c)$. By definition 11.36 this means that given any $\epsilon > 0$, there exists a $\delta_\epsilon > 0$ such that $0 < |x-c| < \delta_\epsilon$ implies $|f(x) - f(c)| < \epsilon$. This already takes care of "most of" the requirements of Definition 11.28. The only thing to worry about is what happens if $0 = |x - c|$: this case is not covered in the definition of limit. But if $0 = |x - c|$, this means $x = c$, in which case $f(x) = f(c)$, so $|f(x) - f(c)| = 0 < \epsilon$ anyway. Thus, we find that if $\lim_{x \mapsto c} f(x)$ exists and equals $f(c)$, then, given any $\epsilon > 0$, there exists a $\delta_\epsilon > 0$ such that $|x-c| < \delta_\epsilon$ implies $|f(x) - f(c)| < \epsilon$. By Definition 11.28, f is continuous at c. □

Remark 11.41. Many textbooks introduce limits first, and then define continuity via limits, as in the statement of Proposition 11.40: A function $f : (a, b) \to \mathbb{R}$ is defined to be continuous at $x = c$, for $c \in (a, b)$, if $\lim_{x \mapsto c} f(x)$ exists, and equals $f(c)$. Whether to define limits first and then define continuity in terms of limit, or to present limits and continuity as separate but closely related concepts (as we have done here) is a matter of taste.

11.6 Series

Associated to any sequence is another mathematical object of vital importance: the *series* that arises from that sequence. We will introduce these objects briefly here, and consider a central example – the geometric series – that is key to the study of other series. We will follow this introduction with a more in-depth discussion in Chapter 12, Section 12.8.

Let (a_n) be a sequence. An expression such as

$$S = a_1 + a_2 + \cdots + a_n + \cdots$$

also written as

$$S = \sum_{i=1}^{\infty} a_i,$$

is known as a *series*. It is a *finite* series if $a_n = 0$ for all $n \geq N$, for some $N \in \mathbb{N}$. Otherwise, the expression is known as an *infinite* series. The terms a_n of the underlying sequence (a_n) are also referred to as the n-th term of the series.

As of now, an infinite series is just an expression. To give meaning to this expression, we consider the following finite sums,

$$\begin{aligned} P_1 &= a_1 \\ P_2 &= a_1 + a_2 \\ \vdots &= \vdots \\ P_n &= a_1 + a_2 + \cdots + a_n \\ \vdots &= \vdots \end{aligned}$$

P_i is known as the *i-th partial sum* of the series. Since each P_i is just a real number, the assemblage $(P_1, P_2, \dots)$ forms a new sequence (P_n), known as the *sequence of partial sums* of the series S. As a sequence, (P_n) can either converge or diverge. If (P_n) converges to a real number L, we say that *the series S converges to L*, or the *the series S sums to L*. Otherwise we say that the series S diverges.

Example 11.42. *Geometric series:* As already pointed out above, this series is key to the study of other series. We have already considered the finite version of geometric series in Example 7.3 in Chapter 7; here we will consider the infinite version.

Let a and r be real numbers. Consider the sequence (a_n), where $a_n = ar^{n-1}$. The series formed from this sequence,

$$G_{a,r} = \sum_{n=1}^{\infty} ar^{n-1}, \tag{11.2}$$

is known as (*an infinite*) *geometric series*. In Example 7.3 in Chapter 7 we considered the finite geometric series $G_{a,r}(n) = a + ar + ar^2 + \cdots + ar^{n-1}$: we now recognize that $G_{a,r}(n)$ is nothing but the n-th partial sum P_n of the infinite geometric series $G_{a,r}$. (Notice that our notation $G_{a,r}$ for the infinite series is a natural extension of the notation for the partial sums used in Chapter 7.)

The series above depends on two parameters a and r. Clearly, if $a = 0$, $G_{a,r}$ reads $0+0+\cdots+\cdots$. It converges to 0, and is of limited interest! So we will assume that $a \neq 0$. If further $r = 0$, then too the series is of limited interest: $G_{a,r}$ reads $a + 0 + 0 + \cdots + \cdots$ and converges to a. So, we will assume that both a and r are nonzero, and consider the convergence and divergence of $G_{a,r}$.

There is one further situation where the series takes on a simple form: when $r = \pm 1$. When $r = 1$, the series reads $a + a + a + \cdots +$, and diverges, since the n-th partial sum has absolute value $n|a|$, which grows without bound as n increases. (At this point, we can accept this last statement that $n|a|$ grows without bound as something that is intuitively clear, based on our experience with the real numbers. However, this fact is actually a deep property of the real numbers known as the Archimedean property, and is a consequence of another deep property of the real numbers known as the least upper bound principle: this is discussed in Chapter 12, in particular in Section 12.3.) When $r = -1$, the series reads $a - a + a - a \pm \cdots$, and the odd partial sums equal a while the even partial sums equal 0. Thus the sequence of partial sums never converges, so $G_{a,r}$ diverges when $r = \pm 1$ (and when $a \neq 0$, as we have assumed).

So we assume now that $a \neq 0$, $r \neq 0$ and $r \neq \pm 1$. The groundwork for studying $G_{a,r}$ with these assumptions has already been laid for us in Example 7.3 in Chapter 7, where we saw that the partial sums P_n (denoted $G_{a,r}(n)$ there) are given by

$$P_n = \frac{a(1 - r^n)}{1 - r} = \frac{a}{1 - r} - \frac{r^n}{1 - r}. \tag{11.3}$$

Working intuitively first, the behavior of P_n as n goes to infinity is thus determined by the behavior of the term $\frac{r^n}{1-r}$ (as the first term is a constant independent of n), and in turn the behavior of $\frac{r^n}{1-r}$ is determined by the behavior of r^n. Again working intuitively, invoking our experience with real numbers, we can see that when $|r| < 1$, we will have $|r| > |r|^2 > |r|^3 > \cdots$, the powers will get smaller and smaller in magnitude as n goes to infinity, and the sequence (r^n) will converge to 0. Thus, when $|r| < 1$, we find that the series $G_{a,r}$ converges to $\frac{a}{1-r}$.

Continuing to work intuitively and invoking our experience with real numbers, when $|r| > 1$, we will have $|r| < |r|^2 < |r|^3 < \cdots$, the powers will get larger and larger in magnitude and will grow without bound as n goes to infinity, and the sequence (r^n) will thus diverge. Thus, the first summand of P_n in equation (11.3) above stays constant, but the second summand diverges, and it follows that the partial sums P_n will diverge. Thus, $G_{a,r}$ diverges if $|r| > 1$.

Putting these results together, we find that $G_{a,r}$ (for $a \neq 0, r \neq 0$) diverges if $|r| \geq 1$ and converges to $\frac{a}{1-r}$ when $|r| < 1$.

The formal justification for our intuitive observations above are provided in Exercises 11.49 through 11.52 and in Exercise 11.54. First, in Exercise 11.54, we show formally that the sequence (r^n) converges to zero when $|r| < 1$ and diverges when $|r| > 1$. Next, we interpret equation (11.3) as saying that the sequence (P_n) is the sum of a sequence whose terms are constant, namely $\frac{a}{1-r}$, and another whose terms are $-\frac{r^n}{1-r}$. In turn, we interpret the sequence $\left(-\frac{r^n}{1-r}\right)$ as the constant $\frac{-1}{1-r}$ times the sequence (r^n). Exercises 11.49 through 11.52 then justify our saying that P_n diverges when (r^n) diverges (that is, when $|r| > 1$) and converges to $\frac{a}{1-r}$ when r^n converges to zero (that is, when $|r| < 1$).

We will consider two other examples of series in Exercises 11.65 and 11.66, and of course, consider series in greater depth in Chapter 12, Section 12.8 as already noted.

11.7 Further Exercises

Exercise 11.43. Practice Exercises:

In Exercises 11.43.1 *through* 11.43.6 *below, prove carefully using* $\epsilon - N_\epsilon$ *arguments like in Examples* 11.15 *through* 11.19 *above that the following sequences converge to their indicated limits:*

Exercise 11.43.1. The sequence (s_n) with $s_n = \frac{3n}{1+n}$ converges to 3.

Exercise 11.43.2. The sequence (s_n) with $s_n = \frac{3n^2+1}{1+4n^2}$ converges to $\frac{3}{4}$.

Exercise 11.43.3. The sequence (s_n) with $s_n = \frac{3n}{1+n^2}$ converges to 0. (Hint: Use estimation to bound the term above by a term of the form constant divided by n.)

Exercise 11.43.4. The sequence (s_n) with $s_n = (\frac{1}{n} - \frac{1}{n+2})$ converges to 0.

Exercise 11.43.5. The sequence (s_n) with $s_n = (\sqrt{4n^2+1} - 2n)$ converges to 0. (Hint: First multiply and divide by the conjugate $(\sqrt{4n^2+1} + 2n)$ and then use estimation.

Exercise 11.43.6. The sequence (s_n) with $s_n = \frac{2n+1}{3n\sqrt{n}+n^2}$ converges to 0. (Hint: Use estimation to bound the terms by a term of the form constant divided by $\sqrt{n}$.)

In Exercises 11.43.7 *through* 11.43.10*, show that the given sequences diverge:*

Exercise 11.43.7. The sequence (s_n) with $s_n = \frac{n^2+1}{1+n}$. (Hint: Use estimation to show the terms are unbounded, and invoke Proposition 11.23.)

Exercise 11.43.8. The sequence (s_n) with $s_n = (-1)^n \frac{n+2}{n+1}$.

Exercise 11.43.9. The sequence (s_n) with $s_n = (-1)^n \frac{n-2}{n+1}$.

Exercise 11.43.10. The sequence (s_n) with $s_n = \tan \frac{(n-1)\pi}{2n}$.

In Exercises 11.43.11 *through* 11.43.13*, show that the following functions are continuous at the indicated values of* x*:*

Exercise 11.43.11. The function $\frac{5x+1}{6x+2}$ defined for $x \neq -1/3$, at 1.

Exercise 11.43.12. The function $\frac{2x+5}{x^2+1}$, at -1.

Exercise 11.43.13. The function $\frac{1}{\sqrt{x+1}}$ defined for $x > -1$, at 0.

Exercise 11.44. Consider the sequence (s_n) with

$$s_n = \frac{a_k n^k + a_{k-1} n^{k-1} + \cdots a_0}{b_k n^k + b_{k-1} n^{k-1} + \cdots b_0}.$$

(Here, $k \geq 0$, and we assume $a_k \neq 0$, $b_k \neq 0$.) We will show that this sequence converges to $\frac{a_k}{b_k}$.

(1) Since $n \geq 1$, we may divide by n^k and write

$$s_n = \frac{a_k + a_{k-1} n^{-1} + a_{k-2} n^{-2} \cdots a_0 n^{-k}}{b_k + b_{k-1} n^{-1} + b_{k-2} n^{-2} \cdots b_0 n^{-k}} = \frac{a_k + u_n}{b_k + v_n},$$

where we have denoted the sum of the terms in the numerator after a_k by u_n and similarly for the denominator. Thus, (u_n) and (v_n) are themselves sequences. Show that (u_n) and (v_n) converge to zero. (Hint: $|a_{k-1} n^{-1} + a_{k-2} n^{-2} \cdots a_0 n^{-k}| \leq |a_{k-1}||n^{-1}| + |a_{k-2}||n^{-2}| \cdots |a_0||n^{-k}|$.)

(2) Part (1) shows, in intuitive terms, that the numerator approaches a_k as n becomes very large, and similarly for the denominator, so it is reasonable to expect that the quotient approaches $\frac{a_k}{b_k}$ (which is the contention of the exercise). Prove this formally. (Hint: $|b_k u_n - a_k v_n| \leq |b_k||u_n| + |a_k||v_n|$.)

Exercise 11.45. Show that if a convergent sequence (s_n) has the property that $s_n \geq 0$ for all $n \in \mathbb{N}$, then the limit of (s_n) cannot be negative. (Hint: Assume to the contrary that (s_n) converges to $-L$, where $L > 0$. Study what happens if you take $\epsilon = L/2$.)

Exercise 11.46. Show that if a convergent sequence (s_n) has the property that $s_n > 0$ for infinitely many n and also $s_n < 0$ for infinitely many n, then the limit of s_n has to be zero. (Hint, first assume that the limit of (s_n) is L for $L > 0$, and then assume the limit is $-L$ for $L > 0$. In each case take ϵ to be $L/2$ and arrive at a contradiction.)

Exercise 11.47. Let (x_n), $n = 1, 2, \ldots$, be a sequence such that $|x_n - c| < d/2^n$ for some c and for some positive real number d. This sequence (x_n) very clearly converges to c, but give a formal $\epsilon - N_\epsilon$ style proof. (Hint: Mimic Example 11.16.)

Exercise 11.48. Give an example of a bounded sequence that is divergent and of a bounded sequence that is convergent. This shows that while convergence implies boundedness by Proposition 11.23, boundedness implies neither convergence nor divergence! Now give an example of a divergent sequence that is bounded and of a divergent sequence that is unbounded. This shows that divergence implies neither boundedness nor unboundedness!

Exercise 11.49. In this and the following three exercises, we will create new sequences from old ones, and study the convergence of the new sequences in terms of the old ones. So, for this exercise, let $r \in \mathbb{R}$ be arbitrary, and let (s_n) be any sequence. Construct the new sequence, denoted (rs_n), whose terms are simply the terms of the original (s_n) multiplied by r. Thus, (rs_n) is the sequence $(rs_1, rs_2, rs_3, \dots)$. Assume that the original sequence (s_n) is convergent, with limit L. Show that (rs_n) also converges and its limit is rL. Show also that if $r \neq 0$ and the original sequence (s_n) diverges, then (rs_n) also diverges. (Hint: If $r = 0$ the proof of convergence should be clear. If $r \neq 0$, note that asking for $|rs_n - rL| < \epsilon$ is equivalent to asking for $|s_n - L| < \epsilon/|r|$. Now invoke the convergence of the original (s_n). For the proof of divergence, argue by contradiction.)

Exercise 11.50. Let (s_n) and $t_n)$ be two sequences. Construct the new sequence, denoted $(s_n + t_n)$, whose terms are simply the sums of the terms of the original two sequences. Thus, $(s_n + t_n)$ is defined to be the sequence $(s_1 + t_1, s_2 + t_2, s_3 + t_3, \dots)$. Show that if both the original sequences (s_n) and (t_n) are convergent, with limits L and M respectively, then $(s_n + t_n)$ also converges, and its limit is $L + M$. (Hint: Note that $|s_n + t_n - L - M| = |(s_n - L) + (t_n - M)|$.)

Combine this result you just proved in this exercise with the result in Exercise 11.49 above to show that if (s_n) converges but (t_n) does not, then $(s_n + t_n)$ cannot be convergent. (Hint: Assume to the contrary and write the sequence (t_n) in terms of (s_n) and the sum sequence.)

Give an example of two *divergent* sequences (s_n) and (t_n) with the property that $(s_n + t_n)$ is convergent.

Exercise 11.51. Let (s_n) and (t_n) be two sequences. Construct the new sequence, denoted $(s_n \cdot t_n)$, whose terms are simply the products of the terms of the original two sequences. Thus, $(s_n \cdot t_n)$ is defined to be the sequence $(s_1 \cdot t_1, s_2 \cdot t_2, s_3 \cdot t_3, \dots)$. Show that if both the original sequences (s_n) and (t_n) are convergent, with limits L and M respectively, then $(s_n \cdot t_n)$ also converges, and its limit is $L \cdot M$. (Hint: Note that by Proposition 11.23 the sequence (t_n) is bounded, so there exists $P > 0$ such $|t_n| < P$ for all n. If $L = 0$, note that $|s_n t_n| \leq |s_n| P$, and find N such that $|s_n| < \epsilon/P$ for all $n \geq N$. Now assume $L \neq 0$. Write $|s_n t_n - LM|$ as $|s_n t_n - t_n L + t_n L - LM|$. Write $(s_n t_n - t_n L)$ as $t_n(s_n - L)$, and write $(t_n L - LM)$ as $L(t_n - M)$. Find N_1 such that $|s_n - L| < \epsilon/2P$ and N_2 such that $|t_n - M| < \epsilon/2|L|$, and take $N_\epsilon = \max(N_1, N_2)$.)

Unlike in the case with sums of sequences, it is possible to have two sequences (s_n) and (t_n) with (s_n) convergent but (t_n) divergent, with the property that the product sequence $(s_n \cdot t_n)$ is convergent! Here is an example: take $s_n = 1/n$ for all $n \in \mathbb{N}$, and take (t_n) to be the sequence $(1, 0, 1, 0, 1, 0, \dots)$. We know that (s_n) converges; this is just an even simpler version of Example 11.9. Clearly (t_n) diverges: it keeps oscillating mindlessly between 0 and 1 and refuses to settle down. Prove that, all the same, $(s_n \cdot t_n)$ converges.

Give an example of two *divergent* sequences (s_n) and (t_n) with the property that $(s_n \cdot t_n)$ is convergent.

Exercise 11.52. Let (s_n) and (t_n) be two sequences. *Assume that $t_n \neq 0$ for any $n \in \mathbb{N}$.* Construct the new sequence, denoted (s_n/t_n), whose terms are simply the quotients of the terms of the original two sequences. Specifically, (s_n/t_n) is defined to be the sequence $(s_1/t_1, s_2/t_2, s_3/t_3, \ldots)$. (Note that our assumption that $t_n \neq 0$ for all $n \in \mathbb{N}$ was necessary for us to be able to consider the quotients s_n/t_n.) Show that if both the original sequences (s_n) and (t_n) are convergent, with limits L and M respectively, and *if we further assume that $M \neq 0$*, then (s_n/t_n) also converges, and its limit is L/M. (Again, note that we need the assumption that $M \neq 0$ to even consider the quotient L/M.) (Hint: Write $|s_n/t_n - L/M|$ as $\frac{|s_nM - t_nL|}{|t_nM|}$. Find N_1 such that $|t_n| > M/2$, so $\frac{|s_nM-t_nL|}{|t_nM|} < |s_nM - t_nL| \cdot (2/M^2)$ for $n > N_1$. Write $|s_nM - t_nL|$ as $|(s_n - L)M + L(M - t_n)|$. Proceed in the same devious manner as in Exercise 11.51.)

Give an example of a convergent sequence (s_n) and a divergent sequence (t_n) with $t_n \neq 0$ for all $n \in \mathbb{N}$, such that the quotient sequence (s_n/t_n) nevertheless converges.

Give an example of two *divergent* sequences (s_n) and (t_n) with the property that (s_n/t_n) is convergent.

Exercise 11.53. *(Squeeze Theorem: Sequences.)* Suppose that (a_n), (b_n), and (c_n) are sequences, and suppose that $a_n \leq b_n \leq c_n$ for all $n \geq N$, where N is some positive integer. Suppose that (a_n) and (c_n) are both convergent, and converge to the same limit l. Show that (b_n) also converges to l. (Hint: We have $b_n = a_n + (b_n - a_n)$, and $0 \leq b_n - a_n \leq c_n - a_n$ for all $n \geq N$. What do Exercises 11.49 and 11.50 above suggest? Note that you cannot apply Exercises 11.49 and 11.50 to sequences unless you know they are convergent!)

Exercise 11.54. We will explore in this exercise the convergence and divergence behavior of the sequences (s_n), with $s_n = r^n$, where r is fixed. Of course, if $r = 0$, the sequence just reads $(0, 0, 0 \ldots)$, and if $r = 1$, the sequence just reads $(1, 1, 1, \ldots)$. Both these sequences clearly converge. If $r = -1$, then the sequence reads $(-1, 1, -1, 1, \ldots)$, and the sequence clearly fails to approach any one real number, so it diverges. Hence, we are left with the cases where $r \neq -1, 0, 1$.

(1) Suppose that $r > 1$. Write $r = 1 + x$, so $x > 0$ and is also fixed. Show that $r^n > 1 + nx$, and use this to conclude that (s_n) diverges when $r > 1$. (Hint: The binomial theorem.)

(2) Extend the arguments in part (1) to show that (s_n) diverges if $|r| > 1$.

(3) Extend the arguments in parts (1) and (2) to show that if $0 < r < 1$ or $-1 < r < 0$, then the sequence (s_n) converges to 0. Give an $\epsilon - N_\epsilon$ proof. (Hint: $1/|r| > 1$.)

Exercise 11.55. Show that the sequence (s_n), given by $s_n = \sqrt[n]{r}$, where $r > 1$ is a fixed real number, converges to 1. (Hint, write $t_n = s_n - 1$, so t_n is positive and $(1 + t_n)^n = r$. Use one or more terms in the binomial expansion of $(1 + t_n)^n$ to argue that t_n is less than some constant divided by n. Finish by appealing to the squeeze theorem, see Exercise 11.53, taking the constant sequence whose terms are all 0 as the lower sequence in that theorem.)

Exercise 11.56. Show that the sequence (s_n), given by $s_n = \sqrt[n]{n}$, converges to 1. (Hint, as in Exercise 11.55 above, write $t_n = s_n - 1$, so t_n is positive and $(1+t_n)^n = n$. Use one or more terms in the binomial expansion of $(1+t_n)^n$ to argue that t_n is less than some constant divided by $\sqrt{n-1}$.)

Exercise 11.57. . Let $f : \mathbb{R} \to \mathbb{R}$ be defined by the rule $f(x) = 1$ if x is rational, and $f(x) = 0$ if x is irrational. Prove that f is not continuous at *any* $y \in \mathbb{R}$. (Hint: Invoke a property of the real number system that you know, namely, that between any two real numbers you can find a rational number, and also, you can find an irrational number. We will formally look at this property later on in Chapter 12, specifically in Exercises 12.63 and 12.64.)

Exercise 11.58. Prove the following analogs of Exercises 11.49 through 11.52 for limits of functions: Here, f and g are two functions from S to $\mathbb{R}$, where $S \subseteq \mathbb{R}$ is of the form (a,b) or $(a,b) \setminus \{c\}$, $c \in (a,b)$. In all cases, it is easy to simply use Definition 11.34 and invoke the results of the analogous results for sequences above in Exercises 11.49 through 11.52, but you can also work with Definition 11.36 and give $\epsilon - \delta$ proofs (in which case, your proofs will be analogous to the proofs in those same exercises).

We assume that $\lim_{x\to c} f(x)$ *and* $\lim_{x\to c} g(x)$ *both exist,* and that $\lim_{x\to c} f(x) = l$ and $\lim_{x\to c} g(x) = m$.

(1) Let $r \in \mathbb{R}$ be arbitrary. Show that $\lim_{x\to c}(rf)(x) = rl$.

(2) Show that $\lim_{x\to c}(f+g)(x) = l + m$.

(3) Show that $\lim_{x\to c}(fg)(x) = lm$.

(4) Assume that $g(x) \neq 0$ for $x \in S$, and that $m \neq 0$. Show that $\lim_{x\to c}(f/g)(x) = l/m$.

Exercise 11.59. Prove the following analogs of Exercises 11.49 through 11.52 for continuity of functions: Here, f and g are two functions from an interval of the form (a,b) to $\mathbb{R}$, and $c \in (a,b)$. You may do it by invoking the results of Exercise 11.58 above and invoking Proposition 11.40, or by using Definition 11.27 and invoking the results of the analogous results for sequences above in Exercises 11.49 through 11.52, but you can also work with Definition 11.28 and give $\epsilon - \delta$ proofs (in which case, your proofs will be analogous to the proofs in Exercises 11.49 through 11.52 above).

We assume that f and g are both continuous at c.

(1) Let $r \in \mathbb{R}$ be arbitrary. Show that the function rf is continuous at c.

(2) Show that the function $f + g$ is continuous at c.

(3) Show that the function fg is continuous at c.

(4) Assume that $g(x) \neq 0$ for $x \in (a,b)$. Show that the function f/g is continuous at c.

Exercise 11.60. *(Squeeze Theorem: Functions.)* Suppose f, g, and h are functions from S to $\mathbb{R}$, where $S \subseteq \mathbb{R}$ of the form (a,b) or $(a,b) \setminus \{c\}$, $c \in (a,b)$.

Suppose that $f(x) \leq g(x) \leq h(x)$ for $x \in S$, and suppose that $\lim_{x \to c} f(x) = \lim_{x \to c} h(x) = l$. Show that $\lim_{x \to c} g(x) = l$. (Hint: You can apply the result of Exercise 11.53 above, along with Definition 11.34. Alternatively, you can work with Definition 11.36. Notice that $g(x) - l = (g(x) - f(x)) + (f(x) - l)$ and that $g(x) - f(x) \leq h(x) - f(x)$. You may need the results in Exercise 11.58.)

Exercise 11.61. Use the squeeze theorem for functions (see Exercise 11.60 above) to establish the following limits:

(1) $\lim_{x \to 0} x \sin(\frac{1}{x}) = 0$. (Hint: Start with the known proprerty of the sine function that $-1 \leq \sin(t) \leq 1$. Use this to bound the given function above by $\pm|x|$. Note that 0 is not part of the domain of $x \sin(\frac{1}{x})$, yet a limit exists as x tends to zero!)

(2) $\lim_{x \to 0} x^2 \sin(\frac{1}{x}) = 0$.

(3) $\lim_{x \to 0} x^2 e^{\sin(\frac{1}{x})} = 0$.

Exercise 11.62. Modify the proof of Proposition 11.33 to provide a proof of Propositon 11.37.

Exercise 11.63. Let $S = \sum_{i=1}^{\infty} a_i$ and $T = \sum_{i=1}^{\infty} b_i$ be two series, and let r be a real number. Show that if S converges, then the series "rS" defined as $\sum_{i=1}^{\infty} r a_i$ also converges, and that if S and T both converge, then the series "$S + T$" defined as $\sum_{i=1}^{\infty}(a_i + b_i)$ also converges. (Hint: Exercises 11.49 and 11.50 above.)

Exercise 11.64. Let $S = \sum_{i=1}^{\infty} a_i$ be a series, and for an integer $t \geq 1$, write S_t for the expression $\sum_{i=t}^{\infty} a_i$. (So, S_t is also a series, whose i-th term is a_{t-1+i}.) Show that S converges if and only if S_t converges. (Hint: Write the sequence of partial sums of S as a sum of two sequences, one whose first $t-1$ terms are the corresponding partial sums of S and all terms after that are equal to $a_1 + a_2 + \cdots + a_{t-1}$, and another whose first $t-1$ terms are zero and all terms after that are suitable partial sums of S_t.)

Exercise 11.65. Show that the series

$$S = \sum_{i=1}^{\infty} \frac{1}{i(i+1)}$$

converges to 1. (Hint: A little trick. $\dfrac{1}{6} = \dfrac{1}{2} - \dfrac{1}{3}$. Can you spot a pattern here? Use this pattern to find a compact representation of the partial sums of this series.)

Exercise 11.66. The series

$$S = \sum_{i=1}^{\infty} \frac{1}{i}$$

is known as the *harmonic series*. It is tempting to think that it should converge, but that is not the case! Prove that S diverges. (Hint: By observing that $\frac{1}{3} + \frac{1}{4} > \frac{1}{4} + \frac{1}{4} = \frac{1}{2}$, show that $P_4 > 1 + \frac{1}{2} + \frac{1}{2}$. By playing a similar game with $\frac{1}{5} + \frac{1}{6} + \frac{1}{7} + \frac{1}{8}$, show that $P_8 > 1 + \frac{1}{2} + \frac{1}{2} + \frac{1}{2}$. Proceed.)

12

The Completeness of $\mathbb{R}$

In this chapter, we will approach mathematics in a way that is quite typical of many advanced areas of the subject: we will focus on some fundamental *structural* properties of mathematical objects, in fact, of the real number system. That is to say, we will study certain properties of the real numbers that at first may seem somewhat abstract and unrelated to the more concrete objects that we associate with the real numbers (such as functions from $\mathbb{R}$ to $\mathbb{R}$), but all the same, have a very significant bearing on these more concrete objects. It will turn out in fact that these properties explain why the real number system is so magnificent an arena in which to do much of mathematics.

To be specific, we will study a property of the real numbers that is crucial to why certain "reasonable" sequences of real numbers converge at all. This is the *Least Upper Bound property* property (abbreviated LUB property) of the real numbers. This has several consequences for the structure of $\mathbb{R}$, many of which are fundamental to the development of calculus. Here are some examples of such consequences, all of which we will study in this chapter: the monotone convergence theorem, which states that an increasing (or decreasing) bounded sequence of real numbers must converge, the Bolzano-Weierstrass theorem, which states that a bounded sequence of real numbers has a convergent subsequence, the Cauchy completeness of the real numbers, which states that every real Cauchy sequence (s_n) must converge, and so on. (Of course, at this stage we have yet to define subsequences, or Cauchy sequences, but we will look at these concepts shortly.) In turn, these properties of $\mathbb{R}$ enable other fundamental results of calculus, such as the Intermediate Value theorem, the Extreme Value theorem, and so on. Thus, in a fundamental sense, the LUB property of $\mathbb{R}$ makes calculus possible!

We will also refer to the LUB property as the *completeness property* of $\mathbb{R}$. Different authors use the term "completeness property" for other, logically equivalent, properties, but for us, completeness will refer to the LUB property.

12.1 Least Upper Bound Property (LUB)

As has been standard all along, we will assume familiarity with the usual arithmetic operations in the set of real numbers. We will also be working with the notion of ordering ("less than" or "$<$" and its companion "greater than" or "$>$") inherent in $\mathbb{R}$.

Remark 12.1. The notion of order in $\mathbb{R}$ can be conceptualized as a relation on $\mathbb{R}$ (see Definition 9.3, Chapter 9). (Indeed, the relation corresponds to the subset of $\mathbb{R} \times \mathbb{R}$ consisting of all (a, b) with $a < b$.) It is convenient to recall from our familiarity with $\mathbb{R}$ that the following properties of the order relation hold:

(1) Trichotomy: For $a, b \in \mathbb{R}$, *exactly one* of the following holds: $a > b$, $a = b$, or $b > a$.

(2) Transitivity: For $a, b, c \in \mathbb{R}$, if $a < b$ and $b < c$, then $a < c$.

(3) Additivity: For $a, b, c \in \mathbb{R}$, if $a < b$, then $a + c < b + c$.

(4) Multiplicativity: For $a, b, c \in \mathbb{R}$, if $a < b$ and $c > 0$, then $ac < bc$.

We first consider the notion of upper and lower bounds of a set $S \subseteq \mathbb{R}$:

Definition 12.2. Let S be a nonempty subset of $\mathbb{R}$. We say $z \in \mathbb{R}$ is an *upper bound* of S if $z \geq x$ for all $x \in S$. Similarly, we say $z \in \mathbb{R}$ is a *lower bound* of S if $z \leq x$ for all $x \in S$.

If S has at least one upper bound, we say S is *bounded above.* Similarly, if S has at least one lower bound, we say S is *bounded below.* If S is bounded above and bounded below, we say S is *bounded.*

Example 12.3. Let us consider some examples:

(1) Let S be the interval $(-1, 1)$. Then any $z \geq 1$ is an upper bound for S. (Note that $1 \geq x$ for every $x \in S$. But also $1.5 \geq x$, as also $100 \geq x$ for every $x \in S$, so 1, 1.5, and 100 are also upper bounds.) Similarly any $z \leq -1$ is a lower bound for S. Notice that there are infinitely many upper bounds and lower bounds for S. Also, note that if $z < 1$, then z is not an upper bound for S. This should be clear; here is the picture: any $z < 1$ must lie in one of the mutually disjoint intervals $(-\infty, -1], (-1, 0.9), [0.9, 0.99), [0.99, 0.999), [0.999, 0.9999), \ldots$. If, for instance, such a z were to lie in $[0.999, 0.9999)$, then already, $z < 0.9999$, and $0.9999 \in S$, so z cannot be an upper bound for S. Similarly, if $z > -1$, then z is not a lower bound for S.

Here is a different technique to show that any $z < 1$ cannot be an upper bound for S, one that will be useful in more general situations as well: Consider any $z < 1$. If already $z < 0$, then z cannot be an upper bound, since 0 is in S and $0 > z$. So assume $z \geq 0$. Write d for $1 - z$ (so $0 < d \leq 1$). Consider the real number $z + d/2$. We have $0 \leq z < z + d/2 < z + d = 1$. This shows that $z + d/2$ is in S, since $z + d/2$ is greater than 0 and less than 1. It now follows immediately from the relation $z < z + d/2$ that z cannot be an upper bound for S. (You should be able to use a similar technique to show that any $z > -1$ is not a lower bound for S.)

(2) Let $S = \mathbb{R}_{>0}$. Then S has no upper bound, since given any $z \in \mathbb{R}$, you can find some $x \in S$ that is "to the right of z" on the number line, that is, such that $z < x$. (For instance, take x to be $z + 1$!) On the other hand, any $z \leq 0$ is a lower bound for S. Thus, S has no upper bounds but infinitely many lower bounds.

Note that no $z > 0$ can be a lower bound for S. (Given $z > 0$, the real number $z/2$ is also positive and is hence in S, and of course, $z > z/2$, so z cannot be a lower bound.)

(3) Let $S = \mathbb{Z}$. Then S has no upper bound, and no lower bound.

(4) Let S be the set $\{3.1, 3.14, 3.141, 3.1415, 3.14159, \dots\}$, where the n-th element in this listing of S is the rational number obtained by truncating the decimal expansion of π to the n-th decimal place. Then any $z \geq \pi$ is an upper bound for S, and any $z \leq 3.1$ is a lower bound. Again, note that if $z < \pi$, then z is not an upper bound for S. (Why?)

Here is a quick exercise:

Exercise 12.4. If a nonempty subset $S \subseteq \mathbb{R}$ is bounded above, show that the set of upper bounds of S is not only infinite but uncountable. State and prove a similar statement for the set of lower bounds.

Now we consider the notions of the least upper bound and the greatest lower bound:

Definition 12.5. Let S be a nonempty subset of $\mathbb{R}$. We say $u \in \mathbb{R}$ is the *least upper bound* of S, written lub(S), if u is an upper bound for S, and if z is any other upper bound for S, then $u < z$. Similarly, we say $l \in \mathbb{R}$ is the *greatest lower bound* of S, written glb(S), if l is a lower bound for s, and if z is any other lower bound for S, then $l > z$.

Remark 12.6. The least upper bound and the greatest lower bound of a set S, if they exist at all, are uniquely determined by S, a fact we will prove in Proposition 12.10. We thus refer to lub(S), if it exists, as *the* least upper bound of S, and similarly for glb(S).

Remark 12.7. Note that the definition does not require that the least upper bound of S actually be a member of S: it could well be a real number that is not in S. Likewise, the definition does not require that the greatest lower bound be a member of S. We will see instances where lub(S) and glb(S) are not in S in Example 12.8 below.

Example 12.8. Let us consider the least upper bound and the greatest lower bound (again, as already remarked above, we will prove ahead that these are unique if they exist) of the sets in Example 12.3 :

(1) For $S = (-1, 1)$, lub(S) $= 1$, and glb(S) $= -1$. We have seen in Example 12.3(1) that 1 is an upper bound, and any $z < 1$ cannot be an upper bound. It follows that any other upper bound must be greater than 1, that is, 1 is indeed the least upper bound of S. Note that neither lub(S) nor glb(S) are members of S.

(2) For $S = \mathbb{R}_{>0}$, 0 is the greatest lower bound of S: We have seen in Example 12.3(2) that $z = 0$ is a lower bound but no $z > 0$ can be a lower bound. S has no least upper bound, since it does not have any upper bound at all. Note that glb(S) is not a member of S.

(3) For $S = \mathbb{Z}$, there is no least upper bound, and no greatest lower bound, as there are no upper bounds at all, and no lower bounds at all.

(4) For $S = \{3.1, 3.14, 3.141, 3.1415, 3.14159, \dots\}$, lub($S$) $= \pi$ and glb(S) $= 3.1$. Note that lub(S) is not a member of S, but glb(S) is indeed a member of S.

It would be a good idea for you to find some least upper bounds and greatest lower bounds yourself, or establish that they do not exist:

Exercise 12.9. Find, with proof, the lub and glb of the following sets (if the lub or glb does not exist, explain why not):

(1) $S = \{x \in \mathbb{R} \mid x^2 < 10\}$.

(2) $S = \{x \in \mathbb{R} \mid x^2 > 10\}$.

(3) $\{x \in \mathbb{R} \mid x^3 > 10\}$.

(4) $\{x \in \mathbb{R} \mid x^2 + 1 < x + 2\}$. (Hint: Bring all terms to one side.)

(5) $\{x \in \mathbb{R} \mid x = \ln(y)$ for some $y \geq 0\}$.

We now show that the greatest lower bounds and least upper bounds of a set $S \subseteq \mathbb{R}$, if they exist, must be unique:

Proposition 12.10. *The least upper bound of a nonempty subset of $\mathbb{R}$, if it exists, is unique. Similarly, the greatest lower bound of a nonempty subset of $\mathbb{R}$, if it exists, is unique.*

Proof. Let S be a nonempty subset of $\mathbb{R}$. Suppose u_1 and u_2 are two least upper bounds of S, and suppose they are distinct. By definition, each of u_1 and u_2 must themselves be upper bounds of S. Focusing on u_1 as a least upper bound, we must have $u_1 < u_2$ since u_2 is another upper bound. Focusing now on u_2 as a least upper bound, we must have $u_2 < u_1$ as u_1 is another upper bound. Thus, we find $u_1 < u_2$ and $u_2 < u_1$, a contradiction. Hence our assumption must be flawed, u_1 must equal u_2 and the least upper bound must be unique. The proof that glb(S) is unique is similar. □

We can turn the idea behind why $z = 1$ is the least upper bound for S in Example 12.3(1) as a formal result that holds more generally:

Proposition 12.11. *Let $S \subseteq \mathbb{R}$ be nonempty, and suppose $u =$ lub(S). Then for any $\delta > 0$, there exists $s \in S$ with $s > u - \delta$. Similarly, if $l =$ glb(S), then for any $\delta > 0$, there exists $s \in S$ with $s < l + \delta$.*

Proof. Given u and δ as in the statement of the theorem, if there were no $s \in S$ with $s > u - \delta$, then $u - \delta$ would already be an upper bound for S, contradicting the fact that u is the *least* upper bound. The proof for the greatest lower bound is similar. □

We now state the basic property of the real numbers that we alluded to at the beginning of the chapter:

Definition 12.12. *Least Upper Bound Property* or *Completeness Property:* Any nonempty subset of $\mathbb{R}$ that is bounded above has a least upper bound in $\mathbb{R}$.

Remark 12.13. Recall that by Proposition 12.10, the least upper bound is unique.

In more formal treatments, the LUB property is taken as one of the fundamental axioms that define the real number system, along with the property that $\mathbb{R}$ is a field, and, further, an ordered field. In fact, one can show that $\mathbb{R}$ is the unique ordered field satisfying the LUB property. Of course, we have not studied fields and ordered fields yet, so we will not go deeper in this direction, but what is salient is that the LUB property is not something that can be deduced from a smaller set of axioms, but is something quite intrinsic to the real numbers.

In the Notes section at the end of the chapter, we list the complete set of axioms that define the real number system.

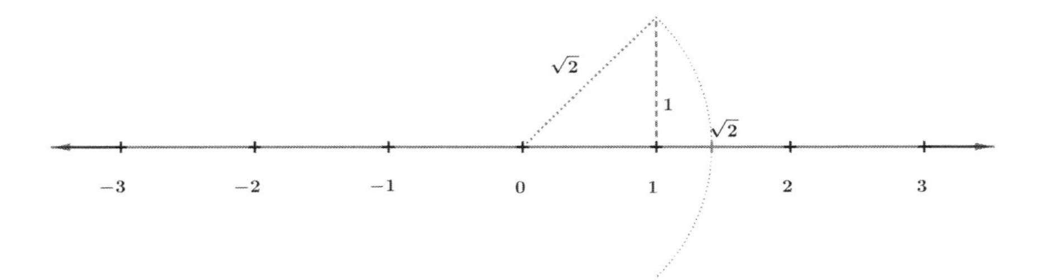

Figure 12.1. The number line, corresponding to positive, negative and zero lengths. A length such as $\sqrt{2}$ can be marked off using the procedure shown: as the hypotenuse of a right triangle with sides 1. But a length such as $\sqrt[5]{11}$? Or for $\sqrt[17]{213}$?

A word here is necessary about the formal development of the real number system. While we intuitively think of real numbers as corresponding to (positive, negative, or zero) lengths along a "number-line" (see Figure 12.1), this intuitive picture does not get us very far in proving in a water-tight fashion many results about the real numbers that are needed for higher mathematics. For instance, let us start with something that is quite elementary: how would we show that there is a real number whose square is 2? If we adopt the intuitive viewpoint that real numbers correspond to (positive, negative, or zero) lengths of line segments, then we could use the Pythagoras theorem: the hypotenuse of a right triangle with sides 1 must have length $\sqrt{2}$, and laying this hypotenuse along the number line, we would have produced the desired line segment with length $\sqrt{2}$. See Figure 12.1. But then, let us carry this forward: what geometric argument would we give for $\sqrt[5]{11}$? Or for $\sqrt[17]{213}$? Here, the more intuitive approach fails us! We need the completeness (LUB) property to deduce that there is indeed a real number whose fifth power is 11 and that there is indeed a real number whose 17-th power is 213. (See Theorem 12.57 ahead.)

There are many other, deeper, properties of $\mathbb{R}$ that are hard to divine from just the geometric picture of $\mathbb{R}$ as lengths. It is necessary therefore to consider $\mathbb{R}$ more formally, and delineate a fundamental set of properties as axioms, and to deduce other properties as logical consequences of these axioms. The LUB property is one of these fundamental properties, along with (as already mentioned above) that of being a field, and, further, an ordered field.

Remark 12.14. Of course, this does not mean we abandon the picture of $\mathbb{R}$ as a set corresponding to (positive, negative, and zero) lengths! Quite the contrary: we must retain our intuitive approach to things as we move towards advanced mathematics. However, this must be supplemented by the awareness that sometimes a purely intuitive approach does not get us all the way, and a more formal approach that dots every i and crosses every t is needed!

Remark 12.15. The completeness (LUB) property is a key structural difference between the rational numbers and the real numbers. Example 12.3(4) gives an instance of a set of rational numbers that is bounded above, but whose least upper bound is actually an irrational number (you will have to accept for now that π is irrational–you will see a proof of this in more advanced courses!). Since the least upper bound of any subset of $\mathbb{R}$ is unique, the fact that the set S of Example 12.3(4) has the least upper bound of π means that the set S does not have a least upper bound *within the rational numbers.*

We will now devote the rest of the chapter to a few deep properties of the real number system that follow from the LUB property.

12.2 Greatest Lower Bound Property

Not surprisingly, there is a property of the real numbers that is analogous to the LUB property, and applies to lower bounds, and that is the *Greatest Lower Bound* property (GLB property). Once the LUB property is postulated as a fundamental axiom of the real number system, the GLB property can be deduced as a logical consequence of the LUB property, and we will do so here:

Theorem 12.16. (*Greatest Lower Bound Property.*) *Every nonempty subset of* $\mathbb{R}$ *that is bounded below has a greatest lower bound.*

Remark 12.17. Recall that by Proposition 12.10, the greatest lower bound is unique.

Proof. Let S be a nonempty subset of $\mathbb{R}$ that is bounded below. Let L be the set of all lower bounds of S, so L is nonempty by the hypothesis that S is bounded below. If $y \in L$ and $s \in S$ are arbitrary, then $y \leq s$ translates to $s \geq y$. Since s and y were arbitrary, we find that every $s \in S$ acts as an upper bound for L! Therefore L is a nonempty subset of $\mathbb{R}$ that is bounded above. Applying the LUB property to L, we find that L has a least upper bound in $\mathbb{R}$, call it l. We claim that $l = \text{glb}(S)$.

First, note that l is indeed a lower bound for S. For, assume to the contrary that there exists an $s \in S$ such that $s < l$. Write d for $l - s$, so $d > 0$. By Proposition 12.11 applied to the set L, there exists a $b \in L$ such that $b > l - d/2$. By virtue of being in L, such a b satisfies $b \leq x$ for all $x \in S$. In particular, this means $b \leq s$. But we also have

$b > l - d/2 > l - d = s$, a contradiction. Hence no such s exists, and l is a lower bound for S.

Next, by the very definition of l as lub(L), $l \geq b$ for all $b \in L$. Since L contains all the lower bounds of S, we find that $l > b$ for any other lower bound b. It follows that $l = \text{glb}(S)$. □

The following exercise will help cement the ideas behind the preceding proof:

Exercise 12.18. Assume that the GLB property had been taken as a fundamental axiom of the real number system instead of the LUB property. Show that the LUB property follows as a logical consequence.

12.3 Archimedean Property

Let x and y be two real numbers, with x positive. Laying off the lengths x, $2x$, $3x$, etc. on the number line, it is geometrically evident that for some n, n a positive integer, the length nx will exceed y. See Figure 12.2. (Of course, if already $x > y$, then we may take $n = 1$.) This intuitively clear result is known as the Archimedean property of $\mathbb{R}$. As we discussed above, for deeper forays into mathematics, the pure geometric picture of real numbers as points on the real line is not sufficient, and it is preferable to set up a body of axioms representing fundamental properties of the real number system, and to derive other results from these axioms. Here, we will prove the Archimedean property starting from the LUB property (one of the fundamental axioms).

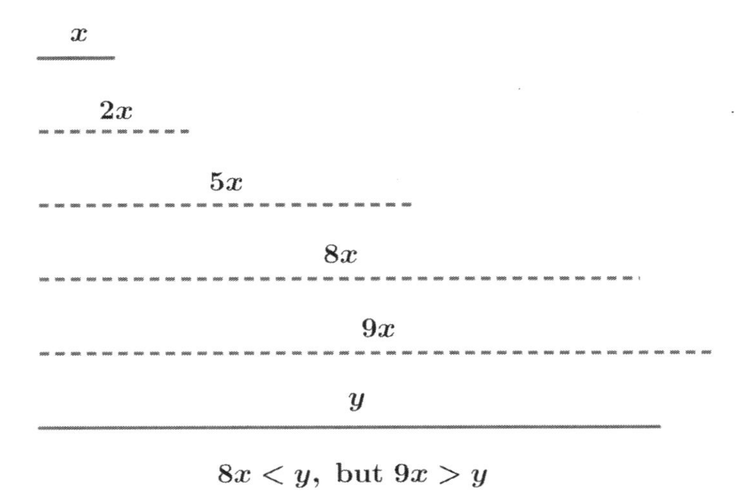

Figure 12.2. Archimedean property: Given $x > 0$ and y, we can find n such that $nx > y$.

The Archimedean property is not so trivial as it might sound. It turns out that there are other number systems that are of importance in mathematics (in particular, ordered fields, but of course, we have not studied what ordered fields are), in which this property does not hold! So, the fact that it holds for $\mathbb{R}$ is quite special.

Theorem 12.19. (*Archimedean Property.*) *Let x and y be real numbers, with x positive. Then there exists a positive integer n such that $nx > y$. More precisely, there exists an integer n such that $(n-1)x \leq y < nx$.*

Proof. We first assume that $x \leq y$. Write S for the set of real numbers of the form mx, such that m is a positive integer and $mx \leq y$. Since x is itself of the form mx with $m = 1$, and since $x \leq y$ by assumption, S contains x and hence is nonempty. S is bounded above by y, by definition. By the LUB property, S has an upper bound, call it l. (Note that we cannot assume that $l \in S$, so we cannot assume that l is already of the form mx for some integer x.) By Proposition 12.11, there exists an element of S that is greater than $l - x$. By the definition of S again, this element must be of the form mx for some suitable positive integer m (and we must therefore have $mx \leq y$). From $mx > l - x$, we find $mx + x > l$, that is, $(m+1)x > l$. Since $l = \text{lub}(S)$, we find $(m+1)x > l \geq s$ for all $s \in S$. In particular, this means that $(m+1)x \notin S$. By the definition of S once more, we must have $(m+1)x > y$. We have thus found $mx \leq y < (m+1)x$. We may hence take $n = m + 1$, establishing both the Archimedean property for this case and the second assertion in the statement of the theorem.

Now assume $x > y$. If we take $n = 1$, we automatically have $nx > y$, which is the Archimedean property that we needed to establish for this (x, y) pair. However, to prove the second assertion of the statement of the theorem for the case $x > y$, we proceed as follows: If $x > y$ and $y \geq 0$, then with $n = 1$, the statement $(n-1)x \leq y < nx$ is certainly true. We are left with the case $x > y$ and $y < 0$, so assume we are in this situation. Then $-y > 0$, and by what we have already proved, there exists an integer m such that $(m-1)x \leq -y < mx$. Multiplying through by -1, we find $(-m)x < y \leq (1-m)x$. If $y < (1-m)x$, then we may take $n = 1 - m$, else if $y = (1-m)x$, we take $n = 1 - m + 1 = 2 - m$ to find that $(n-1)x \leq y < nx$. □

An easy corollary falls out of this, another "obvious" fact if we assume prior knowledge of the real number system, but nonetheless worth establishing as a consequence of our chosen set of axioms:

Corollary 12.20. *Let y be a real number. Then there exists an integer n such that $n-1 \leq y < n$.*

Proof. Simply take $x = 1$ in the statement of Theorem 12.19 above. □

12.4 Monotone Convergence Theorem

We prove the monotone convergence theorem here. This theorem is useful in proving that various sequences converge, and is also useful in proving other theorems about the structure of $\mathbb{R}$. First, some definitions:

Definition 12.21. Let (s_n), $n = 1, 2, \ldots$, $s_n \in \mathbb{R}$, be a sequence. We say (s_n) is *monotonically increasing* if $s_1 \leq s_2 \leq s_3 \ldots$ (so $s_n \leq s_{n+1}$ for all $n = 1, 2, \ldots$), and *monotonically decreasing* if $s_1 \geq s_2 \geq s_3 \ldots$ (so $s_n \geq s_{n+1}$ for all $n = 1, 2, \ldots$). We call a sequence *monotone* if it is either monotonically increasing or monotonically decreasing.

We also say (s_n) is *strictly* monotonically increasing (*strictly* monotonically decreasing) if $s_1 < s_2 < s_2 < \ldots$ (respectively $s_1 > s_2 > s_3 > \ldots$). We say (s_n) is *strictly* monotone if it is either strictly monotonically increasing or strictly monotonically decreasing.

We have the following:

Theorem 12.22. *Monotone Convergence Theorem: Every bounded monotone sequence of real numbers converges.*

Proof. Let (s_n), $n = 1, 2, \dots$, $s_n \in \mathbb{R}$, be a bounded monotone sequence. We will prove the theorem in the case where (s_n) is monotonically increasing; the proof in the other case is very similar. So let (s_n) be a monotonically increasing sequence, bounded above. Consider the *set* $S = \{s_1, s_2, \dots\}$. Since S is bounded above, the LUB property says that $u = \text{lub}(S)$ exists. We show that the sequence (s_n) converges to u. Let $\epsilon > 0$ be given. Then, by Proposition 12.11, there exists some element of S, call it s_{N_ϵ} for some N_ϵ, such that $s_{N_\epsilon} > u - \epsilon$. Since (s_n) is monotonically increasing, $s_n \geq s_{N_\epsilon}$ for all $n \geq N_\epsilon$, and of course, $u \geq s_n$ for all $n \geq N_\epsilon$ since u is an upper bound for S. It follows that $u \geq s_n \geq s_{N_\epsilon} > u - \epsilon$ for all $n \geq N_\epsilon$, so $|u - s_n| < \epsilon$ for all $n \geq N_\epsilon$. Hence (s_n) converges to u. □

Remark 12.23. It is worth remarking that the proof of Theorem 12.22 shows that the limit of a monotonically increasing sequence (s_n) is the *least upper bound* of the set formed by the terms of the sequence, that is, the set $S = \{s_1, s_2, \dots\}$. In a like manner, the limit of a monotonically decreasing sequence is the *greatest lower bound* of the set formed by the terms of the sequence.

Example 12.24. Consider the sequence (s_n), $n = 1, 2, \dots$, where $s_n = \frac{n}{n+1}$. Since s_n can be written as $1 - \frac{1}{n+1}$, the sequence is bounded above by 1. Moreover, $\frac{n}{n+1} < \frac{n+1}{n+2}$, as can be checked by cross multiplication. It follows from the monotone convergence theorem that this sequence converges. Moreover, the expression $s_n = 1 - \frac{1}{n+1}$ shows that the least upper bound of the s_n is exactly 1 (prove this). It follows that the sequence converges to 1.

Example 12.25. The number e is formally defined as the limit of the sequence (s_n), where $s_n = \left(1 + \frac{1}{n}\right)^n$. We can use the monotone convergence theorem to see why this sequence must converge at all.

Using the binomial theorem, we write:

$$\begin{aligned} s_n &= 1 + n\frac{1}{n} + \frac{n(n-1)}{2!}\frac{1}{n^2} + \cdots + \frac{n(n-1)\cdots(n-i+1)}{i!}\frac{1}{n^i} \\ &+ \cdots + \frac{n(n-1)\cdots(n-(n-2))\cdot(n-(n-1))}{n!}\frac{1}{n^n}, \end{aligned} \tag{12.1}$$

$$\begin{aligned} s_{n+1} &= 1 + (n+1)\frac{1}{n+1} + \frac{(n+1)n}{2!}\frac{1}{(n+1)^2} \\ &+ \cdots + \frac{(n+1)n\cdots(n+1-i+1)}{i!}\frac{1}{(n+1)^i} + \cdots \\ &+ \cdots + \frac{(n+1)n\cdots(n+1-(n+1-2))\cdot(n+1-(n+1-1))}{(n+1)!} \\ &\cdot \frac{1}{(n+1)^{n+1}}. \end{aligned} \tag{12.2}$$

We have

$$\frac{n(n-1)\cdots(n-i+1)}{i!}\frac{1}{n^i} = n^i\frac{(1-\frac{1}{n})\cdots(1-\frac{i+1}{n})}{i!}\frac{1}{n^i}.$$

Canceling the n^i, we find

$$\begin{aligned} s_n &= 1+1+\frac{1}{2!}\left(1-\frac{1}{n}\right)+\cdots+\frac{1}{i!}\left(1-\frac{1}{n}\right)\cdots\left(1-\frac{i-1}{n}\right)\\ &\quad+\cdots+\frac{1}{n!}\left(1-\frac{1}{n}\right)\cdots\left(1-\frac{n-2}{n}\right)\cdot\left(1-\frac{n-1}{n}\right), \end{aligned} \tag{12.3}$$

$$\begin{aligned} s_{n+1} &= 1+1+\frac{1}{2!}\left(1-\frac{1}{n+1}\right)+\cdots+\frac{1}{i!}\left(1-\frac{1}{n+1}\right)\cdots\left(1-\frac{i-1}{n+1}\right)\\ &\quad+\cdots+\frac{1}{n!}\left(1-\frac{1}{n+1}\right)\cdots\left(1-\frac{n-2}{n+1}\right)\cdot\left(1-\frac{n-1}{n+1}\right)\\ &\quad+\frac{1}{(n+1)!}\left(1-\frac{1}{n+1}\right)\cdots\left(1-\frac{n-1}{n+1}\right)\cdot\left(1-\frac{n}{n+1}\right). \end{aligned} \tag{12.4}$$

Notice that all summands in both s_n and s_{n+1} are positive. The term s_n has $(n+1)$ summands while s_{n+1} has $(n+2)$ summands: one extra summand compared to s_n. Moreover, from the third summand onward, each summand of s_n is less than the corresponding summand of s_{n+1}. This follows from the fact that $1-\frac{j}{n} < 1-\frac{j}{n+1}$ for $1 \le j \le n-1$. It follows that $s_n < s_{n+1}$, so (s_n) is a strictly increasing sequence. The first two summands show that $2 < s_n$. To find an upper bound for the terms, we note that $1-\frac{j}{n} < 1$ for $1 \le j \le n-1$, while $i! = \underbrace{i(i-1)\cdots 3\cdot 2}_{i-1 \text{ terms}} > \underbrace{2\cdot 2\cdots 2}_{i-1 \text{ terms}} = 2^{i-1}$. It follows that

$$s_n < 1+1+\frac{1}{2}+\frac{1}{2^2}+\cdots+\frac{1}{2^{n-1}} < 1+\frac{1-\frac{1}{2^n}}{1-\frac{1}{2}} < 1+\frac{1}{1-\frac{1}{2}} = 1+2 = 3.$$

(We have used the formula for the sum of a finite geometric series from Example 7.3 in Chapter 7, along with the fact that $1-\frac{1}{2^n} < 1$.)

Thus, $2 \le s_n \le 3$, so (s_n) is bounded (see Remark 11.22 in Chapter 11), and it is (strictly) increasing. By the monotone convergence theorem the sequence converges, and the number it converges to, namely e, lies between 2 and 3.

12.5 Bolzano-Weierstrass Theorem

First, a definition:

Definition 12.26. Given a sequence $(s_n) = (s_1, s_2, s_3, \dots)$ of real numbers, a *subsequence* of (s_n) is the sequence $(s_{n_1}, s_{n_2}, s_{n_3}, \dots)$ formed by the n_ith terms of (s_n), $i = 1, 2, 3, \dots$, for some choice of subscripts n_i, $1 \le n_1 < n_2 < n_3 < \cdots$.

Example 12.27. Let us look at two examples:

(1) Consider the sequence $(s_n) = (2, 4, 6, 8, 10, \dots)$, given by $s_n = 2n$. Then, the sequence $(10, 20, 30, 40, \dots)$ is a subsequence. This sequence is formed by taking the fifth, the tenth, the fifteenth, the twentieth, ..., terms of the original sequence (s_n). Thus, in this example, $n_1 = 5$, $n_2 = 10$, $n_3 = 15$, $n_4 = 20$, and so on.

(2) Consider the sequence $(s_n) = (1, -1/2, 1/4, -1/8, \dots)$, so the terms are given by $s_n = (-1)^{n-1}2^{-(n-1)}$. The sequence formed by taking just the positive terms, namely the sequence $(1, 1/4, 1/16, 1/64, \dots)$, is a subsequence. Here, $n_1 = 1$, $n_2 = 3$, $n_3 = 5$, $n_4 = 7$, and so on.

Remark 12.28. The indices n_k of the terms of a subsequence (in Definition 12.26 above) necessarily satisfy $n_k \geq k$. It is a good exercise for you to prove this formally using induction.

The Bolzano-Weierstrass theorem states that a bounded sequence of real numbers has a convergent subsequence. This is actually a deep theorem about the structure of $\mathbb{R}$, specifically of closed intervals of $\mathbb{R}$: the statement of the theorem leads to a notion called *compactness*. Although we will not study compactness here, it is a central idea in advanced mathematics. The theorem is to be understood as saying that if a sequence is bounded, then, although as in Exercise 11.48, Chapter 11, the sequence need not converge at all, the sequence nevertheless cannot be *too* wild! There has to be some *subsequence*, at least, that must come arbitrarily close to some real number within the bound and stay close, or in other words, must converge to that real number. Here is an example:

Example 12.29. Consider the sequence (s_n) whose terms are defined by $s_n = \cos n\pi$ if n is not a power of 2, and $s_n = 1/(k+1)$ if $n = 2^k$. Then (s_n) is bounded, since $|s_n| \leq 1$, but it does not converge. When n is not a power or 2, s_n equals $\cos n\pi$ which oscillates with values 1 or -1, but when n is a power of 2, the terms show a different behavior. Considering the subsequence formed by the first, second, fourth, eighth, sixteenth, etc. terms, we get the sequence $(1, 1/2, 1/3, 1/4, \dots)$, which converges to zero! (Here, $n_k = 2^k$, $k = 0, 1, 2, \dots$.)

We will first prove the following:

Proposition 12.30. *Every sequence of real numbers has a monotone subsequence.*

Proof. Let (s_n) be a sequence. If (s_n) already has a monotonically increasing subsequence $s_{n_1} \leq s_{n_2} \leq s_{n_3} \leq \dots$, there is nothing to prove, and we are done. So assume that (s_n) has no monotonically increasing subsequences. We claim that (s_n) must then have a monotonically decreasing subsequence.

To create such a subsequence, we first state and prove a lemma:

Lemma 12.31. *Let (s_n) be a sequence. If it contains no monotonically increasing subsequences, then, for any $m \geq 1$, the set $S_m = \{s_n \mid n \geq m\}$ has a maximum element n_m.*

Proof. Assume the contrary. Then $s_m \in S_m$, so it cannot be the maximum, so there exists some $s_{n_2} \in S_m$, $n_2 > m$, such that $s_{n_2} > s_m$. Now s_{n_2} cannot be the maximum of S_m either, so there exists some $s_{n_3} \in S_m$, $n_3 > n_2$, such that $s_{n_3} > s_{n_2} > s_m$. Proceeding thus we create a monotonically increasing subsequence of (s_n), violating the hypothesis. □

Proof of Proposition 12.30 *continued:* We will construct our monotonically decreasing subsequence as follows (See Figure 12.3): Let $S_1 = \{s_n \mid n \geq 1\}$ as in Lemma 12.31.

Then by that lemma and our assumption that (s_n) has no monotonically increasing subsequences, S_1 has a maximum, say at the n_1-th term s_{n_1}. Now take $S_2 = \{s_n \mid n \geq n_1+1\}$. Then S_2 has a maximum by the lemma, say at the n_2-th term s_{n_2}. We must have $n_2 > n_1$ as S_2 only considers terms with index $n_1 + 1$ onward. What is key now is that $s_{n_1} \geq s_{n_2}$. This follows because s_{n_2} is also in S_1 as $n_2 \geq n_1 + 1 \geq 1$, so if to the contrary $s_{n_1} < s_{n_2}$, then s_{n_1} could not have been the maximum of S_1. Now let $S_3 = \{s_n \mid n \geq n_2 + 1\}$. Then S_3 has a maximum s_{n_3}, $n_3 > n_2 > n_1$, and $s_{n_1} \geq s_{n_2} \geq s_{n_3}$. Proceeding, we find our monotonically decreasing subsequence.

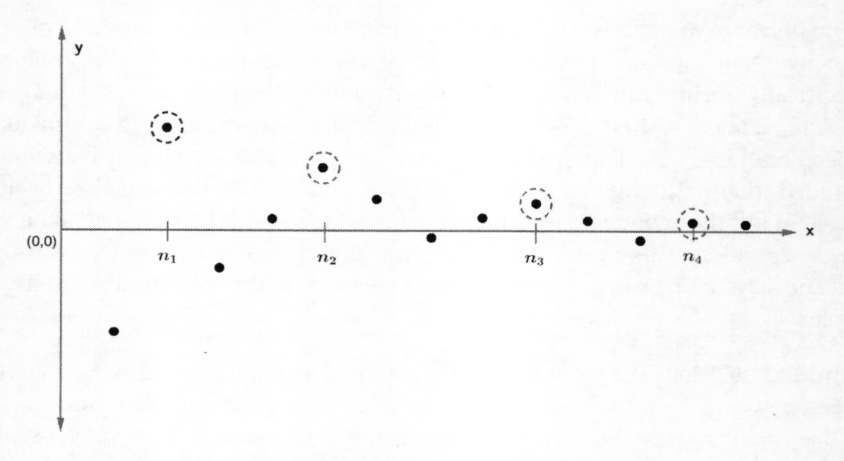

Figure 12.3. Proof of Proposition 12.30. The terms $s_{n_1}, s_{n_2}, s_{n_3}, s_{n_4}$ are circled. They form a decreasing subsequence.

□

Theorem 12.32. (*Bolzano-Weierstrass Theorem.*) *Every bounded infinite sequence of real numbers has a convergent subsequence.*

Proof. With the proof of Proposition 12.30 under our belt, the proof of this theorem is easy. Let (s_n), $n = 1, 2, \ldots$, be a bounded infinite sequence of real numbers. Then (s_n) has a monotone subsequence by the proposition. This subsequence is necessarily bounded, as (s_n) itself is bounded. Then, by the Monotone Convergence Theorem (Theorem 12.22), the subsequence must converge! □

12.6 Nested Intervals Theorem

We prove in this section the Nested Intervals Theorem. It guarantees that a particular intersection of sets is nonempty. This is of use in many situations, including in an alternative (and very imaginative!) proof of the Bolzano-Weierstrass theorem, which we also describe.

In fact, we have already encountered this theorem in our discussion on the Cantor set in Chapter 8, where we used the theorem to prove that the Cantor set is uncountable (Theorem 8.23 of that chapter).

We will consider *closed* intervals here. (Recall that a closed interval is a set of the form $\{x \in \mathbb{R} \mid a \leq x \leq b\}$, where $a < b$ are fixed real numbers. Such a set is denoted

$[a, b]$. See Example 5.40, Chapter 5). We say two closed intervals I and J are nested if $I \supseteq J$ (or if $J \supseteq I$). It is clear that if $I = [a, b]$ and $J = [c, d]$ are two closed intervals, the statement $I \supseteq J$ is equivalent to $a \leq c$ and $b \geq d$ (see Figure 8.9 in Chapter 8).

If $I = [a, b]$, we will define the *width of* I to be $b - a$.

Let $I_1 = [a_1, b_1]$, $I_2 = [a_2, b_2]$, ..., $I_k = [a_k, b_k]$, ..., $k = 1, 2, \ldots$, be closed intervals. Assume they are nested: $I_1 \supseteq I_2 \supseteq I_3 \supseteq \cdots$. We will collectively refer to the I_k as a *sequence of closed nested intervals.* We will refer to $E = \cap_k I_k$ as the *intersection of the closed nested intervals.* If w_k is the width of the k-th interval in the sequence, we will refer to the sequence of real numbers (w_k), $k = 1, 2, \ldots$, as the *width sequence* of the sequence of closed nested intervals.

We note that the definition of being nested can be extended to intervals that are open or half-closed or half-open, but that will not concern us here.

Theorem 12.33. (*Nested Intervals Theorem.*) *The intersection of a sequence of closed nested intervals is nonempty. Moreover, if the corresponding width sequence converges to* 0, *then the intersection of the closed nested intervals consists of precisely one point.*

Proof. Let $I_1 = [a_1, b_1]$, $I_2 = [a_2, b_2]$, ..., $I_k = [a_k, b_k]$, ..., $k = 1, 2, \ldots$, be the closed intervals. Note that the assumption that the intervals are nested is equivalent to $a_1 \leq a_2 \leq a_3 \leq \ldots$ and $b_1 \geq b_2 \geq b_3 \geq \ldots$. Since each $a_i < b_i \leq b_1$, we find that the sequence $(a_1, a_2, a_3, \ldots)$ is a monotonically increasing sequence bounded above by b_1, and therefore, by the Monotone Convergence theorem, converges to a real number a that is actually the least upper bound of the set $A = \{a_1, a_2, a_3, \ldots\}$. (See Theorem 12.22 and Remark 12.23). Similarly, the sequence $(b_1, b_2, b_3, \ldots)$ is a monotonically decreasing sequence bounded below by a_1, so it converges to a real number b, which is the greatest lower bound of the set $B = \{b_1, b_2, b_3, \ldots\}$. Note that $a \leq b$. For, suppose to the contrary that $a > b$. Write d for $a - b$, so $d > 0$. Since a is the least upper bound of the set A, there exists some a_k such that $a - d/4 < a_k \leq a$ (Proposition 12.11). Since the sequence (a_n) is monotonically increasing, we find $a - d/4 < a_k \leq a_n \leq a$ for all $n \geq k$. Similarly, there exists some b_l such that $b + d/4 > b_l \geq b$, and then, for all $n \geq l$, we have $b + d/4 > b_l \geq b_n \geq b$ as the sequence (b_n) is monotonically decreasing. Consider $n = \max(k, l)$. Then we find $b_n < b + d/4 < a - d/4 < a_n$, which is absurd, as $a_n < b_n$ for all n! Hence, $a \leq b$.

It is now easy to see that the set $E = \cap_k I_k$ is precisely the set $S = \{x \mid a \leq x \leq b\}$. (If $a < b$, then this is the closed interval $[a, b]$, else if $a = b$, this is the single element set $\{a\}$.) For, given $x \in S$, then for any k, we have $a_k \leq a \leq x \leq b \leq b_k$ (once again, by virtue of the fact that $a = \text{lub}(A)$ and $b = \text{lub}(B)$). In particular, $a_k \leq x \leq b_k$ for all k, so $x \in I_k$ for all k. Thus, $S \subseteq E$. On the other hand, given $x \in E$, we have $x \in I_k$ for all k by definition of the intersection, so we must have $a_k \leq x \leq b_k$ for all k. Read differently, x is an upper bound for the set A, and a lower bound for the set B. Since a and b are respectively glb(A) and glb(B), we find $a \leq x$ and $x \leq b$, that is, $a \leq x \leq b$. Thus, $x \in S$, so $E \subseteq S$.

Since E is nonempty (it contains a for instance, and if $a \neq b$, then it contains the whole closed interval $[a, b]$), we have proved the first assertion of the theorem. As for the second assertion, since $a_k \leq a \leq b \leq b_k$, we find $b - a \leq w_k = b_k - a_k$ for all k. Since the sequence (w_k) converges to zero by hypothesis, given any $\epsilon > 0$ there exists N_ϵ such that $w_n < \epsilon$ for all $n \geq N_\epsilon$. Since $b - a \leq w_{N_\epsilon}$ in particular, we find $b - a < \epsilon$. Now ϵ was arbitrary. This forces $b - a$ to equal zero (else, if $b - a > 0$, we can choose

$\epsilon = (b-a)/2$ and find $b - a < (b-a)/2$, absurd!). So, when the width sequence converges to 0, we find $E = S = \{x \mid a \le x \le b\} = \{a\}$, a set with one element. □

Remark 12.34. By contrast, if I_k, $k = 1, 2, \ldots$, are nested *open* intervals, then $E = \cap_k I_k$ is no longer guaranteed to be nonempty. For instance, consider the nested open intervals defined by $I_k = (0, 1/k)$, $k = 1, 2, \ldots$. Any $z \in E$ must be in I_1 and therefore must satisfy $0 < z < 1$. Now find a positive integer k such that $1/k < z$. That such a k exists is clear from our intuitive picture of $\mathbb{R}$ as points on the number line, but at any rate, the Archimedean property gives us such a k (simply find a k such that $kz > 1$). Then it is clear that $z \notin I_k$, so $z \notin E$, a contradiction!

If we study the proof of Theorem 12.33 to see where it will fail if instead the nested intervals are open, we will see that the arguments will work till the end, up to the point where we show that $E = \cap_k I_k$ equals the set $S = \{x \mid a \le x \le b\}$. While showing that $E = S$ above, we came to a step where we showed that $a_k \le a \le x \le b \le b_k$ for all k, from which we concluded that $a_k \le x \le b_k$. But this only allows us to conclude that x is in the closed interval $[a_k, b_k]$, not necessarily the open interval (a_k, b_k). Thus, we cannot conclude that $S \subset E$ if our intervals are all open.

We now give an alternative proof of the Bolzano-Weirstrass theorem that is based on the nested intervals theorem above.

Alternative Proof of the Bolzano-Weirstrass Theorem: Recall that the theorem (Theorem 12.32) states that if (s_n), $n = 1, 2, \ldots$, is a bounded infinite sequence of real numbers, then (s_n) has a convergent subsequence.

The proof based on nested intervals is as follows: Since (s_n) is bounded, there exists $l < u$ such that $l \le s_n \le u$ for all $n \ge 1$. Thus, all the s_n lie in the interval $[l, u]$, and we will denote this closed interval by I_0. Note that the "width" of I_0, denoted $|I_0|$, is $(u - l)$.

Consider the two closed intervals $[l, (l+u)/2]$ and $[(l+u)/2, u]$. Since these two intervals form a (not disjoint) partition of I_0 (with $(l+u)/2$ the only common point), and since there are infinitely many terms s_n, at least one of these two intervals, possibly even both, must contain infinitely many of the s_n.

Pick the interval that contains infinitely many terms of (s_n) (if both contain, then pick either one), and call it I_1. Notice that the width of I_1, denoted $|I_1|$, is $|I_0|/2 = (u-l)/2$. Pick an arbitrary term from the infinitely many terms in I_1, say s_{n_1}, for some index n_1. (For example, if I_1 contained $s_5, s_7, s_{15}, s_{57}, s_{102}, \ldots$, and we picked s_{57}, then "n_1" is 57.)

Now divide I_1 into two intervals of equal length by bisecting I_1, just as we did with the initial interval $[l, u]$. If we consider the terms of (s_n) from s_{n_1+1} on, there are infinitely many of them, as there are only finitely many terms with index at most n_1 (namely $s_1, s_2, \ldots, s_{n_1}$). At least one of the two intervals we just created by bisecting I_1, possibly even both, must contain infinitely many of these terms after the n_1-th term. Pick the interval that contains infinitely many terms (if both contain, then pick either one), and call it I_2. Notice that the width of $|I_2$ is $|I_1/2| = |I_0|/2^2$. Pick an arbitrary term from the infinitely many terms that are in I_2, say s_{n_2}, for some index $n_2 \ge n_1 + 1$. (Notice that $n_2 > n_1$ because s_{n_2} has been picked from those terms from the $n_1 + 1$-th term on that are in I_2. Notice also that $I_1 \supseteq I_2$.)

Now proceed in this fashion. Let us say that intervals I_1 containing s_{n_1}, I_2 containing s_{n_2}, ..., I_k containing s_{n_k} have been picked, with $n_1 < n_2 < \cdots < n_k$, $I_1 \supseteq I_2 \supseteq$

$\cdots \supseteq I_k$, and $|I_j| = |I_0|/2^j$, $j = 1, 2, \ldots, k$. Once again, divide the interval I_k into two intervals of equal length by bisecting I_k, and consider the terms of (s_n) from the s_{n_k+1}-th term on. At least one of these intervals, possibly both, must contain infinitely many of these terms. Once again, pick the interval that contains infinitely many of these terms (if both contain then pick either one), and call it I_{k+1}. Pick an arbitrary term from the infinitely many of the terms of (s_n) (from the s_{n_k+1}-th term on) that are in I_{k+1}, say $s_{n_{k+1}}$ for some index $n_{k+1} \geq n_k + 1$. Notice that $I_k \supseteq I_{k+1}$, $|I_{k+1}| = |I_k|/2 = |I_0|/2^{k+1}$, and that $n_{k+1} > n_k$ by the way we have selected $s_{n_{k+1}}$.

Proceeding along, we find a sequence of nested closed intervals $I_1 \supseteq I_2 \supseteq \cdots \supseteq I_k \supseteq \cdots$, one for each positive integer. Since the width of I_k is $|I_0|/2^k$, $k = 1, 2, \ldots$, the sequence $|I_k|$ converges to zero. By the nested intervals theorem (Theorem 12.33), $\cap_k I_k = \{a\}$ for some a. We claim that the subsequence s_{n_j}, $j = 1, 2, \ldots$, that we have selected converges to a. For, given any $\epsilon > 0$, find N_ϵ such that $2^{N_\epsilon} > |I_0|/\epsilon$. (You may use your familiarity with the function 2^x defined on the real numbers to see that such an N_ϵ can be found, or else notice from Exercise 7.11, Chapter 7, that $2^n \geq 1 + n > n$ for positive n, so choose $n > |I_0|/\epsilon$.) Then $|I_0|/2^{N_\epsilon} = |I_{N_\epsilon}| < \epsilon$. Since $s_{n_j} \in I_{n_j} \subseteq I_{N_\epsilon}$ for $n_j \geq N_\epsilon$ (as the intervals are nested), and since $a \in I_n$ for all n, we find $|a - s_{n_j}| \leq |I_{n_j}| \leq |I_{N_\epsilon}| < \epsilon$ for all $n_j \geq N_\epsilon$. This shows that indeed the subsequence s_{n_j}, $j = 1, 2, \ldots$, converges to a.

12.7 Cauchy sequences

We will consider here the central concept of a *Cauchy sequence*. We will see that one consequence of the LUB property is that every Cauchy sequence of elements of $\mathbb{R}$ converges. This is one of the key attributes of the real number system that distinguishes it from the rational numbers. (Not surprisingly, the LUB property, from which the convergence of Cauchy sequences is derived, does not hold for the rational numbers; see Remark 12.15.)

Definition 12.35. Let (s_n), $n = 1, 2, \ldots$, be a sequence of elements in $\mathbb{R}$. We say (s_n) is a *Cauchy sequence* if given any $\epsilon > 0$, there exists a positive integer N_ϵ such that for all n, m with $n, m \geq N_\epsilon$, $|s_n - s_m| < \epsilon$.

Example 12.36. The sequence formed from the set

$$S = \{3.1, 3.14, 3.141, 3.1415, 3.14159, \ldots\}$$

in Example 12.3(4) is a Cauchy sequence. Thus, if we set s_n, $n = 1, 2, \ldots$, to be the n-th element of S in the listing above (so s_n represents π truncated to the n-th decimal place), the sequence (s_n) is Cauchy. To see this, given any $\epsilon > 0$, we first choose N_ϵ such that $10^{-N_\epsilon} < \epsilon$. Then, from the N_ϵ-th term onwards, all terms of the sequence agree to the first N_ϵ decimal places, so for any n, m with $n, m \geq N_\epsilon$, $|s_n - s_m| < 10^{-N_\epsilon} < \epsilon$.

While the convergence of Cauchy sequences can be established directly from the LUB property, we will give a proof based on the Bolzano-Weierstrass theorem (which in turn is based on the LUB property of course!).

Before proving that Cauchy sequences converge, we will establish the converse:

Theorem 12.37. *Every convergent sequence of real numbers is a Cauchy sequence.*

Proof. Let (s_n) be a convergent sequence of real numbers, and let l be its limit. Given any $\epsilon > 0$, by definition of convergence, there exists a positive integer $M_{\epsilon/2}$ such that for all $n \geq M_{\epsilon/2}$, $|l - s_n| < \epsilon/2$. So, for any $n, m \geq M_{\epsilon/2}$, we find $|s_n - s_m| = |s_n - l + s_m - l| \leq |s_n - l| + |s_m - l| < \epsilon/2 + \epsilon/2 = \epsilon$. Hence, taking $N_\epsilon = M_{\epsilon/2}$ we find that any pair of terms from the N_ϵ-th term are within ϵ of each other. Therefore, (s_n) is a Cauchy sequence. □

Theorem 12.37 above shows why one would be interested in Cauchy sequences at all; they are forced on us by the convergence of a sequence!

We start with the following:

Lemma 12.38. *Every Cauchy sequence is bounded.*

Proof. Let (s_n) be the given Cauchy sequence. Taking $\epsilon = 1$, we find that there exists an integer N such that for all $n, m \geq N$, $|s_n - s_m| < 1$. In particular, setting $m = N$, we find $|s_n - s_N| < 1$ for all $n \geq N$, or what is the same thing, $s_N - 1 < s_n < s_N + 1$ for all $n \geq N$. Set $L = \min\{s_1, s_2, \ldots, s_{N-1}, s_N - 1\}$ and $U = \max\{s_1, s_2, \ldots, s_{N-1}, s_N + 1\}$. Then $L \leq s_n$ for all n, since, if $n \geq N$, then $L \leq s_N - 1 < s_n$ (by choice of N and L), else if n is one of $1, 2, \ldots, N-1$, then $L \leq s_n$ (by choice of L). Similarly, $s_n \leq U$ for all n. By Remark 11.22 in Chapter 11, (s_n) is bounded. □

We now prove the convergence of Cauchy sequences:

Theorem 12.39. *Every Cauchy sequence of real numbers is convergent.*

Proof. Let (s_n), $n = 1, 2, \ldots$, be a Cauchy sequence of real numbers. By Lemma 12.38 above (s_n) is bounded, so by the Bolzano-Weierstrass theorem, (s_n) has a convergent subsequence, call it (s_{n_k}). Let L be the limit of this convergent subsequence. We claim that L is also the limit of our original sequence (s_n).

To prove this claim, given $\epsilon > 0$, find N_1 such that for all $k \geq N_1$, $|s_{n_k} - L| < \epsilon/2$, using the definition of convergence of the subsequence (s_{n_k}). (If the double indices are confusing, first define the sequence (t_k) by $t_k = s_{n_k}$. Then, the convergence of (s_{n_k}) implies the convergence of (t_k)—they are both the same sequence!—and the convergence criterion then guarantees that for $\epsilon > 0$ there exists N_1 such for that for all $k \geq N_1$, $|t_k - L| < \epsilon/2$. But t_k is just s_{n_k}, so we find for all $k \geq N_1$, $|s_{n_k} - L| < \epsilon/2$.) Next, find N_2 such that for all $n, m \geq N_2$, $|s_n - s_m| < \epsilon/2$, using the definition of our original sequence being Cauchy. Set $N_\epsilon = \max\{N_1, N_2\}$. Choose any index $l \geq N_\epsilon$, and note that $n_l \geq l \geq N_\epsilon$ (for the first inequality see Remark 12.28). Then for all $n \geq N_\epsilon$, $|L - s_n| = |L - s_{n_l} + s_{n_l} - s_n| \leq |L - s_{n_l}| + |s_{n_l} - s_n| < \epsilon/2 + \epsilon/2 = \epsilon$. (Here, we find $|L - s_{n_l}| < \epsilon/2$ from the convergence of (s_{n_k}), invoking $l \geq N_\epsilon \geq N_1$. Also, we find $|s_{n_l} - s_n| < \epsilon/2$ from the Cauchy condition, invoking the fact that $n_l, n \geq N_\epsilon \geq N_2$.) The sequence (s_n) hence converges to L. □

Remark 12.40. Notice that Theorems 12.37 and 12.39 together show that a sequence converges if and only if it is a Cauchy sequence!

12.8 Convergence of Series

We have already been introduced to series in Section 11.6 in Chapter 11, but with the material on monotone convergence and Cauchy sequences under our belt, we can begin to study some of their convergence and divergence properties. We will give some criteria for series to converge in this section, and use these to develop a few tests of convergence. We will follow this with several examples where these tests are put to use.

First, let us recall from Section 11.6 in Chapter 11 that a series is an expression of the form $S = a_1 + a_2 + \cdots + a_n + \cdots$, also written as $S = \sum_{i=1}^{\infty} a_i$. It is said to converge to a real number L if the corresponding sequence of partial sums (P_n), where $P_n = a_1 + a_2 + \cdots + a_n$, converges to the real number L, and diverge if the sequence of partial sums (P_n) diverges.

Since a series converges, by definition, precisely when the sequence of partial sums converges, we can apply the results of Section 12.7, Theorems 12.37 and 12.39 (see also Remark 12.40) to the sequence of partial sums to obtain the following:

Proposition 12.41. *(Cauchy Criterion for Series.) A series $\sum_{i=1}^{\infty} a_n$ converges if and only if for each $\epsilon > 0$, there exists an integer N_ϵ such that for all $n > m \geq N_\epsilon$, we have*

$$|a_{m+1} + a_{m+2} + \cdots + a_n| < \epsilon.$$

Proof. Theorems 12.37 and 12.39 show that the sequence of partial sums P_n converges if and only if the Cauchy criterion for sequence convergence holds: Given $\epsilon > 0$, there exists an integer N_ϵ such that for all $n, m \geq N_\epsilon$, $|P_n - P_m| < \epsilon$. Here, we may assume first that $n \geq m$ without loss of generality. (The expression *without loss of generality* is used often in mathematics. What it means in this context is that if instead m is less than n, we can simply switch the roles of n and m in what follows, and the proof will go through with no other changes. The meaning of the expression is similar in other contexts.) Further, we can assume that $n > m$ (since trivially $|P_n - P_n| < \epsilon$ for any n). Thus, the series converges if and only if for all $n > m \geq N_\epsilon$, $|P_n - P_m| < \epsilon$. But $|P_n - P_m|$ is precisely $|a_{m+1} + a_{m+2} + \cdots + a_n|$. This proves the proposition. □

We obtain a few results directly from Proposition 12.41:

Corollary 12.42. *(n-th Term Test for Series.) If a series $S = \sum_{i=1}^{\infty} a_n$ converges, then $\lim_{n\to\infty} a_n = 0$. (In particular, reading the contrapositive, if $\lim_{n\to\infty} a_n \neq 0$, then the series must diverge.)*

Proof. Suppose the series S converges. Then by the Cauchy criterion in Proposition 12.41 above, given $\epsilon > 0$, there exists an integer N_ϵ such that for all $n > m \geq N_\epsilon$, $|a_{m+1} + a_{m+2} + \cdots + a_n| < \epsilon$. In particular, taking $n = m + 1$, we find that for all $m \geq N_\epsilon$, or what is the same thing, $m + 1 = n \geq N_\epsilon + 1$, $|a_n| < \epsilon$. By definition, $\lim_{n\to\infty} a_n = 0$. □

Remark 12.43. Note something about the statement of Corollary 12.42 above: it is not worded as an "if and only if" statement! In other words, it is silent about the truth of the converse: "If the n-th term of a series goes to zero, then the series must converge." For good reason: the converse statement is *false!* The harmonic series we have already considered in Exercise 11.66 in Chapter 11 is a perfect example: we have $\lim_{n\to\infty} \frac{1}{n} = 0$, yet, the series $\sum_{i=1}^{\infty} \frac{1}{i}$ diverges!

The utility of the n-th term test lies in the contrapositive: if the n-th term does not go to zero as n goes to infinity, then the series must diverge. For instance, we can see from this that the geometric series $\sum_{n=1}^{\infty} ar^n$ $(a, r \neq 0)$ must diverge when $r > 1$, as $r^n > 1 + n(r-1)$, and thus, r^n grows without bound as $n \to \infty$ (see Exercise 11.54 in Chapter 11).

The following theorem, which is the squeeze theorem for series, also is a consequence of Proposition 12.41:

Theorem 12.44. *Suppose that three series $\sum_{n=1}^{\infty} a_n$, $\sum_{n=1}^{\infty} b_n$, and $\sum_{n=1}^{\infty} c_n$ satisfy the property that $a_n \leq b_n \leq c_n$ for all $n \geq N_0$, for some integer N_0. If the two series $\sum_{n=1}^{\infty} a_n$ and $\sum_{n=1}^{\infty} c_n$ on the outside converge, then the series $\sum_{n=1}^{\infty} b_n$ in the middle also converges.*

Proof. Since $\sum_{n=1}^{\infty} a_n$ converges, the Cauchy criterion above shows that given any $\epsilon > 0$ there exists an integer $N_{1,\epsilon}$ such that for $n > m \geq N_{1,\epsilon}$, we have $|a_{m+1}+a_{m+2}+\cdots+a_n| < \epsilon$. We write this as

$$-\epsilon < a_{m+1} + a_{m+2} + \cdots + a_n < \epsilon.$$

For the same ϵ, the Cauchy criterion applied to the series $\sum_{n=1}^{\infty} c_n$ shows that there exists an integer $N_{2,\epsilon}$ such for all $n > m \geq N_{2,\epsilon}$, we have

$$-\epsilon < c_{m+1} + c_{m+2} + \cdots + c_n < \epsilon.$$

Thus, if $N_\epsilon = \max(N_0, N_{1,\epsilon}, N_{2,\epsilon})$, then, invoking the hypothesis on the a_n, b_n, and c_n and combining with inequalities above, we find that for all $n > m \geq N_\epsilon$,

$$-\epsilon < a_{m+1} + a_{m+2} + \cdots + a_n < b_{m+1} + b_{m+2} + \cdots + b_n < c_{m+1} + c_{m+2} + \cdots + c_n < \epsilon.$$

In particular, this says that for all $n > m \geq N_\epsilon$,

$$-\epsilon < b_{m+1} + b_{m+2} + \cdots + b_n < \epsilon.$$

Put differently, we find that for all $n > m \geq N_\epsilon$, $|b_{m+1} + b_{m+2} + \cdots + b_n| < \epsilon$. Hence, by the Cauchy criterion, the series $\sum_{n=1}^{\infty} b_n$ also converges. □

Here is a quick corollary to the squeeze theorem above, which yields a useful test for convergence and divergence of series:

Corollary 12.45. (*Comparison Test.*) *Suppose that $\sum_{n=1}^{\infty} a_n$ and $\sum_{n=1}^{\infty} b_n$ are two series satisfying $|a_n| \leq b_n$ for all $n \geq N_0$, where N_0 is some integer. Suppose further that the series $\sum_{n=1}^{\infty} b_n$ converges. Then the series $\sum_{n=1}^{\infty} a_n$ also converges. In particular, if the a_n are nonnegative and satisfy $a_n \leq b_n$ for all $n \geq N_0$, then if $\sum_{n=1}^{\infty} a_n$ diverges, then $\sum_{n=1}^{\infty} b_n$ also diverges.*

Proof. The condition $|a_n| \leq b_n$ for all $n \geq N_0$ translates to $-b_n \leq a_n \leq b_n$ for all $n \geq N_0$. It is clear that if the series $\sum_{n=1}^{\infty} b_n$ converges, then so does the series $\sum_{n=1}^{\infty}(-b_n)$, since the partial sums of $\sum_{n=1}^{\infty}(-b_n)$ are simply the negatives of the partial sums of $\sum_{n=1}^{\infty} b_n$. (Recall that if (s_n) is a convergent sequence, so is (rs_n) for any real number r; see Exercise 11.49 in Chapter 11. Apply this to the sequence of partial sums of $\sum_{n=1}^{\infty} b_n$.) By the squeeze theorem (Theorem 12.44), the series $\sum_{n=1}^{\infty} a_n$ must also converge. The last statement is simply the contrapositive, applied to the case where the a_n are nonnegative (so the hypothesis $|a_n| \leq b_n$ for all $n \geq N_0$ is satisfied automatically once we are given $a_n \leq b_n$). □

Remark 12.46. Note that the hypothesis requires that the *absolute value* of the a_n should be bounded by the b_n. It does *not* read "$a_n \leq b_n$." For instance, if we take the series $(-1) + (-2) + (-3) + \cdots$ as our first series, so $a_n = -n$, and take the second series as the geometric series $1 + \frac{1}{2} + \frac{1}{4} + \frac{1}{8} + \cdots$, so $b_n = \frac{1}{2^{n-1}}$, we do have $a_n \leq b_n$, yet, the first series does not converge (for instance, it fails the nth-term test; see Corollary 12.42 above), although the second series converges by Example 11.42 of Chapter 11.

Corollary 12.45 has its own corollary, that then leads to the concept of conditional convergence (Definition 12.48 ahead):

Corollary 12.47. *Let $\sum_{n=1}^{\infty} a_n$ be a series such that the series $\sum_{n=1}^{\infty} |a_n|$ converges. Then $\sum_{n=1}^{\infty} a_n$ also converges.*

Proof. Set $b_n = |a_n|$. Note that we trivially have $|a_n| \leq b_n$. Since $\sum_{n=1}^{\infty} b_n$ converges by hypothesis, the series $\sum_{n=1}^{\infty} a_n$ converges by Corollary 12.45. □

The converse to Corollary 12.47 is false: there are series $\sum_{n=1}^{\infty} a_n$ that converge but for which $\sum_{n=1}^{\infty} |a_n|$ diverges. This leads to the following:

Definition 12.48. If a series $S = \sum_{n=1}^{\infty} a_n$ is such that the series $\sum_{n=1}^{\infty} |a_n|$ converges, then S is said to be *absolutely convergent.* If S converges, but the series $\sum_{n=1}^{\infty} |a_n|$ diverges, S is said to be *conditionally convergent.*

In the language of Definition 12.48, Corollary 12.47 shows that an absolutely convergent series is convergent. Exercise 12.72 ahead shows that there exist conditionally convergent series.

Limit Comparison Test. We can derive from the comparison test (Corollary 12.45) another test that studies limits of ratios. There are situations when this derived test is easier to apply than the original comparison test.

Theorem 12.49. *Suppose that $S = \sum_{n=1}^{\infty} a_n$ and $T = \sum_{n=1}^{\infty} b_n$ are two series. Assume that all terms a_n and b_n are positive. If*

$$\lim_{n\to\infty} \frac{a_n}{b_n} = \begin{cases} r,\ 0 < r < \infty & \text{then } S \text{ converges if and only if } T \text{ converges,} \\ 0 & \text{then if } T \text{ converges } S \text{ also converges,} \\ \infty & \text{then if } T \text{ diverges } S \text{ also diverges.} \end{cases}$$

(See Remark 11.25 *in Chapter* 11 *for what it means for a limit to equal infinity.)*

Proof. First consider the case where the depicted limit r is positive and does not equal ∞.

Assume that T converges. By the definition of limits, there exists an integer N such that $\frac{a_n}{b_n} < 2r$ for all $n \geq N$. (Take $\epsilon = r$ in Definition 11.12 of Chapter 11 to find $-r < \frac{a_n}{b_n} - r < r$ and add r to all terms.) We write this as $a_n < (2r)b_n$ for all $n \geq N$. Since the series $\sum_{n=N}^{\infty}(2r)b_n$ converges (see Exercises 11.63 and 11.64 in Chapter

11), we conclude from the comparison test (Corollary 12.45) that the series $\sum_{n=N}^{\infty} a_n$ converges. By Exercise 11.64 in Chapter 11 again, the series S must converge.

Now assume that S converges. The arguments are similar once we pick a *lower* bound for the ratios. Once again, by the definition of limits, there exists an integer N such that $\frac{a_n}{b_n} > r/2$ for all $n \geq N$. (Take $\epsilon = r/2$ in Definition 11.12 of Chapter 11.) We write this as $b_n < (2/r)a_n$ for all $n \geq N$. Since the series $\sum_{n=N}^{\infty}(2/r)a_n$ converges, the series $\sum_{n=N}^{\infty} b_n$ converges, and hence T converges.

Now assume that the limit $r = 0$, and assume that T converges. Then we can find N such that $\frac{a_n}{b_n} < 1$ for all $n \geq N$. This shows that $a_n < b_n$ for all $n \geq N$. As before, it follows from the fact that the series $\sum_{n=N}^{\infty} b_n$ converges that the series $\sum_{n=N}^{\infty} a_n$ and hence the series S must also converge.

Finally assume that $r = \infty$, and assume that T diverges. By the definition of the limit being infinity (see Remark 11.25 in Chapter 11), there exists an integer N such that $\frac{a_n}{b_n} > 1$ for all $n \geq N$. (Take $L = 1$ in the definition in that remark.) We write this as $b_n < a_n$ for all $n \geq N$. The comparison test (Corollary 12.45) shows that since $\sum_{n=N}^{\infty} b_n$ diverges, $\sum_{n=N}^{\infty} a_n$, and hence S, also diverge. □

Root Test and Ratio Test. We describe two tests that are of much use in determining if series converge. They depend in a crucial way on the results obtained above, particularly Corollary 12.45.

In the first test below, the root test, by $\sqrt[n]{|a_n|}$ we mean the positive n-th root of $|a|$. Note that we will prove formally in Theorem 12.57 below that $\sqrt[n]{|a_n|}$ exists as a real number (that is, there is a real number y such that $y^n = |a|$). Moreover, $|a|$ has a unique positive n-th root (see Remark 12.58 ahead).

Theorem 12.50. (*Root Test.*) *Let $S = \sum_{i=1}^{\infty} a_n$ be a series. If $\lim_{n\to\infty} \sqrt[n]{|a_n|}$ exists and equals r, then, S converges if $r < 1$ and diverges if $r > 1$.*

Proof. Suppose $r < 1$ first. Then, for $\epsilon = (1-r)/2$, there exists an N such that for all $n \geq N$, $|\sqrt[n]{|a_n|} - r| < \epsilon$, so $r - \epsilon < \sqrt[n]{|a_n|} < r + \epsilon$. Write s for $r + \epsilon$, and note that $s < 1$ because of the way we have chosen ϵ. Then $\sqrt[n]{|a_n|} < s$ for all $n \geq N$, or what is the same thing, $|a_n| < s^n$ for all $n \geq N$. Since the geometric series $1 + s + s^2 + \cdots$ converges for $|s| < 1$ (Example 11.42 in Chapter 11), the comparison test (Corollary 12.45 above) shows that $\sum_{i=N}^{\infty} a_n$ converges. By Exercise 11.64 in Chapter 11, S converges.

By a similar argument, if $r > 1$, we can find an M and an $s > 1$ such that for all $n \geq M$, $1 < s < \sqrt[n]{|a_n|}$. Taking n-th powers, we find $1 < s^n < |a_n|$ for all $n \geq M$. By the n-th term test (Corollary 12.42), S diverges. □

Remark 12.51. When $r = 1$, the root test is silent on the convergence of S. This is because both convergence and divergence behaviors can occur when $r = 1$. For instance, the harmonic series $\sum_1^{\infty} \frac{1}{n}$ satisfies $\lim_{n\to\infty} \sqrt[n]{|a_n|} = 1$, by Exercise 11.55 of Chapter 11, and it diverges. As well, the series $\sum_1^{\infty} \frac{1}{n^2}$ satisfies $\lim_{n\to\infty} \sqrt[n]{|a_n|} = 1$, by Exercise 11.51 of Chapter 11. But, by Exercise 12.70 ahead, the series $\sum_1^{\infty} \frac{1}{n^2}$ converges!

Our next test looks at ratios of the terms of a series:

Theorem 12.52. *(Ratio Test.) Let* $S = \sum_{i=1}^{\infty} a_n$ *be a series. (We may assume that no* a_n *is zero, as zero terms will not affect the sum of a series and can be removed.) If* $\lim_{n\to\infty} \frac{|a_{n+1}|}{|a_n|}$ *exists and equals r, then, S converges if* $r < 1$ *and diverges if* $r > 1$*.*

Proof. Suppose $r < 1$. As in the proof of Theorem 12.50 above, we can find an $s < 1$ and an integer N such that for all $n \geq N$, $\frac{|a_{n+1}|}{|a_n|} < s$. Thus, we find successively:

$$\begin{aligned} |a_N| &\leq |a_N| \quad \text{for trivial reasons,} \\ |a_{N+1}| &< |a_N|s, \\ |a_{N+2}| &< |a_{N+1}|s < |a_N|s^2, \\ \vdots &< \vdots \\ |a_n| &< |a_N|s^{n-N} \quad \text{for } n > N. \end{aligned}$$

The right sides of the equalities above are terms of the geometric sequence $|a_N| + |a_N|s + |a_N|s^2 + \cdots$, with $s < 1$. Exactly as in the proof of Theorem 12.50, the comparison test and Exercise 11.64 of Chapter 11 show that S converges.

When $r > 1$, we find, by similar arguments, that there is an $s > 1$ and an integer N such that $|a_n| > |a_N|s^{n-N} > |a_N|$ for all $n \geq N$. By the n-th term test (Corollary 12.42) and Exercise 11.64 of Chapter 11, S diverges. □

Remark 12.53. Just as with the root test, when $r = 1$, the ratio test is silent on the convergence of S. And as with the root test, this is because both convergence and divergence behaviors can occur when $r = 1$. The same two examples apply here too: the harmonic series $\sum_1^{\infty} \frac{1}{n}$ satisfies $\lim_{n\to\infty} \frac{|a_{n+1}|}{|a_n|} = 1$ and diverges, while the series $\sum_1^{\infty} \frac{1}{n^2}$ also satisfies $\lim_{n\to\infty} \frac{|a_{n+1}|}{|a_n|} = 1$ but converges!

Example 12.54. We consider several examples of series now, and show how we can use the various tests developed above – the n-th term test (Corollary 12.42), the comparison test (Corollary 12.45), the limit comparison test (Theorem 12.49), the root test (Theorem 12.50), and the ratio test (Theorem 12.52) – to determine if these series converge or diverge. You are encouraged to attempt Exercise 12.70 ahead before studying these examples, or at least, to familiarize yourself with the statement of that exercise.

(1) The n-th term test readily shows that a series $\sum_{n=1}^{\infty} a_n$ is divergent if a_n does not have a limit of zero as n tends to infinity. Various series of this nature are where $a_n = n$, $a_n = \log n$, $a_n = \cos(n)$ (as also $\cos(n\pi)$ – why?), $a_n = \sin(n)$ (but not $\sin(n\pi)$ – why not?), etc.

(2) The series $\sum_{n=1}^{\infty} \frac{1}{n^2-3n+1}$. Here is the intuition: The trick is to recognize that the denominator "behaves like n^2" for large values of n. What this means precisely is that $\lim_{n\to\infty} \frac{n^2-3n+1}{n^2} = 1$. So, since the convergence of series is unaffected by the first finitely many terms and only depends on the "tail" terms (see Exercise 11.64 in Chapter 11), one would expect that the series $\sum_{n=1}^{\infty} \frac{1}{n^2-3n+1}$ has the same convergence behavior as the series $\sum_{n=1}^{\infty} \frac{1}{n^2}$. Note that by Exercise 12.70, the series $\sum_{n=1}^{\infty} \frac{1}{n^2}$ converges.
Indeed, the limit comparison test (Theorem 12.49) captures *precisely* this intuition! Note, also, that we can show that $\lim_{n\to\infty} \frac{\frac{1}{n^2-3n-1}}{\frac{1}{n^2}} = 1$ quite easily. For instance, we

can rewrite the ratio after dividing by n^2 as $\frac{1}{1-3/n-1/n^2}$. The sequences $(1/n)$ and $(1/n^2)$ converge to 0. Exercises 11.49 through 11.52 of Chapter 11 now show that our desired limit is indeed 1. Now we simply apply the limit comparison test which then says that $\sum_{n=1}^{\infty} \frac{1}{n^2-3n+1}$ converges because the series $\sum_{n=1}^{\infty} \frac{1}{n^2}$ converges!

(3) Here is an instance of the use of the comparison test (Corollary 12.45): we study the behavior of the series $\sum_{n=1}^{\infty} a_n$ where $a_n = \frac{\cos(n)}{n^2}$. We invoke the fact that $|\cos(n)| \leq 1$ and compare with the series $\sum_{n=1}^{\infty} \frac{1}{n^2}$ to establish convergence. (Note that if we used the limit comparison test to compare with the series $\sum_{n=1}^{\infty} \frac{1}{n^2}$ we would come up at a dead end, because the ratio of terms is $\cos(n)$, so the desired limit would not exist.)

(4) Consider the series $1 + 1 + \frac{1}{2!} + \frac{1}{3!} + \frac{1}{4!} + \cdots$. Indexing the terms starting from zero for convenience, we find $a_n = \frac{1}{n!}$. The ratio test shows that the series converges:

$$\lim_{n\to\infty} \frac{|a_{n+1}|}{|a_n|} = \lim_{n\to\infty} \frac{n!}{(n+1)!} = \lim_{n\to\infty} \frac{1}{n+1},$$

and it is clear that this last term above has limit zero. Hence, this series converges. You may recognize this as an alternative representation for *e* from past courses; see Exercise 12.73 ahead.

(5) Consider the series $\sum_{n=1}^{\infty} a_n$ where $a_n = \frac{n2^n}{5^{1+2n}}$. We find $\sqrt[n]{|a_n|} = \frac{2\sqrt[n]{n}}{25\sqrt[n]{5}}$. By Exercises 11.55 and 11.56 of Chapter11, this limit is $\frac{2}{25} < 1$, so the series converges by the root test.

We will see other applications of these tests, particularly the root test and the ratio test, in the exercises at the end of the chapter.

Series with Nonnegative Terms. When all the terms of a series $\sum_{n=1}^{\infty} a_n$ are nonnegative, the partial sums of such series are nondecreasing: $P_n - P_{n-1} = (a_1 + a_2 + \cdots + a_n) - (a_1 + a_2 + \cdots + a_{n-1}) = a_n \geq 0$. Thus, the sequence of partial sums is a monotonically increasing sequence, and the monotone convergence theorem can be applied, yielding further criteria for the convergence or divergence of such series.

We have the following result:

Proposition 12.55. *Let (a_n) be a sequence of nonnegative real numbers, and write S for the series $\sum_{i=1}^{\infty} a_i$. The following are equivalent:*

(1) *S converges.*

(2) *The partial sums are bounded.*

(3) *Some subsequence (P_{n_i}) of (P_n) is bounded.*

Remark 12.56. To say that statements A, B, and C are equivalent is to say that A, B, and C are either simultaneously true or simultaneously false. In other words, when all three of the following statements hold: A is true if and only if B is true, B is true if and only if C is true, and also, C is true if and only if A is true. Of course, if A is true if and only if B is true and B is true if and only if C is true, then automatically, we find that C is true if and only if A is true. Hence, to establish that A, B, and C are equivalent, it is sufficient to show that A is true if and only if B is true and B is true if and only if C is true.

Alternatively, we can also establish the following cyclic implications: $A \implies B \implies C \implies A$! It is easy to see that if this chain of implications holds, then A, B, and C are equivalent.

Proof of Proposition 12.55. We will show part (1) $\iff$ part (2), and part (2) $\iff$ part (3).

Part (1) $\implies$ *part* (2): If S converges, then by definition, the sequence of partial sums converges, so by Theorem 12.37 and Lemma 12.38, the sequence of partial sums is bounded.

Part (2) $\implies$ *part* (1): We have seen in the discussions above the statement of the proposition that the partial sums are monotonically increasing, since the terms $a_n \geq 0$. So, if the partial sums are bounded, then the monotone convergence theorem (Theorem 12.22) shows that the sequence of partials sums must converge, so the series, by definition, converges.

Part (2) $\implies$ *part* (3): This is clear, since the terms of any subsequence are also terms of the original sequence, and the terms of the original sequence are bounded. (In fact, if the partial sums are bounded, then *every* subsequence (P_{n_i}) of (P_n) is bounded!)

Part (3) $\implies$ *part* (2): Given the bounded subsequence (P_{n_i}), let us say that $P_{n_i} \leq B$ for some nonnegative real number B. Given any partial sum P_n, we find i such that $n_i > n$. (The indices n_i of the subsequence are a strictly increasing sequence!) As observed in the discussions above the statement of the proposition, $P_n \leq P_{n_i}$, so we find $P_n \leq B$ for all n. Since the P_n are all nonnegative, we find $|P_n| \leq B$ for all n, so the sequence of partial sums is indeed bounded. □

12.9 n-th roots of positive real numbers

Let us use the LUB property of $\mathbb{R}$ to show that one can always find n-th roots of positive real numbers:

Theorem 12.57. *The n-th root of any positive real number exists as a real number, for any $n \geq 1$.*

Proof. Let a be a positive real number. Consider the set $S = \{y \in \mathbb{R}_{>0} \mid y^n \leq a\}$. This set is nonempty: If $a \geq 1$, then $1^n = 1 \leq a$, so $1 \in S$. Else, if $a < 1$, then multiplying both sides by the positive real number a we find $a^2 < a$, then multiplying again we find $a^3 < a^2 < a$, and so on, to find $a^n < a^{n-1} < \cdots < a$. Hence $a^n \in S$. Moreover, S is bounded above: If $a \geq 1$, then we claim that for any $x \in S$, we must have $x \leq a$. For, if not, we will have $x > a$ for some $x \in S$, and on multiplying both sides by (the positive real number) x, we will find $x^2 > ax$. Now, from $a \geq 1$, we find on multiplying by (again, the positive real number) x, that $ax \geq x$. Thus, putting the inequalities $x^2 > ax$ and $ax \geq x$ together, we find that the assumption $x > a$ yields

$x^2 > x$. Multiplying repeatedly by x we find $x^n > x^{n-1} > \cdots > x$. Coupling with $x > a$, we find $x^n > a$, contradicting the definition of the set S. Thus, when $a \geq 1$, a is an upper bound for S. Now suppose $a < 1$. Then we must have $x < 1$. For if $x \geq 1$, we will find on repeatedly multiplying by x that $x^n \geq x^{n-1} \geq \ldots \geq x \geq 1$. But also, $1 > a$, so we find $x^n > a$, a contradiction. Hence, when $a < 1$, 1 is an upper bound for S.

By the LUB property, S has a least upper bound, call it u. We claim that $u^n = a$, so u is indeed an n-th root of a as desired. We show this by ruling out the two possibilities $u^n < a$ and $u^n > a$.

First assume that $u^n < a$. We claim that we can find a positive real number δ such that $(u + \delta)^n < a$. This means that $u + \delta$ will be in S, contradicting the fact that u is an upper bound for S. Write D for $a - u^n$. For any positive δ, we find by the Binomial theorem that

$$(u + \delta)^n = u^n + \underbrace{\binom{n}{1}u^{n-1}\delta + \binom{n}{2}\delta^2 u^{n-2} + \cdots + \delta^n}_{\text{"}g_\delta\text{"}}.$$

If we select δ so that $g_\delta < D$, then we will find $(u + \delta)^n = u^n + g_\delta < u^n + D = a$, as desired. To do this, first stipulate that δ should be less than 1, so, as we have seen before, $\delta^n < \delta^{n-1} < \cdots < \delta$. It follows that

$$g_\delta < \binom{n}{1}u^{n-1}\delta + \binom{n}{2}\delta u^{n-2} + \cdots + \delta = \delta \underbrace{\left(\binom{n}{1}u^{n-1} + \binom{n}{2}u^{n-2} + \cdots + 1\right)}_{\text{"}h\text{"}}.$$

(Notice that h does not depend on δ and is positive.) It is now sufficient to choose δ so that $\delta \cdot h < D$ since this will automatically guarantee that $g_\delta < D$. But for this, simply select δ so that $0 < \delta < \min(1, D/h)$: the relation $\delta < D/h$ coupled with the fact that h is positive shows that $\delta \cdot h < D$, and of course, the condition $\delta < 1$ was needed for our estimation above. Hence $u^n < a$ is not possible.

Now assume that $u^n > a$. We now claim that we can find a positive real number δ such that $(u - \delta)^n > a$. This will provide us with a contradiction as follows: by Proposition 12.11, there exists $x \in S$ with $x > u-\delta$. Hence, $x^n < a$, while $(u-\delta)^n > a$. So, we have $u-\delta < x$, but $(u-\delta)^n > x^n$, contradicting an elementary fact, whose proof we leave as an exercise:

> **Exercise 12.57.1.** Let $n \geq 1$ be an integer, and a and b positive real numbers. If $a < b$, then $a^n < b^n$. (Hint: multiply $a < b$ first by a, then by b. Repeat appropriately.)

To find such a δ, we proceed as in the case where $u^n < a$, except we drop terms that are negative. Write D for $u^n - a$. Remember we need $(u - \delta)^n > a$. Now, by the Binomial theorem,

$$(u - \delta)^n = u^n - \binom{n}{1}u^{n-1}\delta + \binom{n}{2}\delta^2 u^{n-2} \pm \cdots + (-1)^n\delta^n \underbrace{\geq a}_{\text{needed}}.$$

Thus, we need, after transposing terms appropriately,

$$D \stackrel{\text{def}}{=} u^n - a > \binom{n}{1}u^{n-1}\delta - \binom{n}{2}\delta^2 u^{n-2} \pm \cdots - (-1)^n\delta^n.$$

It is sufficient to choose δ so that

$$D > \binom{n}{1}u^{n-1}\delta + \binom{n}{3}u^{n-3}\delta^3 + \binom{n}{5}u^{n-5}\delta^5 + \cdots,$$

since the term on the right is greater than the term on the right in the previous inequality (we have simply removed all the negative terms).

We stipulate that δ should be less than 1, so that $\delta^n < \delta^{n-1} < \cdots < \delta$. It follows that

$$\binom{n}{1}u^{n-1}\delta + \binom{n}{3}u^{n-3}\delta + \binom{n}{5}u^{n-5}\delta + \cdots > \binom{n}{1}u^{n-1}\delta + \binom{n}{3}u^{n-3}\delta^3 + \binom{n}{5}u^{n-5}\delta^5 + \cdots,$$

so it is sufficient to find $\delta < 1$ such that

$$D > \delta \underbrace{\left(\binom{n}{1}u^{n-1} + \binom{n}{3}u^{n-3} + \binom{n}{5}u^{n-5} + \cdots\right)}_{\text{"}h\text{"}}.$$

Notice that h is positive and does not depend on δ. It is now sufficient to choose any $\delta < \min(1, D/h)$, and our proof is complete! □

Remark 12.58. We can say a bit more about n-th roots of positive real numbers: If n is even there are precisely two real n-th roots, one positive and one negative, and if n is odd, there is precisely one real n-th root, and that is positive. This is the content of Exercise 12.68 ahead.

Note however that a real number a actually has n n-th roots if we allow complex roots. Let $\omega = \cos(2\pi/n) + i\sin(2\pi/n)$. Using DeMoivre's theorem (see Exercise 7.23, Chapter 7, we find $\omega^k = \cos(2k\pi/n) + i\sin(2k\pi/n)$ for $k = 1, 2, \ldots, n-1$ and $\omega^n = 1$. It follows that if y is one n-th root of the positive real number a, then $y\omega, y\omega^2, \ldots, y\omega^{n-1}$ are all distinct and are also n-th roots of a, as $(y\omega^k)^n = y^n\omega^{kn} = y^n(\omega^n)^k = y^n = a$. These elements $y\omega^i$, $i = 0, 1, \ldots, n-1$, already account for n roots of the equation $y^n = a$. It is a theorem that you will study in future courses in abstract algebra that an equation of degree n with real, or even complex-valued, coefficients (such as the equation $y^n = a$ for the unknown y) has exactly n roots in the complex numbers. Thus, these elements $y\omega^i$, $i = 0, 1, \ldots, n-1$, account for *all* n-th roots of a. (See also the factorization of $x^n - y^n$ in Example 13.12 of Chapter 13 ahead.)

12.10 Further Exercises

Exercise 12.59. Practice Exercises:

Exercise 12.59.1. Find, with proof, the lub and glb of the following sets (if the lub or glb does not exist, explain why not):

(1) $S = \{x \in \mathbb{R} | x^2 - 2x \geq 1\}$. (Hint: Bring all terms to one side.)

(2) $S = \{x \in \mathbb{R} | x^2 - 2x < 1\}$.

(3) $S = \{x \in \mathbb{R} | x^3 - 3x^2 + 3x \geq 1\}$. (Hint: Bring all terms to one side.)

(4) $S = \{x \in \mathbb{R} | x^3 - 3x^2 + 3x < 1\}$.

Exercise 12.59.2. Show that the following sequences are Cauchy. Your proofs should find an N_ϵ, given any $\epsilon > 0$, such that $|s_n - s_m| < \epsilon$ for all $n, m \geq N_\epsilon$. (You may find it useful to recall that $|x+y| \leq |x|+|y|$. We can apply this to an expression like $|s_n - s_m|$ by writing it as $|s_n + (-s_m)|$.)

(1) (s_n), where $s_n = \frac{1}{n}$.

(2) (s_n), where $s_n = \frac{(-1)^n}{n}$.

(3) (s_n), where $s_n = 1 + \frac{1}{\sqrt{n}}$.

(4) (s_n), where $s_n = 1 + \frac{1}{n^2+n}$.

(5) (s_n), where $s_n = \frac{\cos(n)}{n}$.

Exercise 12.59.3. Use the tests developed in the chapter to determine if the following series converge or diverge. Study Example 12.54 to get a feel for how to make comparisons. Feel free to use the results of Exercise 12.70. Exercises 11.55 and 11.56 in Chapter 11 may be useful.

(1) $\sum_{n=1}^{\infty} \frac{1}{\sqrt[n]{2}}$.

(2) $\sum_{n=1}^{\infty} \frac{1}{n(n+2)}$.

(3) $\sum_{n=1}^{\infty} \frac{n+2}{n(n+1)}$.

(4) $\sum_{n=1}^{\infty} \frac{2^{-n}}{n}$.

(5) $\sum_{n=1}^{\infty} n!\, e^{-n}$.

(6) $\sum_{n=1}^{\infty} \frac{(n!)^2}{(2n)!}$.

Exercise 12.60. Let $S \subseteq \mathbb{R}$ be nonempty. If $u = \text{lub}(S)$, show that there is a sequence (s_n), $n = 1, 2, \ldots$, with the $s_i \in S$ and satisfying $s_1 \leq s_2 \leq s_3 \leq \ldots$ such that (s_n) converges to u. (Hint: if u is already in S, take (s_n) to be the constant sequence whose elements are all equal to u. Otherwise, use Proposition 12.11 to create such a sequence.)

Exercise 12.61. Here is another result that is obvious if we view that the real numbers are just points on the number line, but in the more formal development of the real number system, and that it is something that needs to be proved: if a and b are real numbers such that $b - a > 1$, then then exists an integer n such that $a < n < b$. (Hint: Use the result of Corollary 12.20.)

Exercise 12.62. We have seen that the least upper bound of a set $S \subseteq \mathbb{R}$ need not be a member of S (see Example 12.8(1)). In this exercise we will show that if S consists only of integers (i.e., $S \subseteq \mathbb{Z} \subseteq \mathbb{R}$), then $\text{lub}(S)$, if it exists, must belong to S. So, let $S \subseteq \mathbb{Z}$ be such that it has a least upper bound, and let $l = \sup(S)$.

(1) By Corollary 12.20, there exists an integer n such that $n \leq l < n + 1$. Show that $l = n$. (Hint: if $l > n$, use Proposition 12.11.)

(2) Now that we have $l = n$, show that l must be in S.

Exercise 12.63. Show that given any two real numbers x and y with $x < y$, there exists a *rational* number q with $x < q < y$. (Hint: Exercise 12.61 if $y - x > 1$. If this condition does not hold, invoke the Archimedean proprety.)

Exercise 12.64. Use the result in Exercise 12.63 to show that given any two real numbers x and y with $x < y$, there exists an *irrational* number r with $x < r < y$. (Hint: we know, for example, that $\sqrt{2}$ is irrational, from Example 10.28 in Chapter 10. Find a rational number between $x/\sqrt{2}$ and $y/\sqrt{2}$ first.)

Exercise 12.65. Let (s_n) be a sequence with $s_n \in \mathbb{Z}$ for all $n \geq 1$. If the sequence converges, show that it must eventually be constant, that is, for all $n \geq N$ for some suitable N, $s_n = p$ for some $p \in \mathbb{Z}$.

Exercise 12.66. Suppose you throw a six-sided die infinitely many times. Let s_n denote the number (from 1 to 6) that shows up on the n-th throw. Show without using the Bolzano-Weierstrass theorem (Theorem 12.32) that the sequence (s_n) has a convergent subsequence. (Hint: Look to Exercise 12.65 for inspiration!)

Exercise 12.67. If a sequence (s_n) converges to some real number s, show that any subsequence of (s_n) must also converge to s.

Exercise 12.68. Prove the assertion in Remark 12.58: If a is a positive real number, and n a positive integer, then the equation $y^n = a$ has precisely two real roots if n is even, one of them negative and one positive, and precisely one real root if n is odd. (Hint: Theorem 12.57 for the existence of at least one root, along with Exercise 12.57.1.)

Exercise 12.69. Use the monotone convergence theorem (Theorem 12.22) to show that the sequence (a_n) defined by $a_1 = 6$ and $a_{n+1} = \frac{a_n}{2} + 6$ for all $n \geq 1$ converges. (Hint: Calculate a few terms of the sequence and guess at an upper bound for the sequence. Use induction to prove that your guess is correct. Show that the sequence is monotone.)

Exercise 12.70. We have seen in Exercise 11.66, Chapter 11, that the harmonic series $S = \sum_{i=1}^{\infty} 1/n$ diverges. We will show here that the series

$$S = \sum_{i=1}^{\infty} \frac{1}{n^p}, \quad p \in \mathbb{Z}, p > 1,$$

converges. (Notice that if we set $p = 1$ we get back the harmonic series, which diverges!)

(1) Show that the partial sums P_n satisfy

$$\begin{aligned} P_1 &\leq \frac{1}{1}, \\ P_3 &\leq 1 + \frac{1}{2^{p-1}}, \\ P_7 &\leq 1 + \frac{1}{2^{p-1}} + \frac{1}{4^{p-1}}, \\ P_{15} &\leq 1 + \frac{1}{2^{p-1}} + \frac{1}{4^{p-1}} + \frac{1}{8^{p-1}}, \\ \vdots \;\; &\vdots \end{aligned}$$

(Hint: Group terms suitably. Use Exercise 12.57.1 in the text, and also use the fact that if $0 < a < b$ then $\frac{1}{a} + \frac{1}{b} < \frac{1}{a} + \frac{1}{a}$.)

(2) Show that the subsequence (P_{2^n-1}) of (P_n) is bounded. (Hint: Finite geometric series! Show that the partial sums are bounded above $\frac{1}{1-\frac{1}{2^{p-1}}}$. The hypothesis that $p > 1$ is crucial here!)

(3) Use Proposition 12.55 to show that the series S above converges.

Remark 12.70.1. The result holds even if p is allowed to be any real number greater than 1. The same proof goes through. Of course this assumes that you are familiar with the function a^x, for $a > 1$ and $x \in \mathbb{R}_{>0}$. For instance, that it is an increasing function of a for a fixed x, or that $a^x b^x = (ab)^x$.

Exercise 12.71. Given a series $S = \sum_{i=1}^{\infty} a_i$ with the property that $\lim_{i\to\infty} a_i = 0$, suppose that the sequence of even partials sums ($P_{2n} = \sum_{i=1}^{2n} a_i$) and the sequence of odd partial sums ($P_{2n+1} = \sum_{i=1}^{2i+1} a_i$) both converge. Show that the series S also converges, and all three of (P_{2n}), (P_{2n+1}), and S converge to the same limit. (Hint: Both sequences (P_{2n}) and (P_{2n+1}) must be Cauchy. Argue that the sequence of partial sums of S must therefore be Cauchy, observing, for instance, that $P_{2j} - P_{2k+1} = a_{2j} + P_{2j-1} - P_{2k+1}$ and similarly for other combinations of P_n and P_m. Invoke the result of Exercise 12.67 above.)

Exercise 12.72. Show that the series $S = 1 - \frac{1}{2} + \frac{1}{3} - \frac{1}{4} \pm \cdots = \sum_{i=1}^{\infty}(-1)^{i-1}\frac{1}{n}$ is conditionally convergent. (Hint: The series obtained by replacing all terms in S by their absolute values is just the harmonic series, which we know to be divergent by Exercise 11.66, Chapter 11. We only need to show that S converges. Study the even partial sums P_{2n} (see Exercise 12.71 above) and the odd partial sums P_{2n+1}. Show that they are both monotonic. Show that they each converge, possibly by finding a relation between P_{2n} and P_{2n+1}, and using this to show that both sequences are bounded by 0 and 1. Note that $\lim_{i\to\infty}\frac{1}{i} = 0$ and invoke the result of Exercise 12.71 above.)

Exercise 12.73. We said in Example 12.25 that the number e is formally defined as $\lim_{n\to\infty}\left(1+\frac{1}{n}\right)^n$. You may have also seen in your previous courses the following definition

$$e = 1 + 1 + \frac{1}{2!} + \frac{1}{3!} + \frac{1}{4!} + \cdots.$$

(Recall that we have already seen in Example 12.54(3), that the series on the right above converges.) We will show in this exercise that both definitions are the same, that is,

$$e = \lim_{n\to\infty}\left(1+\frac{1}{n}\right)^n = 1 + 1 + \frac{1}{2!} + \frac{1}{3!} + \frac{1}{4!} + \cdots. \tag{12.5}$$

Write s_n $(n \geq 1)$ for $(1+\frac{1}{n})^n$ and P_n for the n-th partial sum of the series on the right in equation (12.5) above (with the index starting at zero for convenience – thus, $P_0 = 1$, $P_1 = 1 + 1 = 2$, $P_2 = 1 + 1 + \frac{1}{2} = \frac{5}{2}$, etc.).

(1) Show that $s_n \leq P_n$ for all $n \geq 1$. Conclude by invoking Exercise 11.45 in Chapter 11 that $e \leq \lim_{n\to\infty} P_n$. (Hint: equation (12.3) in Example 12.25. For the second assertion, consider the sequence $(P_n) - (s_n)$.)

(2) Write $s_{n,m}$ for the following:

$$s_{n,m} = 1 + 1 + \frac{1}{2!}\left(1 - \frac{1}{n}\right) + \cdots + \frac{1}{m!}\left(1-\frac{1}{n}\right)\cdots\left(1-\frac{m-1}{n}\right).$$

Use the obvious fact that for $n \geq m$, $s_n \geq s_{n,m}$ to show that $e > P_m$.

(3) Conclude that $e \geq \lim_{n\to\infty} P_n$ and then conclude, using part (1) above, that $e = \lim_{n\to\infty}(1+\frac{1}{n})^n = \lim_{n\to\infty} P_n$.

Exercise 12.74. *(Cauchy Condensation Test.)* We develop a test here for the series $\sum_{n=1}^{\infty} a_n$ in the *special case* where $a_1 \geq a_2 \geq a_3 \geq \cdots \geq 0$. We show here that the series $\sum_{n=1}^{\infty} a_n$ converges if and only if the series

$$\sum_{k=0}^{\infty} 2^k a_{2^k} = a_1 + 2a_2 + 4a_4 + 8a_8 + \cdots$$

converges. (Note that we have indexed the second series $\sum_{k=0}^{\infty} 2^k a_{2^k}$ starting from zero.)

We write P_n $(n \geq 1)$ for the partial sums of $\sum_{n=1}^{\infty} a_n$ and Q_n $(n \geq 0)$ for those of $\sum_{k=0}^{\infty} 2^k a_{2^k}$.

(1) By grouping as follows:

$$P_{2^n-1} = a_1 + (a_2 + a_3) + (a_4 + a_5 + a_6 + a_7) + \cdots$$

show that $P_{2^n-1} \leq Q_{n-1}$ for all $n \geq 1$.

(2) By grouping as follows:

$$P_{2^n} = a_1 + a_2 + (a_3 + a_4) + (a_5 + a_6 + a_7 + a_8) + \cdots$$

show that $P_{2^n} \geq \frac{a_1}{2} + \frac{1}{2}Q_n > \frac{1}{2}Q_n$.

(3) Combine the two inequalities in parts (1) and (2) above into a single chain of inequalities and prove that the sequence P_n converges if and only if the sequence Q_n converges. (Hint: Both series have nonnegative terms. Proposition 12.55 and Exercise 12.67 may be useful.)

Exercise 12.75. Show that the Cauchy condensation test developed in Exercise 12.74 applies to the following and use the test to establish convergence or divergence:

(1) The series $\sum_{n=0}^{\infty} \frac{1}{n^p}$, for $p \in \mathbb{N}$, converges if $p \geq 2$ and diverges if $p = 1$. (This also appears as Exercise 12.70 when $p > 1$, and as Exercise 11.66 in Chapter 11 when $p = 1$. However, if you study those exercises, you will notice that the suggested method of attack in both is really the Cauchy condensation test in disguise!)

(2) The series $\sum_{n=2}^{\infty} \frac{1}{n(\log(n))^p}$, for $p \in \mathbb{N}$, converges if $p \geq 2$ and diverges if $p = 1$. (You may assume that the function $\log(x)$ is the one usually denoted $\log_2(x)$. This function is defined as the inverse of the function $y = e^{x\ln(2)}$. The function $y = e^{x\ln(2)}$ is also denoted "$y = 2^x$," because it reduces to the usual "power of 2" function when x is an integer. Since $e^{x\ln(2)}$ is the composite $g \circ f$ where $f(x) = x\ln(2)$ and $g(x) = e^x$, and since both functions are injective, 2^x is injective. Since f has image all of $\mathbb{R}$ and g has image $(0, \infty)$, 2^x has image $(0, \infty)$. Thus 2^x is a bijection between $\mathbb{R}$ and $(0, \infty)$, and has as inverse $\log_2(x)$. Also note that the series is indexed starting from 2. It would be harmless to take $a_1 = 1$ if you wish, as this does not affect the convergence of the series, by Exercise 11.64 in Chapter 11, and will not affect the applicability of the Cauchy condensation test.)

Exercise 12.76. A *power series* (in the variable x) is an expression of the form $\sum_{n=0}^{\infty} a_n x^n$, where (a_n) is a sequence (that is typically indexed starting from zero). Let us write $S(x)$ for this expression. Just as with the usual series (which we will now refer to as a *numerical* series to distinguish from a power series), a power series is just an expression to begin with. We interpret $S(x)$ as follows: if we substitute any one real number for x, say r, that is, if we consider the expression "$S(r)$" where $S(r) = \sum_{n=0}^{\infty} a_n r^n$, then we get a numerical series. If this numerical series converges (which in turn happens when the sequence of partial sums of $S(r)$ converges), we say that the power series converges for $x = r$, or *at* $x = r$, and if it diverges, we say that the power series diverges for $x = r$, or at $x = r$. Thus, an expression like $S(x)$ is a representation for the entire *family* of numerical series $S(r)$ obtained by substituting various real numbers r in place of x in the expression $\sum_{n=0}^{\infty} a_n x^n$.

Prove the following:

(1) *(Root Test for Power Series.)* Given a power series $S(x)$ as above, suppose that $R' = \lim_{n\to\infty} \sqrt[n]{|a_n|}$ exists. If $R' \neq 0$, set $R = 1/R'$. If $R' = 0$, set $R = \infty$. Show that for x such that $|x| < R$, the series converges, and for x such that $|x| > R$, the series diverges. If $\sqrt[n]{|a_n|}$ is unbounded, show that $S(x)$ converges only for $x = 0$. (Hint: For the last assertion, suppose $x \neq 0$. For any $M > 0$ such that $M|x| > 1$, there exist infinitely many n such that $\sqrt[n]{|a_n|} > M$. What does this say about the terms $a_n x^n$?)

(2) *(Ratio Test for Power Series.)* This time, suppose that $R' = \lim_{n\to\infty} \frac{|a_{n+1}|}{|a_n|}$ exists. (We may assume that no a_n is zero, as zero terms will not affect the sum of a series and can be removed.) If $R' \neq 0$, set $R = 1/R'$. If $R' = 0$, set $R = \infty$. Show that for x such that $|x| < R$, the series converges, and for x such that $|x| > R$, the series diverges.

   Comparing with part (1) above, we see that the limit considered in this exercise and in part (1) are equal when they exist, since they determine the boundary of convergence of the power series.

Exercise 12.77. Show that the power series

$$1 + \frac{x}{1!} + \frac{x^2}{2!} + \cdots + \frac{x^i}{i!} + \cdots$$

converges for all values of x.

The power series above is one definition of the function e^x! Note that setting $x = 1$, we recover the definition of e as $1 + 1 + \frac{1}{2!} + \frac{1}{3!} + \frac{1}{4!} \cdots$ in Exercise 12.73 above.

Exercise 12.78. Show that the power series

$$1 - \frac{x^2}{2!} + \frac{x^4}{4!} - \frac{x^6}{6!} + \cdots + (-1)^i \frac{x^{2i}}{(2i)!} + \cdots$$

converges for all values of x. This power series is one definition of $\cos(x)$!

Exercise 12.79. Show that the power series

$$x - \frac{x^3}{3!} + \frac{x^5}{5!} - \frac{x^7}{7!} + \cdots + (-1)^i \frac{x^{2i+1}}{(2i+1)!} + \cdots$$

converges for all values of x. This power series is one definition of $\sin(x)$!

Notes

Axioms that define the real number system. We list here the axioms that define the real numbers. First, a definition (see also Definition 13.5, Chapter 13):

Definition 12.80. Let S be a set. A function $f : S \times S \to S$ is called a *binary operation* on S.

(1) *Field Axioms:* There exist two binary operations "+" and "·" on $\mathbb{R}$, known, respectively, as addition and multiplication, such that:

 (a) Commutativity of addition: $a + b = b + a$ for all a, b in $\mathbb{R}$.
 (b) Associativity of addition: $a + (b + c) = (a + b) + c$ for all a, b, c in $\mathbb{R}$.
 (c) Existence of additive identity: There exists an element 0 in $\mathbb{R}$, known as the additive identity, such that $a + 0 = a$ for all a in $\mathbb{R}$.
 (d) Existence of additive inverses: For each $a \in \mathbb{R}$, there exists an element denoted "$-a$" such that $a + (-a) = 0$.
 (e) Commutativity of multiplication: $a \cdot b = b \cdot a$ for all a, b in $\mathbb{R}$.
 (f) Associativity of multiplication: $a \cdot (b \cdot c) = (a \cdot b) \cdot c$ for all a, b, c in $\mathbb{R}$.
 (g) Existence of multiplicative identity: There exists an element 1 in $\mathbb{R}$, known as the multiplicative identity, such that $a \cdot 1 = a$ for all a in $\mathbb{R}$.
 (h) Existence of multiplicative inverses: For each *nonzero* $a \in \mathbb{R}$, there exists an element denoted "a^{-1}" such that $a \cdot a^{-1} = 1$.
 (i) Distributivity of multiplication over addition: For all a, b, c in $\mathbb{R}$, $a \cdot (b + c) = a \cdot b + a \cdot c$.

(2) *Ordered Field Axioms:* There exists a relation "<" on $\mathbb{R}$ such that:

 (a) Trichotomy: For $a, b \in \mathbb{R}$, *exactly one* of the following holds: $a < b$, $b < a$, or $a = b$.
 (b) Transitivity: For $a, b, c \in \mathbb{R}$, if $a < b$ and $b < c$, then $a < c$.
 (c) Additivity of Order: For $a, b, c \in \mathbb{R}$, if $a < b$, then $a + c < b + c$.
 (d) Multiplicativity of Order: For $a, b, c \in \mathbb{R}$, if $a < b$ and $c > 0$, then $a \cdot c < b \cdot c$.

(3) *Completeness Axiom:* Every nonempty subset of $\mathbb{R}$ bounded above has a least upper bound.

It is an interesting result that the elements 0 and 1 in the field axioms are unique. In other words, if 0 and, say, $0'$ are two real numbers that satisfy the property that $a+0 = a$ and $a + 0' = a$ for all $a \in \mathbb{R}$, then we must have $0 = 0'$. Similarly for the element 1. Likewise, for any $a \in \mathbb{R}$, the element $-a$ is unique, and if $a \neq 0$, the element a^{-1} is unique. (These properties are actually properties of the "underlying group structure" of $\mathbb{R}$ with the binary operation "+," as also $\mathbb{R} \setminus \{0\}$ with the binary operation "·." You will study these properties in Chapter 13 ahead, Lemmas 13.15 and 13.16.)

It is an interesting fact that $\mathbb{R}$ is the *unique complete ordered field.* In other words, if S is any set that satisfies the axioms listed above, then S is "essentially" the same as $\mathbb{R}$. What this means is the following: there exists a function $f : S \to \mathbb{R}$ such that

(1) f is a bijection.

(2) $f(a+b) = f(a) + f(b)$ and $f(a \cdot b) = f(a) \cdot f(b)$ for all a, b in S.

(3) $f(1_S) = 1$, where we have written 1_S for the multiplicative identity in S.

(4) If $a < b$ for a, b in S, then $f(a) < f(b)$ in $\mathbb{R}$.

This is an example of "sameness" that you will see in several places in advanced mathematics. If such an S exists, then as a set it might be different from $\mathbb{R}$, but there is a way of matching the elements of S with those of $\mathbb{R}$ in such a way that the binary operations of addition and multiplication on S and $\mathbb{R}$ match up, and the order relations on S and $\mathbb{R}$ match up. Thus, up to this matching, S is indistinguishable from $\mathbb{R}$.

13

Groups and Symmetry

In this chapter we will introduce a fascinating branch of mathematics that deals with *symmetry*. At first, this might appear an odd thing to do: much of the mathematics you may have seen may have dealt only with numbers, not something seemingly non-mathematical like symmetry. Even objects like sequences or functions, at least in the version you may have studied so far, may primarily have dealt with numbers (for instance, we ourselves in previous chapters only studied sequences whose terms were real numbers, and mainly worked with functions that took one real number to another).

But after a point, mathematics is no longer about concrete numbers, but about, well, *mathematical objects*. At one level this is of course a perfectly useless statement, but at another level, it gets to the heart of the matter: mathematics is vast, it extends well beyond boundaries that may have been formed in people's minds after an early and brief exposure, and the only way to discover what mathematics studies is to roll up one's sleeves and just study it![1]

It is worth remarking that much of nature exhibits deep symmetries. Sometimes they appear in obvious features like shapes or layouts, for instance, of honeycombs and petal arrangements. But often, they appear in more abstract features, such as in the fact that the laws of motion stay the same under changes of coordinates arising from uniform motion or under more general coordinate changes that account for relativistic motion. From another point of view, various symmetries of a physical system lead to laws of conservation of various associated physical quantities (this is a theorem of mathematics), so symmetry may be viewed as the *cause* of conservation laws in physics.

Groups are the mathematical objects that "measure" symmetry, and it should not be surprising that groups are one of the central objects of mathematics and physics. The way to the study of groups (as is the way to the study of many areas of mathematics) is to first study a number of examples. We will do this now, by considering the symmetries of an equilateral triangle and a square, and then moving on to the symmetries of an

[1]To be fair, the same can be said of many other subjects, such as physics or biology.

n-element set. Along the way, we will introduce the axioms that define groups, and relate these axioms to the examples that we study.

13.1 Symmetries of an equilateral triangle

Let us consider a piece of cardboard cut in the shape of an equilateral triangle lying on a table. To help us keep track of some of the things we will do to the triangle ahead, we will assume that the vertices of the triangle have been marked A, B, C. We will look at all the actions we can perform on this triangular piece of cardboard so that after we are done, *if we were to ignore any labelling we may have done on the triangle,* we would not know that anything has been done to the triangle! To further help us keep track, let us draw the outline of the cardboard on the table by drawing lines on the table where the edges of the triangle lie. Let us mark the corners of the outline on the table: a for the corner where vertex A of the triangle initially lies, b for the corner where vertex B initially lies, and c for the corner where the vertex C initially lies. We will interchangeably use the words *corner* or *spot* for the points a, b, and c on the outline of the triangle on the table, and the word *vertex* for the points A, B, and C on the triangle. See Figure 13.1.

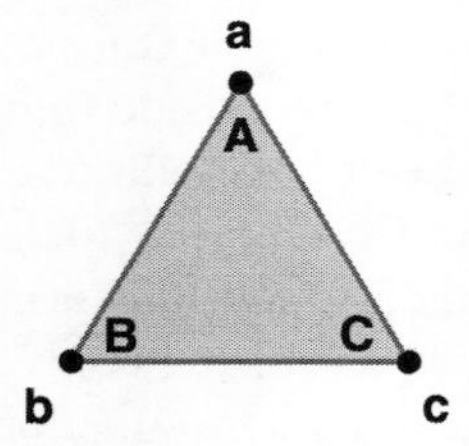

Figure 13.1. Triangle with vertices A, B, C, and corners on the table marked a, b, c.

Now by an "action" on the triangle we will refer to any kind of motion of the triangle on the plane — such as translating the triangle, rotating about some point, flipping about some line, etc. Since we should not be able to tell that anything has been done to the triangle, we find that our actions should not *distort* the triangle. This means that we should not change the distance between any two points of the triangle. (If we changed the distance, then had we labelled the two points and looked at them before and after the action, we would know that an action has occurred!) We are thus looking at *rigid* motions of the triangle.

Moreover, the same requirement that after we are done, we should not be able to tell that anything has been done to the triangle (as long as we ignore the labelling of the vertices), shows that after the motion, the triangle must stay exactly aligned with the outlines initially drawn on the table. Thus, the vertices A, B, and C must land on the corners a, b, and c as marked, although not necessarily on the same corner where they were before the motion. Let P be an arbitrary point of the triangle other than A and B. The triangle PAB must maintain the same shape after the motion, so if P was at a distance d_A from A and d_B from B, then after the motion, P must continue to be at distances d_A and d_B from the new positions of A and B. This determines a *unique* point inside the triangle where P can land. (See Figure 13.2.) Thus, knowing where A and B go under a rigid motion completely determines where every other point of the triangle goes. This immediately allows us to count the number of *possible* such motions: A can land in one of three spots after the motion (either a or b or c), and after determining where A goes, B can occupy one of the remaining two spots, making a total of six possible rigid motions.

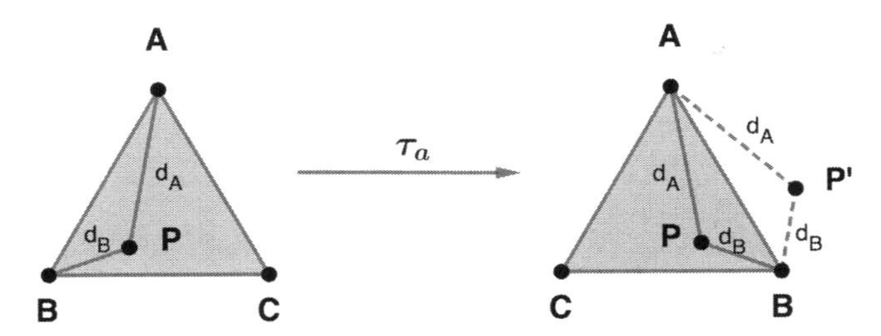

Figure 13.2. After applying the rigid motion τ_a shown which reflects the triangle across a line joining A to the midpoint of BC, the distances d_A and d_B determine two choices for P, with only one inside the triangle.

This count of six was obtained just by looking at the choices for vertices A and B; it is not clear that there actually exists a rigid motion of the triangle that effects each choice of positions of these vertices. As we will now see, all six choices are not just possible, they can actually be effected as rigid motions. Thus, the number of motions of the equilateral triangle that we are seeking: those that are rigid and are such that after the motion, we cannot tell whether a motion has occurred if we ignored the labelling on the vertices, is exactly six.

We will list them below, but first some notation. Since the motion is completely determined by where the vertices are sent (and they have to be sent to another corner and not an arbitrary place on the table), we will write, for example, $\binom{a\,b\,c}{b\,c\,a}$ for the motion that sends whatever vertex lay on the corner a to the corner b, whatever vertex lay on the corner b to the corner c, and whatever vertex lay on the corner c to the corner a. (Initially, these vertices of course are A, B, and C respectively, but after applying several such motions, the configuration of the triangle on the outline could be different, as we will see. Notice also that, as we have seen, it is sufficient to just show what happens at two of the corners; this completely determines what happens on the third corner as well, but this redundancy in notation will help us calculate the effect of combining two motions.) These are the motions (see Figure 13.3):

(1) $\sigma = \binom{a\,b\,c}{b\,c\,a}$. As we can see from the first motion on the left in Figure 13.3, we can effect this by rotating the triangle counterclockwise by 120°. This is indeed a rigid motion (there is no stretching or other distortion involved). Notice that after the rotation is over, if we were to ignore the lettering on the vertices, we would not be able to tell if the cardboard had been moved.

(2) $\sigma^2 = \binom{a\,b\,c}{c\,a\,b}$. This is the second motion on the left in Figure 13.3. We can effect this as a counterclockwise rotation by 240°, or alternatively, as a clockwise rotation by 120°. We will see below that if we were to effect σ first and then another σ, we would arrive at this motion. (Intuitively this should be clear: a counterclockwise rotation by 120° followed by another leads to a counterclockwise rotation by 240°!) It is for this reason that we have denoted this permutation by σ^2.

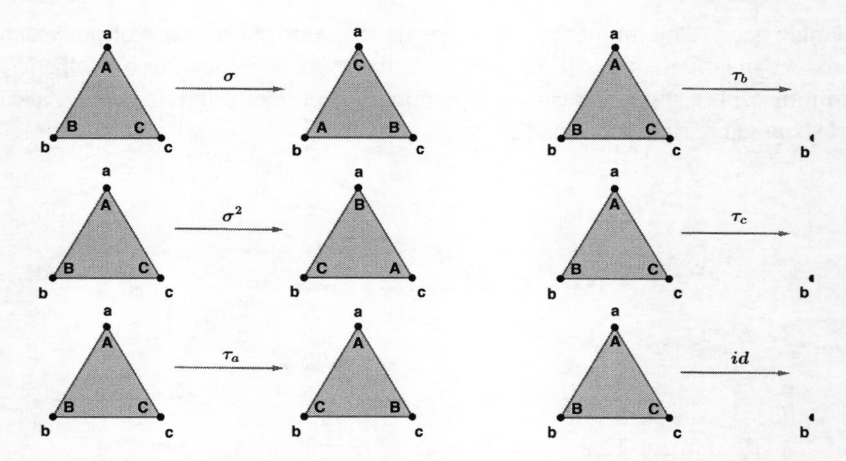

Figure 13.3. D_3: The symmetries of an equilateral triangle.

(3) $\tau_a = \binom{a\ b\ c}{a\ c\ b}$. This is the third motion on the left in Figure 13.3. We can effect this by flipping the triangle about the line joining the point a and the midpoint of the opposite side bc. This too is a rigid motion, and after the flip is over, if we were to ignore the lettering on the vertices, we would not be able to tell if the cardboard had been moved.

(4) $\tau_b = \binom{a\ b\ c}{c\ b\ a}$. This is the first motion on the right in Figure 13.3. We can effect this by flipping the triangle about the line joining the point b and the midpoint of the opposite side ac.

(5) $\tau_c = \binom{a\ b\ c}{b\ a\ c}$. This is the second motion on the right in Figure 13.3. We can effect this by flipping the triangle about the line joining the point c and the midpoint of the opposite side ab.

(6) $id = \binom{a\ b\ c}{a\ b\ c}$. This is the third motion on the right in Figure 13.3. This of course corresponds to *doing nothing* to the triangle! This is a valid operation of the sort that we are seeking: it is clearly a rigid motion of the triangle, and after we have performed this operation (which is to do nothing!), we would not be able to tell whether anybody had disturbed the triangle or not!

We will write D_3 for the set of these six motions of the equilateral triangle.

Now that we have seen how to effect all six possible motions, let us study the most interesting thing about them: it is possible to follow any one of our motions with another, and the result is also one of six rigid motions! Further, it is possible to undo the effect of any of our motions by following it with another of our motions (carefully selected of course!).

Let us consider an example, which we have already alluded to in our description of σ^2 above. What if we follow an application of the motion σ with another application of the same σ? We have already explained this in intuitive terms: a counterclockwise rotation by 120° followed by another leads to a counterclockwise rotation by 240°! But let us work it out carefully in terms of where vertices go:

Let us assume that the triangle was in its initial configuration: A on a, B on b, and C on c. The first application of σ will send the vertex A to the corner b. The second application will take whatever vertex lies on b to the corner c, so it will take A (which was lying on b) to the corner c. Thus, as a result of the combined operation of σ followed by σ, whatever vertex was on the corner a has now landed on corner c. Similarly, the first σ will move B, which was on corner b, to the corner c, and the second σ will send whatever was on the corner c to the corner a, so it will send B to the corner a. Thus, two applications of σ has taken whatever vertex was on corner b to the corner a. Whatever vertex was on corner c now has no choice. Since it must go to a corner, and both a and c are already filled up, it has to to go to b. Thus, the effect of σ followed by σ is the motion denoted $\binom{a\ b\ c}{c\ a\ b}$, in other words, σ^2.

Simillarly, we can see that if we were to follow an application of σ with σ^2, or an application of σ^2 with σ, the two motions will cancel each other, and the net effect is *to do nothing!* For instance, σ sends what was on the corner a to the corner b, and then, σ^2 sends what was on corner b to corner a, so the net effect is to send what was on corner a to corner a. Similarly for the other two corners, and similarly, too, if we go in the other direction and first do σ^2 and then σ. Described differently, the result of following σ with σ^2 (or the other way around) is the motion we have labelled id.

Notice that each of our motions can be thought of as a *function* from the set $C = \{a, b, c\}$ of corners to C. For instance, the motion $\tau_b = \binom{a\ b\ c}{c\ b\ a}$, which sends the vertex on corner a to the corner c, the vertex on corner b to the corner b, and the vertex on corner c to the corner a, can be viewed as the function f that sends a to c, b to b, and c to a. (As well, note that our functions are all bijections between C and C!) The effect of following one motion with another is really just the composition of functions. Thus, to follow, say, σ with τ_a is to perform the function composition $\tau_a \circ \sigma$. (Recall the notion $f \circ g$ for function composition: the function g is applied first and then the function f. Also note from Exercise 5.29 in Chapter 5 that since our functions are bijections between C and C, their composition will also be a bijection between V and V.) Using this notation, we find $\sigma \circ \sigma = \sigma^2$, $\sigma \circ \sigma^2 = id$, etc.

Table 13.1. Composition table for D_3, the symmetries of an equilateral triangle

$\circ$	id	σ	σ^2	τ_a	τ_b	τ_c
id	id	σ	σ^2	τ_a	τ_b	τ_c
σ	σ	σ^2	id	τ_c	τ_a	τ_b
σ^2	σ^2	id	σ	τ_b	τ_c	τ_a
τ_a	τ_a	τ_b	τ_c	id	σ	σ^2
τ_b	τ_b	τ_c	τ_a	σ^2	id	σ
τ_c	τ_c	τ_a	τ_b	σ	σ^2	id

Computing all possible compositions of these six motions of our triangle as we have done above for $\sigma \circ \sigma$ and $\sigma \circ \sigma^2$, we have Table 13.1. In this table, to read the result of $g \circ h$, where g and h are one of our six motions, we find the row labelled g, and in that row, go to the cell in the column labelled h: the motion in that cell represents $g \circ h$.

Thus, the table shows that $\sigma \circ \tau_a$ is τ_c. The table shows clearly that the composition of two of our motions is another motion. Moreover, on each row g the motion id appears in a suitable column h that depends on g, so $g \circ h = id$. The table also shows that for this same g and h, $h \circ g = id$. Thus, for each g, where g is one of our six motions, there is a corresponding h such that performing g first and then h, or performing h first and then g, is to effectively do nothing at all; g and h undo each others' actions!

These rigid motions we have considered are referred to as the *symmetries* of an equilateral triangle. We will discuss the usage of this term later in the chapter.

It would be very helpful to do the following exercise (Exercise 13.2) at this stage to help cement your understanding of these symmetries. But first some remarks that will be helpful for the exercise:

Remark 13.1. The following are some notation and terminology:

(1) For any $g \in D_3$, we will write g^k for $g \circ g \circ \ldots \circ g$ (k times), $k = 2, 3, \ldots$.

(2) Two elements g and h in D_3 are said to *commute* with each other if $g \circ h = h \circ g$. Notice that there are elements in D_3 that do not commute with each other, for instance σ and τ_a!

(3) As discussed above, for each $g \in D_3$, there exists a suitable $h \in D_3$ such that $g \circ h = h \circ g = id$. We will refer to g and h as inverses of each other, and write g^{-1} for h and h^{-1} for g. We will write g^{-k} for $(g^{-1})^k$, for $k = 2, 3, \ldots$. (Note that since each element of D_3 can be thought of as a bijective function from $V = \{a, b, c\}$ to itself, as discussed above, and since under this interpretation the element id is just $\mathbb{1}_V$ in the language of Chapter 5, Definition 5.24, the relations $g \circ h = h \circ g = id$ are tantamount to saying that g and h are inverses of each other as functions from V to V.)

Exercise 13.2. Use Table 13.1 or compute directly with individual motions to answer the following:

(1) Determine the set of elements in D_3 that commute with every other element.

(2) Compute $\sigma \circ \tau_a \circ \sigma^{-1}$ and $\sigma^2 \circ \tau_a \circ \sigma^{-2}$.

(3) Compute $\tau_a \circ \sigma \circ \tau_a^{-1}$ and $\tau_a \circ \sigma^2 \circ \tau_a^{-1}$.

(4) Find the inverses of τ_a, τ_b, and τ_c.

(5) Find the least positive integer k such that $\sigma^k = id$.

(6) Find the least positive k such that $(\sigma^2)^k = id$.

13.2 Symmetries of a square

In this section we will consider the analogous problem of determining the symmetries of a square. Thus, as in the previous section, we have a piece of cardboard cut in the shape of a square lying on a table. To help us track the motions of the square, we will assume that the vertices of the square have been marked A, B, C, and D. To further help us keep track, we will draw the outline of the square on the table, and mark the corners of the outline as a where the vertex A initially lies, b where the vertex B initially lies, and similarly c, and d. Just as with the triangle, we will use the words corner or spot for the points a, b, c, and d on the outline of the square on the table, and the word vertex for the points A, B, C, and D on the square.

We are looking for motions of the square such that after we are done with the motion, we will not be able to tell that anything has been done to the square if we ignore the labelling of the vertices. In particular, just as with the triangle, such motions must be rigid. In the case of the triangle we saw that it is sufficient to know where any two vertices go under such a motion; the position of every other point on the triangle will be uniquely determined (Figure 13.2). This needs to be modified mildly now: for a square, it is sufficient to know where any two *adjacent* vertices go. The proof is the same as for the triangle and we will not repeat it, but it is instructive to see where this will fail if we consider two vertices that are not adjacent. If two vertices are not adjacent, say A and C, they determine a diagonal and not a side of the square, and then, given a pair of distances (d_A, d_C) of a point P from A and C, there is a second point in the square that is at a distance d_A from A and d_C from C. See Figure 13.4. Thus, knowing where A and C go under a rigid motion is not sufficient to determine where a general point P goes.

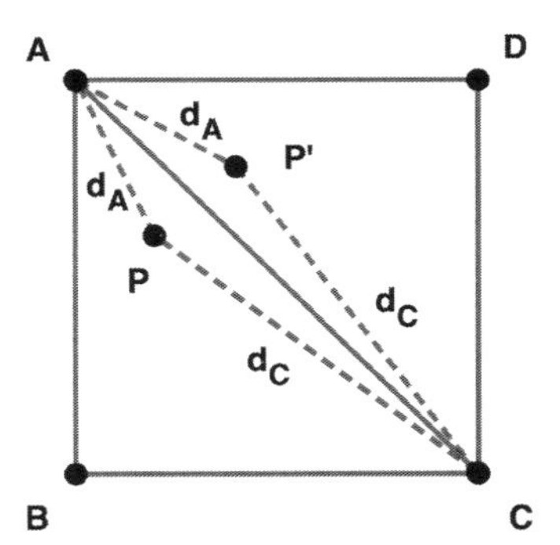

Figure 13.4. There are two points P and P' at the same distances d_A to A and d_C to C.

Arguing as in the case of the triangle, if A and B are two adjacent vertices, then A can go to one of four corners, and then B must go to one of the corners adjacent to where A has gone (else, the distance between A and B will not be preserved after the motion). Thus, there are two choices for B once a destination for A has been determined, making a total of eight *possible* motions. While these are as yet only possibilities, obtained by enumerating where adjacent vertices can go, we will see below that all eight possibilities can be actually realized as rigid motions of the square.

In what follows, just as in the case of the triangle, we will use notation such as $\binom{a\,b\,c\,d}{b\,c\,d\,a}$ to denote a motion that takes whatever vertex was on corner a to corner b, whatever vertex was on corner b to corner c, etc.

The following are all eight of the rigid motions of the square (see Figure 13.5 for a sample):

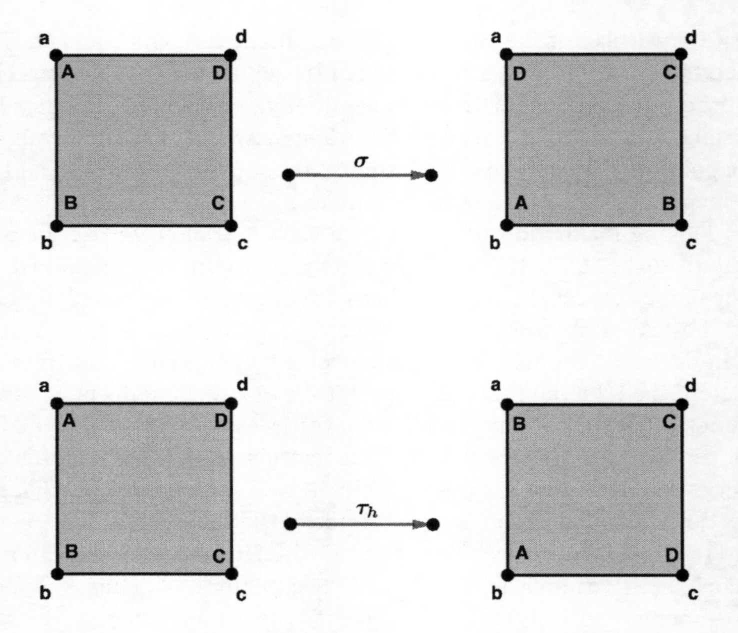

Figure 13.5. D_4: Two symmetries of a square.

(1) $\sigma = \binom{a\ b\ c\ d}{b\ c\ d\ a}$. (See Figure 13.5.) This can be effected by rotating the square counterclockwise by 90°. This is a rigid motion, and after the rotation is over, we would not be able to tell if the cardboard had been moved if we ignore the labelling of the vertices.

(2) $\sigma^2 = \binom{a\ b\ c\ d}{c\ d\ a\ b}$. This is effected by rotating the square counterclockwise (or clockwise) by 180°, and corresponds to the composition $\sigma \circ \sigma$. Hence the name σ^2.

(3) $\sigma^3 = \binom{a\ b\ c\ d}{d\ a\ b\ c}$. This is effected by rotating the square counterclockwise by 270° (or clockwise by 90°), and corresponds to the composition $\sigma \circ \sigma \circ \sigma$.

(4) $\tau_h = \binom{a\ b\ c\ d}{b\ a\ d\ c}$. (See Figure 13.5.) This corresponds to flipping the square about its horizontal axis, or what is the same thing, about the line joining the midpoints of the sides *ab* and *cd* of the outline. This is indeed a rigid motion and of course, after the rotation is over, we would not be able to tell if the cardboard had been moved if we ignore the labelling of the vertices.

(5) $\tau_v = \binom{a\ b\ c\ d}{d\ c\ b\ a}$. This corresponds to flipping the square about its vertical axis, or what is the same thing, about the line joining the midpoints of the sides *ad* and *bc* of the outline.

(6) $\tau_{ac} = \binom{a\ b\ c\ d}{a\ d\ c\ b}$. This corresponds to flipping the square about the top-left to bottom-right diagonal, or what is the same thing, about the line joining the points a and c of the outline.

(7) $\tau_{bd} = \binom{a\ b\ c\ d}{c\ b\ a\ d}$. This corresponds to flipping the square about the bottom-left to top-right diagonal, or what is the same thing, about the line joining the points b and d of the outline.

(8) $id = \begin{pmatrix} a & b & c & d \\ a & b & c & d \end{pmatrix}$. This of course corresponds to doing nothing to the square. As in the case of the equilateral triangle, this operation on the square is a rigid motion of the square, and after we have performed this operation, we would not be able to tell whether anybody had disturbed the square or not.

We will write D_4 for the set of these eight motions of the square.

We can compute the effect of composing two of these rigid motions just as we did with the triangle, and we will find that the composition of any two of these motions from this list is a third motion in this list! We will also find that given any motion on the list, there is another motion on this list which undoes the effect of this motion. (Also, just as with D_3, the elements of D_4 may be viewed as bijections between the set of corners $C = \{a, b, c, d\}$ and itself.) The composition table is partly filled in Table 13.2: it is a good exercise for you to complete the table.

Table 13.2. Partial composition table of D_4, the symmetries of a square. Complete this table!

$\circ$	id	σ	σ^2	σ^3	τ_h	τ_v	τ_{ac}	τ_{bd}
id	id	σ	σ^2	σ^3	τ_h	τ_v	τ_{ac}	τ_{bd}
σ	σ	σ^2	σ^3	id	τ_{bd}	τ_{ac}	τ_h	τ_v
σ^2								
σ^3								
τ_h								
τ_v								
τ_{ac}								
τ_{bd}								

As with the equilateral triangle, these rigid motions we have considered are referred to as the *symmetries* of a square.

It would be helpful to do the following exercise at this stage. (Remark 13.1 may be useful for notation and terminology.)

Exercise 13.3. Answer the following, using the completed Table 13.2 or by direct computations:

(1) Determine the set of elements in D_4 that commute with every other element.

(2) Determine $\sigma \circ \tau_h \circ \sigma^{-1}$, $\sigma^2 \circ \tau_h \circ \sigma^{-2}$, and $\sigma^3 \circ \tau_h \circ \sigma^{-3}$.

(3) Determine $\sigma \circ \tau_{ac} \circ \sigma^{-1}$, $\sigma^2 \circ \tau_{ac} \circ \sigma^{-2}$, and $\sigma^3 \circ \tau_{ac} \circ \sigma^{-3}$.

(4) Determine $\tau_v \circ \tau_h \circ \tau_v^{-1}$.

(5) Determine $\tau_{ac} \circ \tau_h \circ \tau_{ac}^{-1}$.

(6) Determine $\tau_{bd} \circ \tau_h \circ \tau_{bd}^{-1}$.

(7) Determine $\tau_h \circ \tau_{ac} \circ \tau_h^{-1}$.

(8) Determine $\tau_v \circ \tau_{ac} \circ \tau_v^{-1}$.

(9) Determine $\tau_{bd} \circ \tau_{ac} \circ \tau_{bd}^{-1}$.

(10) Determine $\tau_{ac} \circ \sigma \circ \tau_{ac}^{-1}$, $\tau_{ac} \circ \sigma^2 \circ \tau_{ac}^{-1}$, and $\tau_{ac} \circ \sigma^3 \circ \tau_{ac}^{-1}$.

(11) Determine $\tau_h \circ \sigma \circ \tau_h^{-1}$, $\tau_h \circ \sigma^2 \circ \tau_h^{-1}$, and $\tau_h \circ \sigma^3 \circ \tau_h^{-1}$.

13.3 Symmetries of an n-element set

Let n be a positive integer, and let Σ_n denote the set $\{1, 2, \dots, n\}$. In this section we will consider the set of all bijections between Σ_n and itself. We will write S_n for this set.

Recall from Definition 5.17, Chapter 5, that a bijection between Σ_n and Σ_n is just a function $f : \Sigma_n \to \Sigma_n$ that is both injective and surjective. We will use the same notation to describe such a function that we used for D_3 and D_4 (after all, the elements of D_3 and D_4 were themselves bijections between the set of corners and itself). So, if such a bijection sends 1 to the integer $a_1 \in \Sigma_n$, 2 to $a_2, \dots,$ and n to a_n, we will write

$$f = \begin{pmatrix} 1 & 2 & \dots & n \\ a_1 & a_2 & \dots & a_n \end{pmatrix}.$$

Since we will introduce another notation later in this section we will informally refer to the notation above as the *stack notation*.

Note that any bijection between Σ_n and itself is nothing but a permutation of 1, 2, $\dots, n$. We have already counted the number of such permutations in Chapter 4, Proposition 4.7. There are $n!$ such permutations. We thus find that S_n has $n!$ elements.

Note too that the composition of two bijections between Σ_n and itself is another bijection. (We have noted this fact before in this chapter when we composed two motions in D_3. See Exercise 5.29, Chapter 5.)

It will be instructive to write down explicitly the elements in Σ_n for small values of n. When $n = 1$, Σ_1 has only one element in it, the bijection "*id*" that sends 1 to 1! This bijection *id* when composed with itself gives back *id*: the composition also sends 1 to 1! Thus there is not a whole lot to say about Σ_1.

When $n = 2$ there are two: $f = \begin{pmatrix} 1 & 2 \\ 2 & 1 \end{pmatrix}$ and $id = \begin{pmatrix} 1 & 2 \\ 1 & 2 \end{pmatrix}$. We can compose these: for instance, $f \circ f$ sends 1 to 2 under the first f, and then 2 back to 1 under the second f, so $f \circ f$ sends 1 to 1. By default, 2 has to go to 2 under $f \circ f$ (it has nowhere else to go!). Thus, $f \circ f$ equals id. This is depicted in Table 13.3.

Table 13.3. Composition table for S_2, the symmetries of $\Sigma_2 = \{1, 2\}$.

$\circ$	id	f
id	id	f
f	f	id

When $n = 3$ there are $3! = 6$ bijections. We list them systematically: there are permutations that fix no element (that is, send no element to itself), so they move all three elements, there are permutations that fix just one element and therefore flip the other two, and there is the one permutation that fixes all three elements, in other words, it does nothing! (Note that if a permutation fixes two elements, it must fix the third as well, as the third element has nowhere else to go!) These are as follows:

Fix none, move all three: $r_1 = \binom{1\ 2\ 3}{2\ 3\ 1}$, $r_2 = \binom{1\ 2\ 3}{3\ 1\ 2}$

Fix exactly one, flip the other two: $f_1 = \binom{1\ 2\ 3}{1\ 3\ 2}$, $f_2 = \binom{1\ 2\ 3}{3\ 2\ 1}$, $f_3 = \binom{1\ 2\ 3}{2\ 1\ 3}$

Fix all three: $id = \binom{1\ 2\ 3}{1\ 2\ 3}$

We can compute out their various compositions. For instance, $f_2 \circ r_1$ sends 1 to 2 under r_1, and then 2 to 2 under f_2, so $f_2 \circ r_1$ sends 1 to 2. Similarly, $f_2 \circ r_1$ sends 2 to 3 under r_1, and then 3 to 1 under f_2, so $f_2 \circ r_1$ sends 2 to 1. By default, $f_2 \circ r_1$ must send 3 to 3. Thus, $f_2 \circ r_1$ equals f_3.

Computing out in full, we have Table 13.4.

Table 13.4. Composition table for S_3, the symmetries of $\Sigma_3 = \{1, 2, 3\}$.

$\circ$	id	r_1	r_2	f_1	f_2	f_3
id	id	r_1	r_2	f_1	f_2	f_3
r_1	r_1	r_2	id	f_3	f_1	f_2
r_2	r_2	id	r_1	f_2	f_3	f_1
f_1	f_1	f_2	f_3	id	r_1	r_2
f_2	f_2	f_3	f_1	r_2	id	r_1
f_3	f_3	f_1	f_2	r_1	r_2	id

Just as with D_3 and D_4, we see that the composition of any two bijections of Σ_3 is another bijection, but we knew this already from Chapter 5. Further, for every bijection f, there is another bijection g such that $f \circ g = g \circ f = id$. This too is something we knew would happen from Chapter 5. (See Proposition 5.32 in that chapter, along with the definition of invertibility of a permutation in Definition 5.31. Note that $\mathbb{1}_{\Sigma_3}$, in the notation of Chapter 5, is the same as the permutation that we have labelled id.)

Note the similarity between the D_3 table (Table 13.1) and the S_3 table above. (We will have more to say on the similarity in Example 13.41 ahead, where we show that the similarity is actually an *isomorphism*.) It is easy to see that this similarity will not extend to a similarity between D_4 and S_4, even though we have not yet computed out the composition table for S_4. This is because D_4 has 8 elements, but S_4 has $4! = 24$, so we cannot have this tidy match between their composition tables — the sizes will not match!

These bijections between Σ_n and itself are known as the symmetries of the n-element set. The collection of symmetries S_n, for each n, is known as the *symmetric group* on n letters. We will see in the next section why S_n is called a group.

Example 13.4. *Cycle Decomposition of a Permutation:* Let us use the notation $(1\ 4\ 6)$ for the permutation in S_6 that sends 1 to 4, 4 to 6, and then 6 back to 1, and that acts as the identity on the remaining elements 2, 3, and 5. Thus, $(1\ 4\ 6)$ is notation for $\binom{1\,2\,3\,4\,5\,6}{4\,2\,3\,6\,5\,1}$. We call a permutation such as $(1\ 4\ 6)$ a 3-*cycle*: "cycle" because it moves the elements cyclically, and "3" because there are 3 elements that are being moved cyclically.

Now consider the permutation $g = \binom{1\,2\,3\,4\,5\,6}{3\,6\,1\,5\,2\,4}$. We find that 1 goes to 3 and then 3 goes to 1, and next, 2 goes to 6 which goes to 4 which goes to 5 which goes back to 2. Thus there are two cycles present here: the 2 cycle $(1\ 3)$ and the 4-cycle $(2\ 6\ 4\ 5)$.

Exercise 13.4.1. Verify that $g = (1\ 3) \circ (2\ 6\ 4\ 5) = (2\ 6\ 4\ 5) \circ (1\ 3)$.

We call the decomposition of g into products using the algorithm above the *cycle decomposition* of g. It is clear that any $\sigma \in S_n$ (for any $n \geq 2$) has such a cycle decomposition: We start with 1 and follow where it goes, exactly as we did for g above, to obtain one cycle. Next, we take the smallest k that does not appear in the first cycle, and follow that where it goes, to get the second cycle. And so on. The various cycles that arise this way for a given σ commute among themselves, exactly as in Exercise 13.4.1, since the cycles each move disjoint sets of elements. (For instance, in the decomposition of g above, $(1\ 3)$ only affects 1 and 3, and leaves the others unchanged. On the other hand, $(2\ 6\ 4\ 5)$ affects the others and leaves 1 and 3 unchanged. As a result, it does not matter whether you perform $(1\ 3)$ first and $(2\ 6\ 4\ 5)$ next or the other way around.)

Groups

With the examples of D_3, D_4 and S_n behind us, we are ready to consider the concept of a group.

First, a quick definition, which expands on Definition 12.80, Chapter 12:

Definition 13.5. Let S be a set. A *binary operation* on S is a function $f : S \times S \to S$. The binary operation is said to be *commutative* if for all $s, t \in S$, $f(s,t) = f(t,s)$. The binary operation is said to be *associative* if for all $s, t, u \in S$, $f(f(s,t),u) = f(s, f(t,u))$.

Notice something common to D_3, D_4 and S_n. Each is a set, with a *binary operation* defined on it, namely the composition of two motions in the case of D_3 and D_4, and the composition of bijections, in the case of S_n. For example, in D_3, $\sigma \circ \tau_a = \tau_c$. Thus, $\circ$ may be viewed as the binary operation on D_3 that takes in (σ, τ_a) and yields τ_c. This binary operation is not commutative (for instance, in D_3, $\sigma \circ \tau_a = \tau_c$ but $\tau_a \circ \sigma = \tau_b$, and in D_4, $\sigma \circ \tau_h = \tau_{bd}$ but $\tau_h \circ \sigma = \tau_{ac}$). However, the binary operation is definitely associative, that is, $(g \circ h) \circ k = g \circ (h \circ k)$ for all $g, h, k \in D_3$, as well as for all $g, h, k \in D_4$ or in S_n. This is because, as we have observed, the elements of D_3 may be interpreted as functions on $C = \{a, b, c\}$, and function composition is certainly associative (see Exercise 5.30, Chapter 5), and similarly for D_4. In the case of S_n, the elements are already functions (more precisely bijective functions, but at root, functions), so their composition is certainly associative.

Here is another fact common to all three of D_3, D_4, and S_n, namely they each have a distinguished element, denoted *id* in all three sets, that satisfy the property that $g \circ id = id \circ g = g$ for all g in their respective sets. (You can see these relations by examining the composition tables for D_3 and D_4 and S_2 and S_3, or directly for all of D_3, D_4 and S_n by noting that *id* is the do nothing operation and thus does not affect any motion or bijection that precedes it or follows it!)

And a last fact common to all three of D_3, D_4 and S_n that will be of interest to us: for each $g \in D_3$ there exists an element $h \in D_3$ such that $g \circ h = h \circ g = id$, and similarly for D_4 and for S_n.

As it turns out, there are many objects that show up in many places in mathematics (as well as in other subjects, such as physics and chemistry) that have exactly these properties that we have focused on above. We give such objects a special name: *groups*. And as is common in mathematics, when different objects appear in many places with the same properties, these objects with their common properties are studied for their own sake. Such study often reveals rich depths to these objects, and usually, the results of the study can be applied back to the original situation in which these objects arose, to provide new and fruitful insight. Such is definitely the case with groups!

The formal definition is as follows:

Definition 13.6. A *group* is a set G with a binary operation " $\cdot$ ": $G \times G \to G$ with the properties:

(1) The binary operation $\cdot$ is associative.

(2) There exists a special element in G, the "identity element," denoted id_G or usually just id for simplicity, that satisfies the property that $g \cdot id = id \cdot g = g$ for all $g \in G$.

(3) For each $g \in G$, there is an element $h \in G$ such that $g \cdot h = h \cdot g = id$. The element h is also denoted g^{-1} and is known as the "inverse of g."

An *abelian* group is a group in which the binary operation is also commutative.

If G is a finite set we say that G is a finite group. Otherwise, we say that G is an infinite group.

The binary operation in G is sometimes referred to as the *composition* in G.

We sometimes write $(G, \cdot)$ for the group, to emphasize that there are two ingredients to the definition, the set and the binary operation. Also, when the context is clear, we often just write gh instead of $g \cdot h$. For $k = 2, 3, \ldots$, we will write g^k for $g \cdot g \cdots g$ (k times). (Of course, g^1 will denote just g.) For $k = -2, -3, \ldots$, we will write g^k for

$g^{-1} \cdot g^{-1} \cdots g^{-1}$ ($|k|$ times). We will let g^0 denote the identity element. (All this is just like in Remark 13.1.) We will see later that the identity element, as well as the inverse of a given $g \in G$, are unique.

All three of D_3, D_4 and S_n are clearly groups under the binary operations defined by the composition of motions or the composition of bijections in the case of S_n. D_3, D_4 and S_3 are not abelian as we can see from their composition tables. S_2 (and trivially, S_1) is abelian. As for S_n for $n \geq 4$, we have the following easy exercise:

Exercise 13.7. Show that S_n is not abelian for $n \geq 4$. (Hint: from Table 13.4, we know that r_1 and f_1 in S_3 do not commute. Consider two elements of S_n whose restrictions to Σ_3 are r_1 and f_1.)

Here are some other examples of groups:

Example 13.8. What is common to the equilateral triangle and the square is that they are regular 3-gons and regular 4-gons respectively. Just as with D_3 and D_4, we can consider the symmetries of a regular 5-gon, also known as a regular pentagon. The set of these symmetries is known as D_5. Prove that there are 10 elements in D_5, by copying the arguments used for D_3 or D_4. Play with these symmetries. These will form a group. While you do not have to develop the full composition table to show formally that these form a group, you should play with a few products and inverses of motions to get a feel for D_5.

In a similar fashion, we get groups D_n, $n = 6, 7, \ldots$, by considering the symmetries of a regular hexagon, a regular heptagon, etc. These groups are known as the *dihedral groups.* Each D_n contains $2n$ elements.

Example 13.9. $(\mathbb{Z}, +)$, the integers with the usual addition taken as the binary operation. Note that addition may indeed be viewed as a binary operation "$+$" $: \mathbb{Z} \times \mathbb{Z} \to \mathbb{Z}$ that takes the pair (m, n) to $m + n$. (We have considered this situation before, where familiar addition or multiplication processes are described as binary operations: for instance in Chapter 9, in Remark 9.41, and Exercises 9.43, and 9.46.) As it turns out, this is operation is both commutative and associative. The integer 0 acts as the identity element, since $m + 0 = 0 + m = m$ for all $m \in \mathbb{Z}$. The integer $-m$ is the inverse of m, since $m + (-m) = (-m) + m = 0$. Thus, $(\mathbb{Z}, +)$ is a group, and in fact, it is an abelian group. Note that $(\mathbb{Z}, +)$ is an infinite group.

We say that $\mathbb{Z}$ admits a *group structure with respect to addition.* We refer to $-m$ as the *additive inverse* of m.

In a similar vein, $(\mathbb{Q}, +), (\mathbb{R}, +), (\mathbb{C}, +), (\mathbb{Z}/n\mathbb{Z}, +), (M_2(\mathbb{C}), +)$ are all abelian groups, or put differently, $\mathbb{Q}, \mathbb{R}, \mathbb{C}, \mathbb{Z}/n\mathbb{Z}, M_2(\mathbb{C})$ all admit a group structure with respect to addition. $(\mathbb{Z}/n\mathbb{Z}, +)$ is a finite group, and the rest are infinite groups. (To see that addition in $M_2(\mathbb{C})$ is associative, see Exercise 13.46 ahead.)

Example 13.10. Write $\mathbb{R}^*$ for the set $\mathbb{R} \setminus \{0\}$. The usual multiplication of real numbers, $(a, b) \mapsto a \cdot b$, is a binary operation on $\mathbb{R}^*$. (It is also a binary operation on $\mathbb{R}$, but we will see in a moment why we restrict our attention to $\mathbb{R}^*$.) This operation is both commutative and associative. The identity element for this operation is the number 1, since $r \cdot 1 = 1 \cdot r = r$ for all $r \in \mathbb{R}^*$. Every *nonzero* element r admits a companion number $1/r$ that satisfies $r \cdot 1/r = 1/r \cdot r = 1$. Hence $(\mathbb{R}^*, \cdot)$ is an abelian group.

We say that $\mathbb{R}^*$ admits a *group structure with respect to multiplication.* We refer to $1/r$ as the *multiplicative inverse of* r.

The last fact that every element has a multiplicative inverse will fail to hold if we were to consider all of $\mathbb{R}$ as a candidate for a group under multiplication, since 0 does not have a multiplicative inverse. It is for this reason that we restrict our attention to $\mathbb{R}^*$. (Although, as we have observed in Example 13.9 above, if we consider addition instead as our binary operation, then all of $\mathbb{R}$ forms a group.)

In a similar vein, $(\mathbb{Q}^*, \cdot)$ and $(\mathbb{C}^*, \cdot)$ also are abelian groups. So also are $(\mathbb{R}_{>0}, \cdot)$ and $(\mathbb{Q}_{>0}, \cdot)$. Can you see that when n is a prime, $(\mathbb{Z}/n\mathbb{Z} \setminus \{0\}, \cdot)$ is an abelian group, but fails to be a group if n is not prime? (See Exercise 10.48, Chapter 10 and Exercise 1.16, Chapter 1.)

Here is a quick exercise that builds further on the example of $(\mathbb{Z}/n\mathbb{Z} \setminus \{0\}, \cdot)$ in the paragraph above:

> **Exercise 13.10.1.** Use Exercise 10.49, Chapter 10, to show that the subset $\mathbb{Z}/n\mathbb{Z}^* \stackrel{\text{def}}{=} \{[a]_n \in \mathbb{Z}/n\mathbb{Z} \mid \gcd(a, n) = 1\}$ of $\mathbb{Z}/n\mathbb{Z}$ forms a group with respect to the multiplication operation in $\mathbb{Z}/n\mathbb{Z}$. (Note that by the same Exercise 10.49, Chapter 10, $\mathbb{Z}/n\mathbb{Z}^*$ is well defined, since if $\tilde{a}$ and $\tilde{\tilde{a}}$ are two representatives of $[a]_n$, then $\gcd(\tilde{a}, n) = 1$ if and only if $\gcd(\tilde{\tilde{a}}, n) = 1$.)

Example 13.11. Our next example will deal with $M_2(\mathbb{C})$, the set of 2×2 matrices with entries in $\mathbb{C}$ that we have already considered in Chapter 5, Example 5.44. We have seen above that $(M_2(\mathbb{C}), +)$ is a group: the identity element of $(M_2(\mathbb{C}), +)$ is the zero matrix

$$\mathbf{0}_2 = \begin{pmatrix} 0 & 0 \\ 0 & 0 \end{pmatrix}.$$

One can verify that $\mathbf{0}_2 + M = M + \mathbf{0}_2 = \mathbf{0}_2$ for all matrices M.

One can also verify that $\mathbf{0}_2 \cdot M = M \cdot \mathbf{0}_2 = \mathbf{0}_2$ for all matrices M.

The identity matrix for multiplication is the matrix

$$\mathbf{I}_2 = \begin{pmatrix} 1 & 0 \\ 0 & 1 \end{pmatrix}.$$

One can similarly check that $\mathbf{I}_2 \cdot M = M \cdot \mathbf{I}_2 = M$ for all matrices M.

In particular this means that there cannot exist a matrix M such that $\mathbf{0}_2 \cdot M = M \cdot \mathbf{0}_2 = \mathbf{I}_2$, since $\mathbf{0}_2 \cdot M = \mathbf{0}_2$. Thus, the zero matrix has no multiplicative inverse. It follows that $M_2(\mathbb{C}, \cdot)$ cannot be a group. (In fact, as we will see in Exercise 13.47, there are many other matrices in $M_2(\mathbb{C})$ that do not have multiplicative inverses.)

Arguing by analogy from the redeeming example of $(\mathbb{C}^*, \cdot)$ being a group even though $(\mathbb{C}, +)$ is not (and also, from the equally redeeming example of $(\mathbb{Z}/n\mathbb{Z}^*, \cdot)$ being a group even though $(\mathbb{Z}/n\mathbb{Z}, +)$ is not; see Exercise 13.10.1), if we were to restrict our attention to the *multiplicatively invertible* elements of $M_2(\mathbb{C})$, that is, those matrices M for which there is a matrix N such that $M \cdot N = N \cdot M = \mathbf{I}_2$, one should expect to get a group structure with respect to matrix multiplication. But we need to study this further:

Given the matrix

$$M = \begin{pmatrix} a & b \\ c & d \end{pmatrix},$$

we define the *determinant of* M, denoted $\det(M)$, as the complex number $\det(M) \stackrel{\text{def}}{=} ad - bc$. We can verify by direct multiplication that if $\det(M)$ is nonzero, then the matrix

$$N = \begin{pmatrix} d/\det(M) & -b/\det(M) \\ -c/\det(M) & a/\det(M) \end{pmatrix}$$

satisfies $M \cdot N = N \cdot M = \mathbf{I}_2$. (We have seen these constructions of the determinant and the inverse in Exercises 5.62 and 5.63 in Chapter 5 already, except that we restricted our attention there to matrices over $\mathbb{R}$.) Thus, $\det(M) \neq 0$ is a sufficient condition for M to have a multiplicative inverse. As it turns out, this condition is also necessary (see Exercise 13.47 ahead). Thus, we need to focus our attention on the subset $Gl_2(\mathbb{C}) \stackrel{\text{def}}{=} \{M \in M_2(\mathbb{C}) \mid \det(M) \neq 0\}$.

We need to show that multiplication is a binary operation on $Gl_2(\mathbb{C})$. Thus, if M and N are two invertible matrices, we need to show that their product $M \cdot N$ (which certainly is another matrix in $M_2(\mathbb{C})$) actually lives in $Gl_2(\mathbb{C})$, that is, $M \cdot N$ is also invertible. This is an easy and instructive computation, which we will leave as an exercise:

> **Exercise 13.11.1.** Show that if M and N are matrices in $M_2(\mathbb{C})$, then $\det(M \cdot N) = \det(M)\det(N)$. Conclude that if M and N are both invertible, then MN is also invertible.

It is now easy to verify that $(Gl_2(\mathbb{C}), \cdot)$ is a group. The associativity of matrix multiplication follows from the associativity of addition and multiplication of complex numbers (see Exercise 13.46 ahead). The multiplicative identity, as already noted, is $\mathbf{I}_2$, and by our choice of sets, every matrix in $Gl_2(\mathbb{C})$ has a multiplicative inverse.

By analogy, one can define a corresponding determinant function $\det : M_n(\mathbb{C}) \to \mathbb{C}$ on the set of $n \times n$ matrices with entries in $\mathbb{C}$ (but we will not do so here). It turns out that $M \in M_n(\mathbb{C})$ is invertible if and only if $\det(M) \neq 0$, and we may consider the set $Gl_n(\mathbb{C}) \stackrel{\text{def}}{=} \{M \in M_n(\mathbb{C}) \mid \det(M) \neq 0\}$. This will analogously be a group with respect to multiplication.

Example 13.12. Fix an integer $n \geq 2$, and let ω_n denote the complex number $\cos(2\pi/n) + i\sin(2\pi/n)$. This is the complex number of modulus 1, at a counterclockwise angle of $2\pi/n$ radians with respect to the positive x-axis. By DeMoivre's theorem (see Exercise 7.23, Chapter 7, for example), the various powers ω_n^k, $k = 1, 2, \ldots$, are the complex numbers $\cos(2k\pi/n) + i\sin(2k\pi/n)$, which are also of modulus 1 but at a counterclockwise angle of $2k\pi/n$ radians with respect to the positive x-axis. Thus, the various powers $\omega_n, \omega_n^2, \ldots$ are evenly spaced about the unit circle appearing every $2\pi/n$ radians. Of course, this means that ω_n^n is of modulus 1 and is at an angle of $2n\pi/n = 2\pi$ with respect to the positive x-axis, or in other words, $\omega_n^n = 1$. Play with these numbers for various n, and prove for yourselves that for each n, the set of complex numbers $\boldsymbol{\mu}_n \stackrel{\text{def}}{=} \{\omega_n^k, \mid k = 0, 1, \ldots, n-1\}$ forms a group with respect to complex number multiplication. Can you write the inverse of ω_n^k as ω_n^l for some nonnegative l?

The groups $\boldsymbol{\mu}_n$ are known as the *cyclic groups of order n*.

Notice that $(\omega_n^k)^n = \omega_n^{kn} = (\omega_n^n)^k = 1^k = 1$. Thus, the members of $\boldsymbol{\mu}_n$ are all solutions of $x^n = 1$. Hence, they are known as the *n-th roots of unity*. When $n = 2$, $\omega_2 = -1$, so $\boldsymbol{\mu}_2 = \{-1, 1\}$. When $n = 3$, $\omega_3 = \dfrac{-1 + i\sqrt{3}}{2}$, so $\boldsymbol{\mu}_3 = \left\{\dfrac{-1 + i\sqrt{3}}{2}, \dfrac{-1 - i\sqrt{3}}{2}, 1\right\}$. When $n = 4$, $\omega_4 = i$, so $\boldsymbol{\mu}_4 = \{i, -1, -i, 1\}$. And so on.

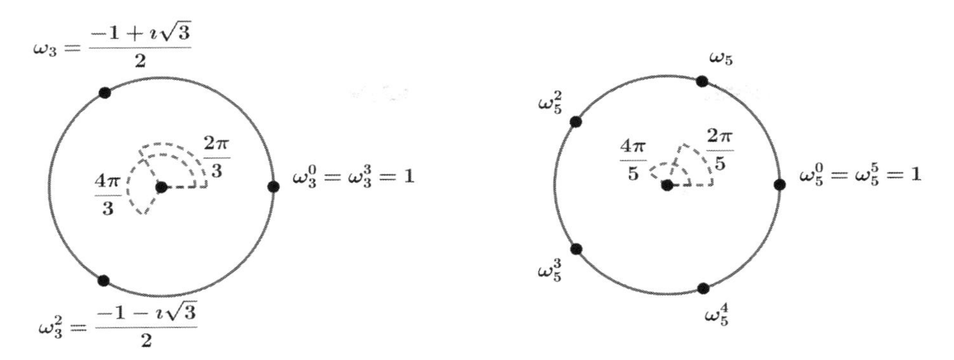

Figure 13.6. The third roots of unity $\boldsymbol{\mu}_3$ on the left and the fifth roots of unity $\boldsymbol{\mu}_5$ on the right.

See Figure 13.6 for the cases $n = 3$ and $n = 5$. Incidentally, the well-known factorization $x^2 - y^2 = (x-y)(x+y)$ uses the values ± 1 in front of the "y," which are just the elements of $\boldsymbol{\mu}_2$. Analogously, we have the factorizations

$$\begin{aligned} x^3 - y^3 &= (x-y)(x-\omega_3 y)(x-\omega_3^2 y), \\ x^4 - y^4 &= (x-y)(x-\omega_4 y)(x-\omega_4^2 y)(x-\omega_4^3 y) \\ \vdots &= \quad \vdots \\ x^n - y^n &= \prod_{i=0}^{n-1}(x-\omega_n^i y). \end{aligned}$$

Example 13.13. Let $(G, \cdot)$ and $(H, *)$ be groups, where we have used "$*$" to denote the composition in H to distinguish it from the composition in G. Consider the direct product set $G \times H = \{(g,h) \,|g \in G \text{ and } h \in H\}$ (see Definition 5.12, Chapter 5). Although initially defined to be just a set, there is a natural binary operation on $G \times H$ that arises from the individual binary operations in G and H: we define $(g,h)\cdot(g',h')$ to be $(g \cdot g', h * h')$, where, as indicated, the composition in the first slot is happening in G and the one in the second slot is happening in H. It is an illustrative exercise to show that $G \times H$ becomes a group, with identity element the order pair (id_G, id_H).

When both groups are abelian, we often denote $G \times H$ as $G \oplus H$. $G \oplus H$ will also be abelian, as can readily be checked.

A particularly useful example of this is when we take $G = H = (\mathbb{R}, +)$. We write $\mathbb{R}^2$ for $\mathbb{R} \oplus \mathbb{R}$. This is the group underlying ordinary 2-dimensional space. After choosing a coordinate system on the plane, we may view the ordered pair (a,b) of $\mathbb{R}^2$ as the point with x-coordinate a and y-coordinate b.

Remark 13.14. If one were to think about it, it should be clear that it really does not matter what name we give to the binary operation that defines our group, all that matters is that it should have the properties listed in Definition 13.6. In the case of $(\mathbb{Z}, +)$ for example, we have historically called the process of taking 2 and 3 and getting back 5 "addition" and denoted this symbolically with the "+" sign. But human history would likely not have changed if we had called this something else, concatenation for example, and chosen some other symbol, "!!" for example. (Thus, we could just as easily have written 2! ! 3 = 5 and read this as "2 concatenated with 3 gives 5.")

Many of the groups we have considered have arisen in concrete settings where a name and a symbol have already been used historically, exactly as in $(\mathbb{Z}, +)$ above. In such situations, mathematicians typically continue to use the historically accepted name and symbol for the operation. When talking of groups in the abstract, conventionally, mathematicians just refer to the binary operation as the *composition* that defines the group, and typically use the multiplicative symbol $\cdot$ to denote this composition and the suggestive notation 1 to denote the identity element. Moreover, just as with variables, where we do not bother writing $x \cdot y$ every time but shorten it to xy, composition in abstract groups is often shown without the "$\cdot$" symbol: thus, instead of writing $g \cdot h$ for elements g and h in the group, we often write gh.

There is one deviation from this convention: when we know ahead of time that our group is abelian. In that case, we conventionally think of our group composition as addition (even though there may be no actual addition in the conventional sense taking place!), and write $g + h$ for the composition instead of gh, and 0 instead of 1 for the identity element.

We will take care of a few housekeeping details now: these are results that can be derived very quickly from the axioms. They enable us to talk about "the" identity element, "the" inverse of an element, to cancel across two sides of an equality, to write out the inverse of a product, and so on.

Lemma 13.15. *The identity element in a group is unique.*

Proof. Let id and id' be two identity elements in a group G. Then, viewing id as the identity element and id' as just another element of G, we find $id \cdot id' = id'$. But then, viewing id' as the identity element and id as just another element, we find $id \cdot id' = id$. Thus, $id' = id \cdot id' = id$, proving that the identity element is unique. □

Lemma 13.16. *The inverse of any element in a group is unique.*

Proof. Let h and k be two inverses of an element g of a group G, so $gh = hg = id$ and $gk = kg = id$. Hence, $gh = gk$. Multiplying on the left by, say h, we find $h(gh) = h(gk)$. By associativity, we find $(hg)h = (hg)k$. Since $hg = id$, we find $id \cdot h = id \cdot k$, and since id is just the identity element, this last equality yields $h = k$. □

The argument used in Lemma 13.16 yields a more general statement, namely, that left and right cancellation is possible in groups:

Lemma 13.17. *Let G be a group, and g, h, k elements of G. Then*

(1) *Left cancellation: If $gh = gk$ then $h = k$.*

(2) *Right cancellation: If $hg = kg$ then $h = k$.*

Proof. The proofs of both statements are similar; we will prove just the right cancellation property: Given $hg = kg$, multiply on the *right* by g^{-1} (which by Lemma 13.16 is uniquely specified!). Doing so, we find $(hg)g^{-1} = (kg)g^{-1}$, so invoking associativity of composition, $h(gg^{-1}) = k(gg^{-1})$, so $h \cdot id = k \cdot id$, so $h = k$. □

Here is an easy result:

Lemma 13.18. *Let G be a group and let $g_1, g_2, \ldots, g_k$ be elements of G. Then $(g_1 g_2 \cdots g_k)^{-1} = g_k^{-1} g_{k-1}^{-1} \cdots g_1^{-1}$. In particular, for $j = 2, 3, \ldots$, $(g^j)^{-1} = (g^{-1})^j = g^{-j}$.*

The proof is elementary and we leave it as an exercise. Note that the last equality $(g^{-1})^j = g^{-j}$ above is simply a matter of convention: we have *agreed* to write g^{-j} for $(g^{-1})^j$ (see the discussion after Definition 13.6, where we also agreed that g^0 will mean the identity).

The following is an exercise:

> **Exercise 13.19.** Let G be a group, and g an element of G. Show that $g^m g^n = g^{m+n}$ for m and n in $\mathbb{Z}$. (Hint: divide your proof according to whether m and n are individually either positive or zero or negative, and consider each case separately.)

13.4 Subgroups

We will consider the notion of subgroups in this section. Informally, a subgroup is a subset of a group that itself forms a group with respect to the same binary operation that defines the larger group. Groups very often arise as subgroups of larger groups, and the relationship between the subgroup and the larger group often carries information on related mathematical objects.

First we study the notion of a subset of a group being closed with respect to the binary operation:

Definition 13.20. Let $(G, \cdot)$ be a group, and let H be a nonempty subset of G. We say H is *closed with respect to "$\cdot$"* if $h_1 \cdot h_2 \in H$ for all h_1, h_2 in H.

Let us consider an example. Consider the group $(\mathbb{Z}, +)$. The subset $H = \{2n \mid n \in \mathbb{Z}\}$ is closed with respect to "$+$." Indeed, H consists of all the even integers, and the sum of any two even integers is definitely even, and thus, an element of H. By contrast, the subsets $H = \{2n + 1 \mid n \in \mathbb{Z}\}$ and $H' = \{n^2 \mid n \in \mathbb{Z}\}$ are not closed with respect to $+$: the sum of two odd integers is never odd, and the sum of two squares is not always a square, for instance, $2^2 + 3^2$ is not a square. (Although, as would no doubt be familiar to you, the sum of two squares can sometimes very pleasingly be another square!)

When a subset $H \subseteq (G, \cdot)$ is closed with respect to $\cdot$, then $\cdot$ becomes a binary operation on H, since it takes in two elements of H and yields another element in H.

We are now ready for the definition of a subgroup:

Definition 13.21. Let $(G, \cdot)$ be a group. A (nonempty) subset H of G is called a *subgroup* of G if H is closed with respect to $\cdot$, and $(H, \cdot)$ is itself a group.

Let us consider an example, one we looked at just before this definition. Take $G = (\mathbb{Z}, +)$ and let H be the subset $\{2n \mid n \in \mathbb{Z}\}$, which we will now denote $2\mathbb{Z}$. As we just observed, $2\mathbb{Z}$ is closed with respect to $+$. Now $(2\mathbb{Z}, +)$ itself forms a group: the associativity of $+$ as a binary operation on H is a given because it is associative as a binary operation on G already (in other words, the relation $m + (n + r) = (m + n) + r$ holds for *all* integers, and in particular for the even integers!). The integer 0 is even, and therefore is an element of $2\mathbb{Z}$, and of course, $0 + m = m + 0 = m$ holds for *all* integers, and in particular for the even integers. Finally, given any even integer $2n$, its additive inverse is $-2n = 2(-n)$, which is also even. Thus, every element of $2\mathbb{Z}$ has an inverse in $2\mathbb{Z}$ with respect to $+$. $(2\mathbb{Z}, +)$ is therefore a group in its own right, so it is a subgroup of $(\mathbb{Z}, +)$.

It is instructive to look at a nonexample! For the same group $G = (\mathbb{Z}, +)$, if we were to consider $H = \{2n \mid n \in \mathbb{N} \text{ and } n \geq 0\}$, we would find that H is indeed closed with respect to addition, and contains an identity element, namely 0, but still fails to be a subgroup of G because it does not contain inverses of all its elements! For instance, $-2 \notin H$!

A small detail to take care of: Definition 13.21 only requires that H be a group in its own right with respect to $\cdot$. As such, as a group in its own right, the identity element of H could conceivably be different from the identity element in G. Thus, if id_G is the identity element of G, it is conceivable that H contain its own identity element $id_H \neq id_G$. (Lemma 13.15 only shows that if two elements in G act as the identity element *for all of* G then they must be equal. In our case id_H is only the identity element of H, so Lemma 13.15 does not apply.) Reassuringly, this does not happen! We will leave this as an exercise:

> **Exercise 13.22.** Let H be a subgroup of G. Show that the identity elements of G and H are the same. (Hint: Consider $id_G id_H$ and also $id_H id_H$.)

Before we consider further examples of subgroups, it is useful to have a test to easily determine if a given nonempty subset H of a group G is a subgroup, something easier to apply than working through all axioms. The following is one such test:

Lemma 13.23. (*One-step subgroup test.*) *Let H be a nonempty subset of a group G. If $ab^{-1} \in H$ for all $a, b \in H$, then H is a subgroup of G.*

Proof. Since H is nonempty, there is an element $h \in H$. Taking $a = b = h$ in the given hypothesis, we find $hh^{-1} = id \in H$. Thus, the identity element of G lives inside H, where it will therefore also serve as the identity element for H.

Now, for any $h \in H$, we take $a = id$ and $b = h$ in the hypothesis to find $id \cdot h^{-1} = h^{-1} \in H$. Thus, H contains the inverses of all its elements. To check for closure with respect to the composition in G, note that given h, k in H, $k^{-1} \in H$ as we have just seen. Hence, taking $a = h$ and $b = k^{-1}$, we find $h\left(k^{-1}\right)^{-1} = hk \in H$, so indeed H is closed with respect to the composition in G. This composition continues to be associative in H as it is already associative in G. Hence, H satisfies all the requirements for being a group with respect to the composition in G, and is thus a subgroup of G. □

This test can be further simplified when H is finite:

Lemma 13.24. (*Finite subgroup test.*) *Let H be a nonempty and finite subset of a group G. If H is closed with respect to the composition in G, then H is a subgroup of G.*

Proof. Note that the associativity of the composition in H is inherited from its associativity in G. Now, given any $h \in H$ (remember H is nonempty!), the elements $h, h^2, \ldots$

all belong to H as H is closed with respect to the composition in G. These elements h^i, $i = 1, 2, \ldots$, cannot all be distinct as H is finite. Hence, there exist $1 \leq i < j$ such that $h^i = h^j$. We write this as $h^i \cdot id = h^i h^{j-i}$. By left cancellation (Lemma 13.17), we find $id = h^{j-i}$. Since $j-i \geq 1$, h^{j-i} is the product of h with itself a suitable number of times, so it belongs to H by closure under the composition in G. Thus we find $id \in H$, and it will of course continue to act as the identity in H. We also wish to show that the inverse of h is in H. We saw above that $id = h^{j-i}$, for some $1 \leq i < j$. If $j - i = 1$, this relation reads $h^1 = id$, that is, h is the identity, so it is its own inverse. Otherwise, $j - i > 1$, so $j - i - 1 \geq 1$. We write $id = h^{j-i}$ as $id = h \cdot h^{j-i-1} = h^{j-i-1} \cdot h$. Thus $h^{-1} = h^{j-i-1}$. Now since $j - i - 1 \geq 1$, h^{j-i-1} is the product of h with itself a suitable number of times, so it belongs to H by closure under the composition in G. Hence $h^{-1} \in H$. We find that H satisfies all the axioms for being a group, and is thus a subgroup of G. □

Example 13.25. We will now consider examples of subgroups. We will encounter many new examples of groups in the process:

(1) Here are two subsets of a group G that are trivially subgroups of G: the set $\{id\}$ consisting of just the identity, as well G itself! Go through the definitions, or use Lemmas 13.23 or 13.24 (for the set $\{id\}$), and verify this.

(2) For each $k = 3, 4, \ldots$, the set $k\mathbb{Z} = \{kn \mid n \in \mathbb{Z}\}$ are subgroups of $(\mathbb{Z}, +)$, exactly like the set $2\mathbb{Z}$ considered above. See Exercise 13.51 ahead.

(3) Let G be a group, and let g be an element in G. Consider the set $\{g^n \mid n \in \mathbb{Z}\}$, which we will denote $\langle g \rangle$. Thus, $\langle g \rangle$ is nonempty, and consists of $g, g^2, g^3, \ldots$, along with $g^0 = id$, as well as $g^{-1}, g^{-2} = (g^2)^{-1}, g^{-3} = (g^3)^{-1}$, etc. (See Lemma 13.18 as well as Exercise 13.19.) For any g^n and g^m in this set, we have $g^n (g^m)^{-1} = g^n g^{-m} = g^{n-m}$, another element in the same set. By Lemma 13.23, $\langle g \rangle$ is a subgroup of G.

$\langle g \rangle$ is known as the *subgroup generated by* g. The reason for this name is the following: Let H be any subgroup of G that contains g. We claim that H must contain $\langle g \rangle$. This is because if H contains g, then first, by virtue of H being a subgroup and therefore being closed under the composition in G, it must contain all positive powers g^n, $n > 0$. Next it must contain $id_H = id_G = g^0$. It must also contain the inverse of g, i.e., g^{-1}. Finally, by closure again, it must contain all powers $(g^{-1})^n$, $n > 0$, i.e., all g^{-n}, $n > 0$. This proves our claim. In a sense therefore, $\langle g \rangle$ is the *smallest* subgroup of G that contains g: because $\langle g \rangle$ is already contained in every subgroup that contains G. It is for this reason that we call $\langle g \rangle$ the subgroup generated by G.

See Exercise 13.48 ahead.

In $(\mathbb{Z}, +)$, for instance, the subgroup $\langle 2 \rangle$ is just the subgroup $2\mathbb{Z}$ that we have already considered. Note that it is an infinite group. Whereas, in $M_2(\mathbb{C})$, for instance,

$$\left\langle \begin{pmatrix} 0 & 1 \\ 1 & 0 \end{pmatrix} \right\rangle = \left\{ \begin{pmatrix} 0 & 1 \\ 1 & 0 \end{pmatrix}, \begin{pmatrix} 1 & 0 \\ 0 & 1 \end{pmatrix} \right\},$$

a finite group!

Definition 13.25.1. The cardinality of $\langle g \rangle$, considered as just a set, is known as the *order of* g. It is a positive integer if $\langle g \rangle$ is finite, else it is ∞.

(4) Very interesting groups arise as subgroups of $Gl_2(\mathbb{C})$ (and more generally as subgroups of $Gl_n(\mathbb{C})$ for $n \geq 2$). Here is one, denoted $U(2)$:

Consider the subset of $Gl_2(\mathbb{C})$ consisting of matrices M with the property that $M^*M = I_2$. Here, M^* denotes the *conjugate transpose* of M, that is,

$$M^* = \begin{pmatrix} \overline{a} & \overline{c} \\ \overline{b} & \overline{d} \end{pmatrix}, \text{ if } M = \begin{pmatrix} a & b \\ c & d \end{pmatrix},$$

where $\overline{z}$ denotes the complex conjugate of the complex number z. This will be our set $U(2)$.

Recall that for M as above,

$$\overline{M} = \begin{pmatrix} \overline{a} & \overline{b} \\ \overline{c} & \overline{d} \end{pmatrix} \text{ and } M^t = \begin{pmatrix} a & c \\ c & d \end{pmatrix},$$

where M^t denotes the *transpose* of M.

Exercise 13.25.1. We will explore the properties of this subset in the following:

(a) Verify that $\overline{(M^t)} = \left(\overline{M}\right)^t = M^*$.

(b) If N is another matrix in $M_2(\mathbb{C})$, verify that $(MN)^t = N^tM^t$, $\overline{MN} = \overline{M}\,\overline{N}$, and $(MN)^* = N^*M^*$.

(c) Show that for $M \in U(2)$, $\det(M)$ lies on the unit circle. In particular, it is nonzero.

(d) Show that $M^*M = I_2$ if and only if $MM^* = I_2$. (Hint: Use the fact that M and M^* are invertible, see the previous part, and multiply the hypotheses on the right and left by suitable matrices.)

(e) Conclude that if $M \in U(2)$ then $M^{-1} = M^*$.

(f) Verify that for $M, N \in U(2)$, MN^{-1} is also in $U(2)$.

(g) Produce a concrete matrix in $U(2)$ to show it is nonempty, and conclude using Lemma 13.23 that $U(2)$ is a subgroup of $Gl_2(\mathbb{C})$.

$U(2)$ is known as the *unitary group of complex* 2×2 *matrices.* It and its further subgroups $SU(2)$, the subgroup of complex unitary 2×2 matrices with determinant exactly 1, have immense application in diverse fields, particularly in physics.

13.5 Cosets, Lagrange's Theorem

In this section we will prove a very pretty theorem for finite groups known as Lagrange's theorem. Recall from Example 13.25(3) (see Definition 13.25.1) that we defined the order of an element g in a group G to be the cardinality of the subgroup $\langle g \rangle$, considered as just a set. This definition is actually a special case of a more general definition:

Definition 13.26. If H is any subgroup of a group G, we define the *order* of H to be the cardinality of H considered as just a set. (In particular, the order of G is the cardinality of G considered as just a set.)

This theorem says that for a finite group G, the order of any subgroup H must divide the order of G. The route to this theorem is via the concept of *cosets* of a subgroup, a concept of independent interest.

So let G be a group, as of now finite or infinite, and let H be a subgroup. For any $g \in G$, we construct the following two subsets of G:

Definition 13.27. With G, H, and g as above:

(1) The set $gH \stackrel{\text{def}}{=} \{gh \mid h \in H\}$ is known as the *left coset of* H *with respect to* g.

(2) The set $Hg \stackrel{\text{def}}{=} \{hg \mid h \in H\}$ is known as the *right coset of* H *with respect to* g.

Note that the left and right cosets gH and Hg are defined as simply *subsets* of G. They will not in general be subgroups of G (although they will for special values of g, as we will see in Exercise 13.29 ahead).

Example 13.28. Here are some examples of cosets:

(1) Take $G = D_3$, and $H = \langle\tau_a\rangle$. Note that $H = \{id, \tau_a\}$, as $\tau_a^2 = id$ already (and so $\tau_a^{-1} = \tau_a$). Take $g = \sigma$. Then the left coset of H with respect to σ, σH, is the set $\{\sigma \cdot id, \sigma\tau_a\} = \{\sigma, \tau_c\}$. On the other hand, the right coset of H with respect to σ, $H\sigma$, is the set $\{id \cdot \sigma, \tau_a\sigma\} = \{\sigma, \tau_b\}$. Note that the left and right cosets are not the same!

In this example, notice that H, σH, and $H\sigma$ all have the same number of elements, namely 2. Notice that $|D_3| = 6$, and notice that 2 divides 6, presaging Lagranges' theorem! Notice the "obvious" one-to-one correspondence between the sets H and σH that sends id to σ and τ_a to $\sigma\tau_a$. There is a similar correspondence between the sets H and $H\sigma$ that sends id to σ and τ_a to $\tau_a\sigma$.

Notice too that neither $H\sigma$ nor σH are subgroups of G in this example: for instance, neither of these two sets contains the identity!

(2) In a like manner, the left coset $\tau_c H = \{\tau_c \cdot id, \tau_c \cdot \tau_a\} = \{\tau_c, \sigma\}$, while, by a similar calculation, $H\tau_c = \{\tau_c, \sigma^2\}$. Exactly as in part (1), the left and right cosets are not the same, $|H|$ divides $|G|$, there are obvious one-to-one correspondences between H and $\tau_c H$ and between H and $H\tau_c$, and neither $\tau_c H$ nor $H\tau_c$ are subgroups of G.

(3) Now take $G = D_3$ as above, but $H = \langle\sigma\rangle$. This is the subgroup $\{id, \sigma, \sigma^2\}$. Take $g = \tau_a$. Then $\tau_a H = \{\tau_a \cdot id, \tau_a\sigma, \tau_a\sigma^2\} = \{\tau_a, \tau_b, \tau_c\}$. As for the right coset, $H\tau_a = \{id \cdot \tau_a, \sigma\tau_a, \sigma^2\tau_a\} = \{\tau_a, \tau_c, \tau_b\}$. Notice that in this example, the left and right cosets are the same sets!

We have $|H| = 3$, and as in parts (1) and (2), $|H|$ divides $|G| = 6$. Also, we have an obvious one-to-one correspondence between H and $\tau_a H$ that sends id to τ_a, σ to $\tau_a\sigma$, and σ^2 to $\tau_a\sigma^2$. And just as in parts (1) and (2), $\tau_a H$ is not a subgroup of G, it does not contain id.

(4) In the previous two examples, the various elements g which we used to create the cosets of H were not in the subgroup H. Let us see what happens when g happens

already to be in H. Take $G = D_3$ and $H = \langle\tau_a\rangle$, and $g = \tau_a$. Then $\tau_a H = \{\tau_a, \tau_a\tau_a\} = \{\tau_a, id\} = \langle\tau_a\rangle$, and similarly $H\tau_a = H$ as well. Thus, in this example, our left and right cosets turned out to be *subgroups* of G! See Exercise 13.29 ahead.

(5) Here is an example where the groups G and H are infinite, and where the cosets have a particularly pleasing pictorial interpretation! See Figure 13.7.

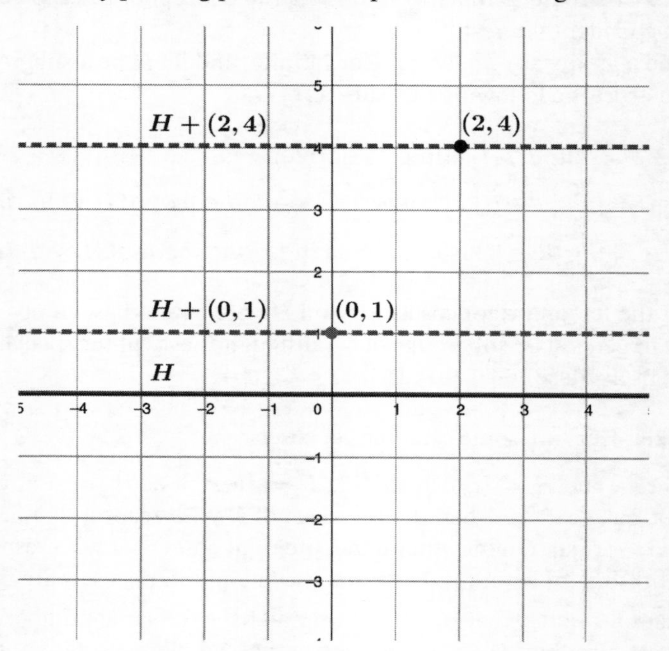

Figure 13.7. Cosets of of the subgroup consisting of the points on the x-axis. The cosets are all parallel to the x-axis.

Let $G = \mathbb{R}^2$, as in Example 13.13. As explained in that example, we may, by choosing a coordinate system on the plane, view $\mathbb{R}^2$ as points on the plane, with the group element (a, b) representing the point with x-coordinate a and y-coordinate b. Let H be the subgroup consisting of ordered pairs of the form $(a, 0)$ as a varies in $\mathbb{R}$ (verify that H is a subgroup). Thus, H consists of all points on the x-axis. Take for example $g = (0, 1)$. What do the points corresponding to Hg, which we now write as $H + g$, look like? They will be all points of the form $(a, 0) + (0, 1)$, that is, all points of the form $(a, 1)$, as a varies in $\mathbb{R}$. But this is just the line parallel to the x-axis through the point $(0, 1)$ on the y-axis! Similarly, other cosets of H will be lines parallel to the x-axis:

Exercise 13.28.1. Show that for a general (r, s), the coset $H + (r, s)$ is the line parallel to the x-axis through the point (r, s), and thus, the line parallel to the x-axis through the point $(0, s)$ on the y-axis. (Hint: Show that the set $\{a + r \mid a \in \mathbb{R}\} = \mathbb{R}$ first.)

Based on these examples, you should do the following exercise to firm up your understanding of cosets.

Exercise 13.29. Let G be a group and H a subgroup. We will study conditions on an element g that will determine whether Hg (or gH) is a subgroup of G or not.

(1) Show that if $g \notin H$, neither gH nor Hg can contain the identity element of G and therefore cannot be subgroups of G.

(2) Show that if $g \in H$, then $gH = Hg = H$, and therefore gH $(= Hg)$ is a subgroup of G. (Hint: To show $H \subseteq gH$, for instance, note that $g^{-1} \in H$, and consider $g(g^{-1}h)$.)

In particular, note that H is itself a coset of H. It is both a left coset and a right coset with respect to any element $h \in H$.

We will now prove:

Theorem 13.30. *(Lagrange's Theorem.) The order of any subgroup of a finite group G divides the order of G.*

Proof. Let H be a subgroup of G. The proof involves an analysis of either the left or the right cosets of H. We will work with left cosets here. There are three ingredients to the proof:

(1) We will show that any two cosets of H have the same number of elements. Since H is itself a coset (see Exercise 13.29 above), it is sufficient to prove that $|gH| = |H|$ for any $g \in G$. We have seen instances of this in our examples above (see Example 13.28(1) for instance). The idea is to consider the obvious correspondence between H and gH obtained as follows:

$$\begin{aligned} l_g : H &\to gH, \\ h &\mapsto gh. \end{aligned}$$

This map is injective because of left cancellation: if $gh = gh'$ then we can cancel on the left to get $h = h'$. It is clearly surjective: any $gh \in gH$ is $l_g(h)$. Thus l_g provides a bijection between H and gH, showing that $|H| = |gH|$.

(2) We will now show that any two left cosets are either equal or disjoint as sets. For this, let us prove something else of interest: if $k \in gH$, then $gH = kH$. To show this, note that $k \in gH$ means that $k = gh'$ for some $h' \in H$. Then $kH = \{kh \mid h \in H\} = \{(gh')h \mid h \in H\} = \{g(h'h) \mid h \in H\}$. But the set $\{h'h \mid h \in H\}$ is just the left coset $h'H$, and as you would have seen in Exercise 13.29 above, $h'H$ is just H, so $\{g(h'h) \mid h \in H\} = \{gh \mid h \in H\} = gH$. (Although we do not need it here, note that the reverse direction is also true: If $gH = kH$, then $k \in gH$. This is because $k = k \cdot id$ is an element of $kH = gH$.)

Now assume that we have two cosets gH and $g'H$. If they are disjoint there is nothing to prove. So assume they have nonempty intersection, and let k be an element of $gH \cap g'H$. By what we just saw above, $gH = kH$, and as well, $g'H = kH$. Hence, if gH and $g'H$ are not disjoint, then $gH = g'H$.

(3) Now we will observe that the union of the cosets gH as g varies in G is all of G. This is clear because for any $g \in G$, $g = g \cdot id$ is an element of the coset gH.

From these three results proved above, Lagrange's theorem follows almost immediately: Since G is a disjoint union of the cosets gH, and since any two cosets have the same number of elements (namely the number of elements in H), we find

$$|G| = |H| \text{ times the number of left cosets of } H \text{ in } G.$$

In particular, $|H|$ divides $|G|$. □

Remark 13.31. Although we will not do it here, we can repeat this proof with right instead of left cosets, and the arguments will be essentially identical, once the obvious switches have been made (for example, we consider a map $r_g : H \to Hg$ that sends h to hg, we show that $Hg = Hk$ if and only if $k \in Hg$, etc.) We would then find that $|G| = |H|$ times the number of *right* cosets of H in G. It follows by comparing the two proofs of Lagrange's theorem that the number of left cosets of H in G is the same as the number of right cosets of H in G. This common number, namely $|G|/|H|$, is known as the *index* of H in G, and is often denoted $[G : H]$.

Remark 13.32. Exercise 13.61 ahead shows that (left and right) cosets can be modeled as equivalence classes under suitable equivalence relations in G, and that parts (2) and (3) of the proof of Lagrange's theorem above are immediate consequences of properties of equivalence classes (Theorem 9.19 in Chapter 9). We have chosen to compute concretely with cosets (particularly in part (2)) and reprove results that have already been proven in the context of equivalence classes just so that you can "get your hands dirty" working with cosets! Concrete calculations help build intuition (although there is no doubt that appealing to the equivalence relation is cleaner). Likewise, Proposition 13.34 below is a direct consequence of properties of equivalence relations.

Remark 13.33. It is worth highlighting something we proved in the course of proving Lagrange's theorem as a separate result (the proofs for right cosets is very similar as pointed out in Remark 13.31 above):

Proposition 13.34. *If H is a subgroup of a group G, and $g, k \in G$, then $gH = kH$ if and only if $k \in gH$. Similarly, $Hg = Hk$ if and only if $k \in Hg$.*

We get an immediate corollary to Lagrange's theorem. We have the following:

Corollary 13.35. *The order of any element of a finite group divides the order of the group.*

Proof. The proof follows from Langrange's theorem and the definition of the order of an element of g of a group G as the order of the subgroup $\langle g \rangle$ (Definition 13.25.1). □

Remark 13.36. If say $|\langle g \rangle| = n$, then n can also be characterized as the *least positive integer t such that $g^t = id$*. This follows from two considerations:

(1) There has to be a positive integer t such that $g^t = id$. This is because, as we have seen in the proof of Lemma 13.24, the element g^m and g^n cannot all be distinct for $1 \leq m < n$, so as in the proof of that lemma, there is a positive t such that $g^t = 1$.

(2) Now let n be the least positive t such that $g^t = id$. Then the elements of the set $\{id, g, g^2, \ldots, g^{n-1}\}$ are all distinct. This is clear if $g = id$, since $n = 1$ and the set only consists of id. Otherwise, if $g^i = g^j$ for $0 \leq i < j \leq n-1$, then as in the proof

of Lemma 13.24, $g^{j-i} = id$. Since $1 \leq j - i < n$, this violates the minimality of n. Now, the set $\{id, g, g^2, \dots, g^{n-1}\}$ is already a subgroup of G, as is readily seen by the finite subgroup test (Lemma 13.24) and the following:

$$g^i g^j = \begin{cases} g^{i+j} & \text{if } 0 \leq i, j < n, \text{ and } i + j < n, \\ g^{i+j-n} & \text{if } 0 \leq i, j < n, \text{ and } i + j \geq n. \end{cases}$$

Since $\langle g \rangle$ must be contained in any subgroup of G that contains g, we find $\langle g \rangle \subseteq \{id, g, g^2, \dots, g^{n-1}\}$. Since the reverse inclusion is clear by the definition of $\langle g \rangle$, we find $\langle g \rangle = \{id, g, g^2, \dots, g^{n-1}\}$. Since the elements of $\{id, g, g^2, \dots, g^{n-1}\}$ are all distinct, indeed $|\langle g \rangle| = n$.

13.6 Symmetry

We have invoked symmetry many times in this chapter and alluded to how groups quantify symmetry, but we have not formally defined the term. In this section, we will try to quantify symmetry, and the precise role that groups have to play in symmetry.

To begin with, let S be any set. We denote by $\mathrm{Bij}(S)$ the collection of all functions $f : S \to S$ that are bijections between S and itself. For instance, when $S = \{1, 2, \dots, n\}$ for a positive integer n, then $\mathrm{Bij}(S)$ is nothing but the group S_n that we have considered. If S is a finite set, with say n elements in it, then exactly as for the specific n element set $\{1, 2, \dots, n\}$, we can see that $\mathrm{Bij}(S)$ will have $n!$ elements in it. But when S is an infinite set, $\mathrm{Bij}(S)$ will also be infinite.

The first thing to note is that $\mathrm{Bij}(S)$ is a group under function composition. Indeed, we have seen this already quite explicitly in the case when $S = \{1, 2, \dots, n\}$, but the ideas for why $\mathrm{Bij}(S)$ is a group for any set S have been implicit all along in what we studied in Chapter 5: We have seen that the composition of two bijective functions is bijective (see Exercise 5.29 of that chapter) and that function composition is associative (see Exercise 5.30 of that chapter), we have seen the definition of the identity function $\mathbb{1}_S$ (Definition 5.24 of that chapter), we know that for trivial reasons, $f \circ \mathbb{1}_S = \mathbb{1}_S \circ f = f$ for all $f \in \mathrm{Bij}(S)$, and finally, we know that every element of $\mathrm{Bij}(S)$ is invertible (Proposition 5.32 of that chapter). Thus, $\mathrm{Bij}(S)$ under the binary operation of function composition satisfies all the axioms for a group.

We will quantify symmetry in the following way:

Definition 13.37. Let S be a set. Then by *symmetries of S* we will mean the elements of some distinguished subgroup H of $\mathrm{Bij}(S)$. The subgroup H is then referred to as the *symmetry group* of S.

This definition might seem surprising: it does not pin down which subgroup we have in mind. Also, the connection between the day-to-day use of the expression "symmetry" and bijection is not immediately clear.

Let us tackle both of these issues by considering the case of the "symmetries" of the equilateral triangle that we studied earlier in this chapter (see Section 13.1). Here, the set S is a cardboard in the shape of an equilateral triangle, located somewhere on the xy plane, with the location determined say by the outline of its edges. As we have seen, the motions that we like to think of as symmetries are those that end up preserving the outline of its edges (else we would know that the triangle has been moved). Hence, we can think of our motions as simply moving the points of S to other points of S, without changing the overall position of S. Thus, our motions are functions $f : S \to S$. But more, we want our f to be rigid, to preserve distances. This shows that f must be injective—otherwise the nonzero distance between two distinct points P and Q will collapse to zero if $f(P) = f(Q)$. Moreover, using the ideas inherent in Figure 13.2, we can see that for the motion to be rigid, our shuffling must be surjective as well (if Q is at distances d_1 and d_2 from $f(A)$ and $f(B)$ respectively, consider the unique point P that is at distances d_1 and d_2 from A and B respectively: $f(P)$ has to be Q). Thus, our motions are bijections of S. We have determined our allowable motions to be the group D_3: we thus find that D_3 lives inside $\text{Bij}(S)$ as a subgroup.

Note that D_3 is a *proper* subgroup of $\text{Bij}(S)$, that is, $D_3 \subsetneq \text{Bij}(S)$. For instance, in Figure 13.8, there is a bijection of S in which all points on and inside the inner circle are rotated counterclockwise by 90°, but all other points on the triangle stay in the same location. In theory this bijection could also constitute a symmetry of S, but we have restricted ourselves to motions that preserve distances between points, so we do not consider this a symmetry for our purposes.

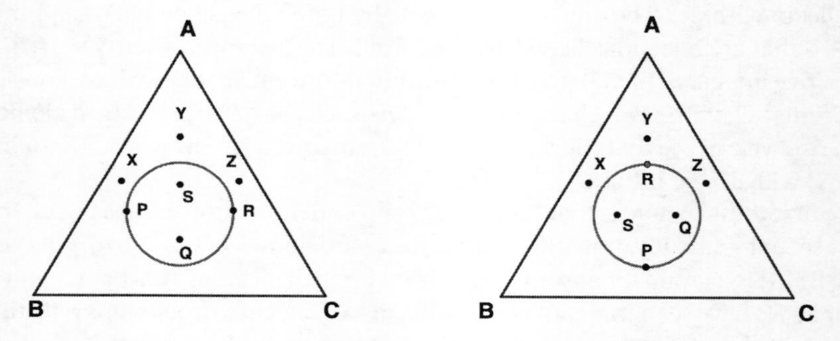

Figure 13.8. A symmetry of an equilateral triangle T that is not rigid. Here, every point on and inside the inner circle is rotated counterclockwise by 90°, but all other points on the triangle stay in the same location. This is certainly a bijection between the points of T, but it shears the interior of the triangle and is not a rigid symmetry.

In general, this is what happens when we talk of "symmetries" of an object S. We typically do not mean all symmetries (which would correspond to elements of the entire group $\text{Bij}(S)$), but only those that typically, "preserve a property." In the cases of interest, the elements of $\text{Bij}(S)$ that preserve a given property usually form a subgroup of $\text{Bij}(S)$. That is, if bijections f and g preserve a property, then $f \circ g$ and f^{-1} and g^{-1} also preserve the property. With this interpretation, the subgroup itself becomes a measure of the symmetry of S. The structure of this subgroup: the composition table that shows how elements compose, its own further subgroups and their relation to one another, all give us information on the symmetry inherent in S. (Thus, this is an instance where we "measure" something not by a real number, but by an abstract object like a group!)

For instance, in the case of the equilateral triangle S at a given location on the xy plane, we only considered those symmetries that preserved the rigidity of the triangle. These symmetries naturally formed a subgroup of $\text{Bij}(S)$, which we called D_3. Similarly, the symmetries of a square S at a given location on the xy plane that preserved its rigidity formed a subgroup of $\mathrm{Bij}(S)$, which we called D_4. As an instance of a very quick and dirty inference we can make about the symmetries of T and S from its groups, we can say that the square is "more symmetric" than a triangle since its symmetry group D_4 has more elements in it than the symmetry group D_3 of the triangle.

Note that for the same set S, we may in different contexts consider symmetries that preserve different properties. Accordingly, we will get different subgroups of $\mathrm{Bij}(S)$ that we may refer to as "the" symmetry group of S. Although this is an abuse of terminology, it is an acceptable practice as long as we keep in mind what the context is and what property is being preserved in the definition of the symmetry group. For instance, while considering the symmetries of the equilateral triangle we could require for various reasons that in addition to the rigidity, the orientation of the vertices should also be maintained. That is, if the vertices are labelled A, B, C in the counterclockwise direction as in Figure 13.3, then we could require that after the motion, A, B, and C must end up still oriented counter clockwise. This immediately rules out the motions τ_a, τ_b and τ_c, as they flip the orientation to clockwise. The symmetry group now is just the subgroup $\{id, \sigma, \sigma^2\}$.

The advantage of this very general definition of symmetry is that it can be applied to a vast number of situations even where there is no obvious *geometric* notion of symmetry. For instance, we can apply it to the symmetries of a polynomial equation. Consider for example the polynomial $x^4 - 2$, a polynomial with rational (in fact integer) coefficients. It has no roots in the rational numbers, but in the complex numbers, it has four roots: $\sqrt[4]{2}$, $-\sqrt[4]{2}$, $i\sqrt[4]{2}$ and $-i\sqrt[4]{2}$ (see also the factorization of $x^4 - y^4$ in Example 13.12). Write R for the set of these roots. By the term "symmetries of the polynomial $x^4 - 2$" we actually mean the symmetries of R, the set of roots of $x^4 - 2$. But more: we refer not to all bijections between the set of roots and itself, but only those that *preserve arithmetic relations.* By this, we mean that if for some polynomial $p(x, y, z, w)$ with coefficients in $\mathbb{Q}$ we have

$$p(\sqrt[4]{2}, i\sqrt[4]{2}, -\sqrt[4]{2}, -i\sqrt[4]{2}) = 0,$$

then for a bijection $f \in \mathrm{Bij}(R)$ to be allowable, we must also have

$$p\Big(f(\sqrt[4]{2}), f(i\sqrt[4]{2}), f(-\sqrt[4]{2}), f(-i\sqrt[4]{2})\Big) = 0.$$

For instance, if an allowable symmetry f sends $\sqrt[4]{2}$ to $i\sqrt[4]{2}$, then $-\sqrt[4]{2}$ must go to $-i\sqrt[4]{2}$. This is because, taking $p(x, y, z, w)$ to $x+z$, we find p is satisfied by plugging in $x = i\sqrt[4]{2}$, $z = -i\sqrt[4]{2}$. Hence, $f(i\sqrt[4]{2})$ and $f(-i\sqrt[4]{2})$ must also satisfy p, that is, we must have $i\sqrt[4]{2} + f(-i\sqrt[4]{2}) = 0$. Hence, $f(-i\sqrt[4]{2})$ must necessarily be $-i\sqrt[4]{2}$.

There is a vast theory that describes the theory of symmetries of equations such as the one in the example above. This theory allows us to calculate, for example, that the set of symmetries of the roots of $x^4 - 2$ must have exactly 8 elements, and allows us to determine all eight symmetries explicitly. These naturally form a group. (We will have more to say about the symmetries of $x^4 - 2$ in Section 13.7 ahead, where we will show that these symmetries are naturally related to the group D_4.)

Group theory was essentially born from the example of the symmetries of the roots of a polynomial equation with coefficients in $\mathbb{Q}$. The theory behind these symmetries was laid out by Evariste Galois, who used the structure of the group of symmetries of an equation, including its list of subgroups, to determine if there would be a formula for the roots of an equation that involved only the coefficients, the usual operations of addition, subtraction, multiplication, and division, along with the process of taking n-th roots. (Such a formula would be akin to the quadratic formula: it gives the roots of the equation $ax^2 + bx + c = 0$ as $\dfrac{-b \pm \sqrt{b^2 - 4ac}}{2a}$. This is a formula for the roots in terms the coefficients, arithmetic operations of addition, subtraction, multiplication, and division, and a square root.) He showed using his theory that for equations of degree 5 or higher, in general, there is no such formula!

Physics, in particular, uses the formulation for symmetry in Definition 13.37 to describe various patterns that arise in the physical world. For instance, when physicists talk of symmetries of a physical law, what they are referring to is coordinate changes that preserve the form of the law. These are, first of all, coordinate changes of four-dimensional x, y, z, t space (where t stands for time). We denote this space by $\mathbb{R}^4$, as a set $\mathbb{R}^4$ is just the four fold direct product $\mathbb{R} \times \mathbb{R} \times \mathbb{R} \times \mathbb{R}$. Thus, a coordinate change of $\mathbb{R}^4$ is a bijective map f from $\mathbb{R}^4$ to $\mathbb{R}^4$ that maps (x, y, z, t) to

$$\left(x' = f_x(x, y, z, t), y' = f_y(x, y, z, t), z' = f_z(x, y, z, t), t' = f_t(x, y, z, t)\right),$$

with inverse g that maps (x', y', z', t') to

$$\left(x = g_{x'}(x', y', z', t'), y = g_{y'}(x', y', z', t'), z = g_{z'}(x', y', z', t'), t = g_{t,}(x', y', z', t')\right).$$

Here, f_x, f_y, $g_{x'}$, $g_{y'}$ etc. are functions from $\mathbb{R}^4$ to $\mathbb{R}$. Usually, these coordinate changes are restricted to affine changes, that is, those where the f_x, etc. are of the form $ax + by + cz + dt + e$ for suitable constants $a, \dots, e$. The symmetries of the physical law are then those elements of $\mathrm{Bij}(\mathbb{R}^4)$ above such that after replacing x with $g_{x'}(x', y', z', t')$, y with $g_{y'}(x', y', z', t')$, etc., the physical law reads the same except with x replaced by x', y by y', etc. Such bijective maps form a subgroup of the group of all bijective maps from $\mathbb{R}^4$ to $\mathbb{R}^4$.

The well-known crisis in the late nineteenth century that led to Einstein's formulation of the theory of relativity occurred because Maxwell's equations showed that the speed of light was constant, independent of which coordinate system was used, among any from an "inertial" family that were all moving with uniform velocity with respect to one another. But according to the principle of Galilean relativity, which states that the equations of motion should remain unchanged in a second coordinate system that is moving with uniform relative velocity with respect to the first coordinate system, the speed of light should be different in these different coordinate systems. Einstein developed his theory of relativity that accepted as a given the constancy of the speed of light, and showed that the correct symmetry group for the equations of motion under his hypotheses is not the set of *Galilean transformations* (for say motion just along the x-axis)

$$x' = x - vt, \quad y' = y, \quad z' = z, \quad t' = t,$$

but the *Lorenzian transformations* (discovered earlier by the physicist Lorenz)

$$x' = \gamma(x - vt), \quad y' = y, \quad z' = z, \quad t' = \gamma\left(t - \frac{vx}{c^2}\right).$$

Here, v is the uniform velocity with which the second coordinate system (x', y', etc.) moves with respect to the first (x, y, etc.); as measured in the first, c is the (constant in all coordinate systems) speed of light, and $\gamma = \left(\sqrt{1 - \frac{v^2}{c^2}}\right)^{-1}$.

Exercise 13.38. Suppose you were at rest on Earth, and a photon of light sped past you, and you observed its speed as c. Now suppose instead that you had been in a spaceship moving in the same direction as the photon of light, at a speed v ($0 < v < c$). Show that had you used used the Galilean transformation, you would have calculated the speed of light from your new point of view to be $c - v$, which goes against what Maxwell had predicted. Show instead that had you used the Lorenzian transformation; you would indeed calculate the speed of light to still be c. (Hint: Relative to the first coordinate axes on Earth, aligned so that the x axis points in the direction of travel of the photon, the coordinate axes attached to the spaceship are moving at the uniform speed v in the x direction. Use the formulas for the two transformations to calculate the speed of light in the spaceship coordinate system. What is critical is that the speed of light as measured in the spaceship is not dx'/dt but instead dx'/dt'.)

13.7 Isomorphisms Between Groups

In this section we will introduce a more general notion of symmetry, the notion of isomorphism. We will do this in the context of groups, but this is a notion that you will see in many different contexts in mathematics in your courses ahead.

As we saw in the previous section, a symmetry of a set S is, at root, a bijection of S. (Typically we require that it belong to some subgroup of $\mathrm{Bij}(S)$, but that is not critical for now.) Thus, a symmetry of a set S captures some "sameness" between S and itself. Now suppose we loosened this a bit, and considered along with S a different set T. If there exists a bijection f between S and T, then, at the rather loose level of just being sets, there is indeed some sameness to S and T: there is a way to pair the elements of S and T, and f captures that sameness. Loosely, we may think of f as providing a symmetry between S and T.

But just as in many contexts we do not consider an arbitrary element of $\mathrm{Bij}(S)$ to be a symmetry of S but require that the element be in some distinguished subgroup (typically obtained as the subgroup preserving some property of S), the symmetries between different sets that we find interesting are those that preserve some property of the two sets, or preserve some structure in the two sets. In different contexts this will have a different meaning specific to that context, but let us now consider this idea in the context of groups:

Let $(G, \cdot)$ and $(H, *)$ be groups, where we have denoted the group operation in G by $\cdot$ and in H by $*$. We are in the context of groups, and what makes G and H interesting to us is that they are more than mere sets: they are sets with respective binary operations defined on them satisfying the three group axioms. Thus, to a mathematician, G and H have *structure*; they are groups. Accordingly, the bijections between S and T that we will find interesting are not arbitrary ones, but those that *preserve the group structure.* What this means is captured in the following definition:

Definition 13.39. An *isomorphism* between two groups $(G, \cdot)$ and $(H, *)$ is a function

$$f : G \mapsto H$$

that is one-to-one and onto, and satisfies

$$f(g \cdot h) = f(g) * f(h).$$

If $f : G \to H$ is an isomorphism, then we say that G and H are *isomorphic*, or that f *provides an isomorphism between* G *and* H.

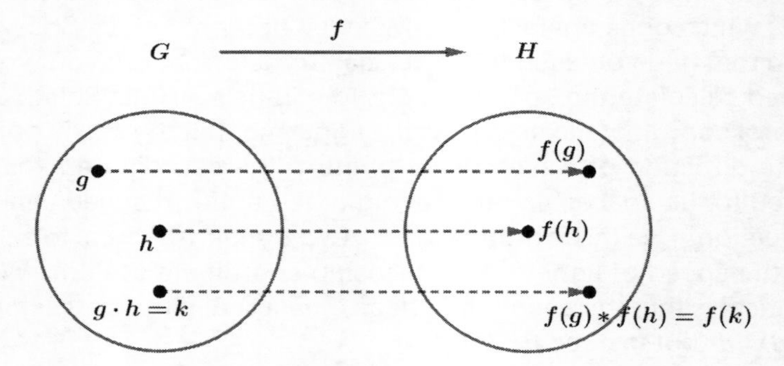

Figure 13.9. An isomorphism f between groups G and H. The figure shows that f preserves the group structure in G and H, in the sense that $f(g \cdot h) = f(g) * f(h)$.

Let us examine this definition. The requirement that f be a bijection (or pairing) is something we already considered above. What is new is the requirement $f(g \cdot h) = f(g) * f(h)$. See Figure 13.9. The operation on the left side of this equality is the one in G, while the one on the right is in H. Thus, what this requirement is saying is that f must not only match the elements of G and H, but must be such that if g and h multiply in G to give, say k, then, $f(g)$, which pairs with g, and $f(h)$, which pairs with h, must multiply in H to give $f(k)$, the element that pairs with k! We see that the pairing then respects the group operations in the respective group: we can multiply g and h in G and then go to the match of gh in H, or go to the matches of g and h in H and then multiply those in H! This requirement quantifies exactly what it means for f to preserve the group structure of G and H. The group operations in G are the same as those in H once the elements have been renamed according to f.

When two groups G and H are isomorphic, then any true statement about the structure of G as a group must also be true for H, after making any necessary labelling changes. For instance, if G is abelian, H must also be abelian. If G has exactly 3 elements of order 2, H must also have exactly three elements of order 2. If there exists a chain of "nested" subgroups of G of the form $G_0 = G \supset G_1 \supset \cdots \supset G_k = \langle id_G \rangle$, then there must also exist in H a chain of nested subgroups of the form $H_0 = H \supset H_1 \supset \cdots \supset H_k = \langle id_H \rangle$, and so on.

When two very different sets turn out to have isomorphic symmetry groups, it is a matter of considerable joy to mathematicians! For instance, the group of symmetries

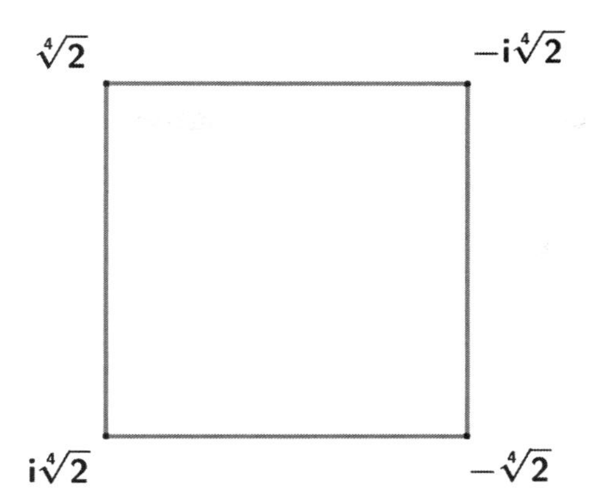

Figure 13.10. The group of symmetries of $x^4 - 2$. This group is isomorphic to D_4. By placing the roots of $x^4 - 2$ on the vertices of a square as shown and letting the elements of D_4 act on the square, we get the symmetries of this equation.

of the equation $x^4 - 2$ considered in the previous section turns out be isomorphic to the dihedral group D_4! The correspondence may be pictured as follows: the group of symmetries of $x^4 - 2$ are gotten by imagining a square with vertices labelled in the counterclockwise direction $\sqrt[4]{2}$, $i\sqrt[4]{2}$, $-\sqrt[4]{2}$ and $-i\sqrt[4]{2}$. See Figure 13.10. The images of these vertices under the symmetries in D_4 is exactly where the roots go under the various symmetries of this equation! For instance, corresponding to the element σ (see Section 13.1 where the elements of D_4 have been described), there is a symmetry that sends $\sqrt[4]{2}$ to $i\sqrt[4]{2}$, $i\sqrt[4]{2}$ to $-\sqrt[4]{2}$, $-\sqrt[4]{2}$ to $-i\sqrt[4]{2}$, and $-i\sqrt[4]{2}$ back to $\sqrt[4]{2}$. Similarly, corresponding to the element τ_{bd}, there is a symmetry that sends $\sqrt[4]{2}$ to $-\sqrt[4]{2}$ and leaves $i\sqrt[4]{2}$ and $-i\sqrt[4]{2}$ unchanged. Etc.

Here is a quick exercise to make you think through the definition of an isomorphism:

> **Exercise 13.40.** Let G and H be groups and $f : G \to H$ be an isomorphism. Since f is a bijection between the two sets G and H, it has an inverse $f^{-1} : H \to G$. (Proposition 5.32, Chapter 5.) Show that f^{-1} is an isomorphism between H and G.

We consider some examples of isomorphism groups:

Example 13.41. The groups D_3 and S_3 we have considered are isomorphic. We can see this clearly from their composition tables developed earlier, which we display side by side in Table 13.5 for easy comparison. They are identical except for the names of the different elements.

Specifically, the function $f : D_3 \to S_3$ that provides this isomorphism must satisfy the following: f must match σ to r_1, σ^2 to r_2, τ_a to f_1, τ_b to f_2, τ_c to f_3, and id_{D_3} to id_{S_3}. With this pairing, the tables match up exactly: for instance, σ and τ_a multiply to τ_c in D_3, and correspondingly, r_1 and f_1 multiply to f_3 in S_3.

Table 13.5. Composition tables for D_3 and S_3 for comparison

$\mathbf{D_3}$	id	σ	σ^2	τ_a	τ_b	τ_c
id	id	σ	σ^2	τ_a	τ_b	τ_c
σ	σ	σ^2	id	τ_c	τ_a	τ_b
σ^2	σ^2	id	σ	τ_b	τ_c	τ_a
τ_a	τ_a	τ_b	τ_c	id	σ	σ^2
τ_b	τ_b	τ_c	τ_a	σ^2	id	σ
τ_c	τ_c	τ_a	τ_b	σ	σ^2	id

$\mathbf{S_3}$	id	r_1	r_2	f_1	f_2	f_3
id	id	r_1	r_2	f_1	f_2	f_3
r_1	r_1	r_2	id	f_3	f_1	f_2
r_2	r_2	id	r_1	f_2	f_3	f_1
f_1	f_1	f_2	f_3	id	r_1	r_2
f_2	f_2	f_3	f_1	r_2	id	r_1
f_3	f_3	f_1	f_2	r_1	r_2	id

Example 13.42. The subgroup $\langle\sigma\rangle$ of D_3 is isomorphic to the group $\boldsymbol{\mu}_3$. Work through this, consider the function $f\colon D_3 \to \boldsymbol{\mu}_3$ that sends σ to ω_3, σ^2 to ω^2 and id to 1, and check that it satisfies the conditions for it to be an isomorphism.

Example 13.43. The group $(\mathbb{R}, +)$ is isomorphic to $(\mathbb{R}_{>0}, \cdot)$ via the map

$$f\colon (\mathbb{R}, +) \to (\mathbb{R}_{>0}, \cdot),$$
$$x \mapsto e^x.$$

We saw this function in Example 5.37 in Chapter 5. There is a well-known formula $e^{a+b} = e^a \cdot e^b$ for all $a, b \in \mathbb{R}$, which you may have seen already, or else, is something you will see in a course on analysis. This formula is just the statement that the map f preserves the group structure in $(\mathbb{R}, +)$ and $(\mathbb{R}_{>0}, \cdot)$ respectively! Its inverse is the map

$$f^{-1}\colon (\mathbb{R}_{>0}, \cdot) \to (\mathbb{R}, +),$$
$$x \mapsto \ln(x),$$

where ln stands for the natural logarithm (that is, logarithm to the base e). The equally well-known formula $\ln(ab) = \ln(a) + \ln(b)$ is just the statement that the map f^{-1} preserves the group structure in $(\mathbb{R}_{>0}, \cdot)$ and $(\mathbb{R}, +)$ respectively!

We will end with one lemma:

Lemma 13.44. *Let $(G, \cdot)$ and $(H, *)$ be groups, and let $f\colon G \to H$ be an isomorphism between them. Then $f(id_G) = id_H$.*

Proof. Starting with the relation $id_G \cdot id_G = id_G$, we find, from the definition of an isomorphism, that $f(id_G) * f(id_G) = f(id_G)$. On the other hand, we have in H the relation $f(id_G) * id_H = f(id_G)$. Thus, $f(id_G) * f(id_G) = f(id_G) * id_H$. Left cancellation in H yields $f(id_G) = id_H$. □

In your future courses you will see this idea of structure preserving bijections in other contexts. For instances, a *homeomorphism* between two topological spaces X and Y is a bijection $f\colon X \mapsto Y$ such that f and f^{-1} are both continuous. A *diffeomorphism* between two C^∞ manifolds M and N is a bijection $f\colon M \to N$ such that f and f^{-1} are C^∞. (Note that in the case of a group isomorphism, thanks to Exercise 13.40 above, we did not have to impose the condition on f^{-1} that it preserve the group operation; that condition came for free!) In the same vein, we have the notion of isomorphisms (structure preserving bijections) between rings, between vector spaces, between algebraic varieties, between Banach spaces, and on and on. You have much deep mathematics awaiting you!

13.8 Further Exercises

Exercise 13.45. Practice Exercises:

Exercise 13.45.1. Compute the following in S_6:

(1) gh, where

$$g = \begin{pmatrix} 1 & 2 & 3 & 4 & 5 & 6 \\ 3 & 1 & 6 & 5 & 2 & 4 \end{pmatrix} \text{ and } h = \begin{pmatrix} 1 & 2 & 3 & 4 & 5 & 6 \\ 4 & 2 & 3 & 5 & 6 & 1 \end{pmatrix}.$$

(2) $(gh)^{-1}$, where gh is the answer you got to part (1) above. (Remember, if say $\sigma \in S_6$ sends 1 to 4, then σ^{-1} has to send 4 to 1. If σ sends 2 to 3, then σ^{-1} has to send 3 to 2. And so on. This will allow you to write down explicitly the inverse of any element of S_6.)

(3) g^{-1} and h^{-1}, where g and h are as in part (1) above.

(4) $h^{-1} \cdot g^{-1}$, where h^{-1} and g^{-1} are the answers you got in part (3) above.

(5) Confirm that the answers to parts (2) and (4) are the same (see Lemma 13.18).

Exercise 13.45.2. Compute the cycle decomposition (see Example 13.4) of the following in S_6:

(1) The element h in Exercise 13.45.1 above.

(2) $h = \begin{pmatrix} 1 & 2 & 3 & 4 & 5 & 6 \\ 3 & 2 & 1 & 5 & 6 & 4 \end{pmatrix}$.

(3) $h = \begin{pmatrix} 1 & 2 & 3 & 4 & 5 & 6 \\ 6 & 3 & 5 & 4 & 2 & 1 \end{pmatrix}$.

Exercise 13.45.3. A 2-cycle is also known as a *transposition.* We will see that every k-cycle can be expressed as a product of $k-1$ transpositions:

(1) Verify that $(1\ 2\ 3) = (1\ 3)(1\ 2)$.

(2) Verify that $(1\ 2\ 3\ 4) = (1\ 4)(1\ 3)(1\ 2)$.

(3) Write a k-cycle $(a_1\ a_2\ \dots\ a_k)$ as a product of $k-1$ transpositions, based on parts (1) and (2).

(4) Verify that $(1\ 2\ 3) = (2\ 1)(2\ 3)$. Hence, the representation as a product of transpositions is not unique. (Remark: you should compute this out directly, but it is worth noticing that the cycle $(1\ 2\ 3)$ can also be written as the cycle $(2\ 3\ 1)$. The general representation as a product of transpositions you obtained in part (3) can be applied to this form.)

Exercise 13.45.4. We will compute some conjugation actions on S_n by k-cycles:

(1) Let $g = (1\ 2\ 3)$ and $h = (1\ 2\ 3\ 4\ 5\ 6)$ in S_6. Compute ghg^{-1}: you should find it to be another 6-cycle. Write it in cycle notation.

(2) Let $g = (3\ 6\ 2\ 4)$ and $h = (6\ 3\ 4\ 1\ 5\ 7)$ in S_7. Compute ghg^{-1}: you should find it to be another 6-cycle. Write it in cycle notation.

Exercise 13.45.5. $Gl_2(\mathbb{C})$ (see Example 13.11): Compute the following:

(1) $\begin{pmatrix} 1 & 2 \\ 3 & 4 \end{pmatrix}^{-1}$.

(2) $\begin{pmatrix} 2 & 1 \\ -4 & -3 \end{pmatrix}^{-1}$.

(3) $\begin{pmatrix} 1 & 2 \\ 3 & 4 \end{pmatrix} \cdot \begin{pmatrix} 2 & 1 \\ -4 & -3 \end{pmatrix}$.

(4) $\begin{pmatrix} 2 & 1 \\ -4 & -3 \end{pmatrix} \cdot \begin{pmatrix} 1 & 2 \\ 3 & 4 \end{pmatrix}$.

Exercise 13.45.6. Just like $Gl_2(\mathbb{C})$, we can consider $Gl_2(\mathbb{R})$, the set of 2×2 matrices with entries in $\mathbb{R}$ of nonzero determinant.

(1) Show that $Gl_2(\mathbb{R})$ is a subgroup of $Gl_2(\mathbb{C})$. (Hint: Lemma 13.23.)

(2) Let $SO_2(\mathbb{R})$ denote the set of matrices in $Gl_2(\mathbb{R})$ of determinant 1. Show that $SO_2(\mathbb{R})$ is a subgroup of $Gl_2(\mathbb{R})$.

(3) Let $SL_2(\mathbb{Z})$ denote the set of matrices in $SO_2(\mathbb{R})$ whose entries are all integers. Show that $SL_2(\mathbb{Z})$ is a subgroup of $SO_2(\mathbb{R})$.

(4) For any integer $n \geq 2$, let $\Gamma(n)$ denote the set of matrices in $SL_2(\mathbb{Z})$ whose $(1,1)$ and $(2,2)$ entries are each congruent to 1 mod n and whose $(1,2)$ and $(2,1)$ slots are each congruent to 0 mod n. (See Remark 1.5 in Chapter 1 for the meaning of congruence mod n.) Show that $\Gamma(n)$ is a subgroup of $SL_2(\mathbb{Z})$.

Exercise 13.45.7. Let G stand for the subgroup generated by the rotation $\sigma \in D_4$. Show that G and $(\mathbb{Z}/4\mathbb{Z}, +)$ are isomorphic. Do the same for G and μ_4. (Study their composition tables to see how to match them up, and show that the conditions in Definition 13.39 are met.)

Exercise 13.46. Use the known fact that addition and multiplication of complex numbers is associative to prove that addition and multiplication of matrices in $M_2(\mathbb{C})$ is associative. (Hint: If M, N, P are in $M_2(\mathbb{C})$, compute the entries of both $(M+N)+P$ and $M+(N+P)$. Compare the (i,j) entries for $i,j=1,2$ for both sums, and invoke the associativity of addition in $\mathbb{C}$. Adopt the same strategy of comparing the entries of $(M\cdot N)\cdot P$ and $M\cdot(N\cdot P)$ to establish the associativity of multiplication.)

Exercise 13.47. We have seen in Example 13.11 that if $\det(M)$ is nonzero for some $M\in M_2(\mathbb{C})$, then M is invertible. We will show here the converse: if $M\in M_2(\mathbb{C})$ is invertible then $\det(M)$ is nonzero. (We will establish this result by proving its contrapositive.)

(1) Show that if $M\cdot N=\mathbf{0}_2$ for two nonzero matrices M and N, then neither M nor N could have a multiplicative inverse. (Hint, assume to the contrary and multiply by the inverse either on the right or on the left as suitable.)

(2) Suppose that $\det(M)=0$ for a nonzero matrix M. Find a nonzero matrix N such that $M\cdot N=\mathbf{0}_2$. (Hint: Construct N by choosing as its entries some permutation and/or slight alteration of the entries of M, and invoking the fact that $\det(M)=0$. This is classic mathematics: you need to discover your N by *playing!*)

(3) Conclude that if $\det(M)=0$, then M cannot be invertible.

Exercise 13.48. Let I be some index set, and let H_α, $\alpha\in I$, be a family of subgroups of a group G. Show that $\cap_\alpha H_\alpha$ is a subgroup of G. Now let $S\subseteq G$ be an arbitrary subset of G. We define the *subgroup generated by* S, denoted $\langle S\rangle$, to be the intersection of all subgroups of G that contain S. Compute and list the elements of $\langle S\rangle$ in the following cases:

(1) G is any group, and $S\subseteq G$ is the empty set.

(2) $G=S_3$, $S=\{\sigma,\tau_a\}$.

(3) $G=S_4$, $S=\{\sigma^2,\tau_h\}$.

(4) $G=Gl_2(\mathbb{C})$, $S=\left\{\begin{pmatrix} i & 0\\ 0 & -i\end{pmatrix},\begin{pmatrix} 0 & -1\\ 1 & 0\end{pmatrix}\right\}$. The group generated by this set S is a very well-known group and is known as the *quaternion group.* It has 8 elements.

Remark: If $S\subseteq G$ and $\langle S\rangle=G$, we say that G *is generated by* S.

Exercise 13.49. Show that for all $n\geq 2$, S_n is generated by transpositions. (See Exercise 13.48 for what it means to be generated by transpositions. Example 13.4 as well as Exercise 13.45.3 will be helpful.)

Exercise 13.50. Let g be some k-cycle in S_n. Let $h=(a_1\ a_2\ \cdots\ a_l)$ be an l-cycle in S_n. Compute ghg^{-1}: prove that it is the l-cycle $(g(a_1)\ g(a_2)\ \cdots\ g(a_l))$. (Hint: Compute ghg^{-1} acting on $g(i)$ for various i. See also Exercise 13.45.4.)

Exercise 13.51. We will show here that the only subgroups of $(\mathbb{Z},+)$ are $\mathbb{Z}$ itself, $\{0\}$, and the groups $k\mathbb{Z}$, $k=2,3,\ldots$, considered in Example 13.25(2). Proceed as follows:

(1) Let H be a subgroup not equal to $\mathbb{Z}$ or $\{0\}$. Show that H must contain at least one positive integer.

(2) Expain why H must contain at least positive integer h.

(3) Show that H must contain the subgroup $h\mathbb{Z}$.

(4) Show that H must equal $h\mathbb{Z}$ by showing that if $m \in H \setminus h\mathbb{Z}$, then H must contain a positive integer *less than* h, violating the minimality of h. (Hint: the division algorithm, Chapter 10, Theorem 10.2.)

Exercise 13.52. Fix an integer $n \geq 2$. In the group $\boldsymbol{\mu}_n$ considered in Example 13.12, show that for any integer $k \geq 1$, $\langle \omega_n^k \rangle = \langle \omega_n^d \rangle$, where $d = \gcd(n, k)$. Show that $|\langle \omega_n^k \rangle| = n/d$. (Hint: Write d as an integer linear combination of k and n.)

Exercise 13.53. Let g be an element in a group G. Assume that g has finite order n. Repeat the calculations of Exercise 13.52 to show that for any integer $k \geq 1$, g^k has order $d = \gcd(n, k)$. Use this to show the following: Let G be any group whose order is prime. Then $G = \langle g \rangle$ for any $g \neq id$ in G.

Exercise 13.54. Show that when n is prime, a group of order n has no subgroups other than itself and $\langle id \rangle$.

Exercise 13.55. An element $\pi \in S_n$ is said to be a *derangement* if $\pi(i) \neq i$ for any $i \in \{1, 2, \ldots, n\}$. (By way of terminology, if $\pi(i) = i$ for any $i \in \{1, 2, \ldots, n\}$, then i is known as a *fixed point* of π, so a derangement is also known as a *fixed point free* permutation.) We will count the number of derangements in S_n in this exercise:

(1) Let A_i, $i \in \{1, 2, \ldots, n\}$, denote the subset of all $\pi \in S_n$ that fix i. Show that A_i is a subgroup of S_n that has $(n-1)!$ elements, and that more generally, $A_{i_1} \cap A_{i_2} \cap \cdots \cap A_{i_k}$ is a subgroup of S_n that has $(n-k)!$ elements.

(2) Use the principle of Inclusion Exclusion (Exercise 7.27, Chapter 7) to show that the number of permutations that fix *at least* one element in $\{1, 2, \ldots, n\}$ is

$$n!\left(\frac{1}{1!} - \frac{1}{2!} + \frac{1}{3!} \pm \cdots + (-1)^{n-1}\frac{1}{n!}\right).$$

(3) Conclude that the number of derangements in S_n is

$$n!\sum_{i=0}^{n}(-1)^n\frac{1}{i!}.$$

(4) (For those who are familiar with series from previous studies of calculus or from Chapter 12:) Conclude that as $n \to \infty$, the probability that a randomly selected permutation will be a derangement is $1/e$. (Hint: Exercise 12.77 in Chapter 12.)

Exercise 13.56. Let G be a group. Show that G cannot be written as the set-theoretic union $H \cup K$ of two proper subgroups H and K. (A subgroup H of G is called proper if $H \neq G$.)

Exercise 13.57. Suppose a finite group G contains an element $g \neq id$ such that $g = g^{-1}$. Show that $|G|$ must be even.

Exercise 13.58. The converse to the statement in Exercise 13.57 also is true: Let G be a finite group G. Show that if $|G|$ is even, then G contains an element $g \neq id$ such that $g = g^{-1}$. (Hint: Match elements of G with their inverses.)

Exercise 13.59. Let G be a group, and write $Z(G)$ for the set $\{h \in G \mid hg = gh \text{ for all } g \in G\}$. Clearly, the identity element of G belongs to $Z(G)$.

(1) Show that $Z(G)$ is a subgroup of G. This is a distinguished subgroup of G known as the *center* of G.

(2) Determine $Z(D_4)$.

Exercise 13.60. If G is a group and g and h are in G, we say g is conjugate to h if there is a $k \in G$ such that $g = khk^{-1}$. Write $g \sim h$ if g is conjugate to h.

(1) Show that $\sim$ is an equivalence relation on G.

(2) The equivalence classes under $\sim$ are referred to as the *conjugacy classes* of G. Determine the conjugacy classes of D_4. (Hint: Exercise 13.3 may be helpful.)

Exercise 13.61. Let G be a group and H a subgroup. Define a relation (suggestively) labeled L_H by $(g, k) \in L_H$ if $k^{-1}h \in H$. Define another relation (also suggestively) labeled R_H by $(g, k) \in R_H$ if $gk^{-1} \in H$.

(1) Show that L_H and R_H are both equivalence relations on G.

(2) Show that the equivalence class of g under L_H is the left coset gH and that the equivalence class of g under R_H is the right coset Hg.

Thus, as alluded to in Remark 13.32, once we view cosets (left or right) as the equivalence classes under a suitable equivalence relation, we see that the proofs of parts (2) and (3) of Lagrage's theorem (Theorem 13.30) as well the proof of Proposition 13.34 are immediate consequences of the properties of equivalence relations we have already considered in Theorem 9.19 in Chapter 9.

Exercise 13.62. Recall that we defined an isomorphism between two groups $(G, \cdot)$ and $(H, *)$ as a map $f : G \to H$ that is bijective and satisfies $f(g \cdot h) = f(g) * f(h)$ for all $g, h \in G$. Very often, we find in practice a weaker version of such a phenomenon: we find maps $f : G \to H$ such that f is not necessarily bijective, but all the same satisfies the relation $f(g \cdot h) = f(g) * f(h)$ for all $g, h \in G$. We call such a map a *homomorphism* from G to H. While clearly not as strong as an isomorphism, the presence of a homomorphism from G to H nevertheless exhibits at least some similarity between the group structure in G and that in H, or at least, that in the range $f(G) \subseteq H$ of f. We will study a few properties of such an f here.

So let $f : G \to H$ be a homomorphism from G to H.

(1) Show that $f(id_G) = id_H$. (Hint: this should be the same proof as for the case where f is an isomorphism!)

(2) Show that the image of f, i.e., $f(G)$, is a subgroup of H.

(3) We define the *kernel of* f, denoted ker(f), to be the set $f^{-1}(\{id_H\})$, that is, the set $\{g \in G \mid f(g) = id_H\}$. Show that ker($f$) is a subgroup of G.

(4) Show that for $g \in G$, $ghg^{-1} \in$ ker(f) for all $h \in$ ker(f).

When $f : G \to H$ is a group homomorphism, then there is some similarity between the group structure in $f(G)$ and that in G, "if we forget ker(f)." This is of course a very loose statement, but one that will be made precise in courses ahead.

Exercise 13.63. Let G be a group. We discuss in this exercise the centralizer of an element $h \in G$ and its relation to the conjugacy class of h (Exercise 13.60 above).

(1) Denote by C_h the set of elements $\{g \in G \mid gh = hg\}$. These are the elements in G that commute with h, and is known as the *centralizer of* h. Show that C_h is a subgroup of G.

(2) Let S denote the set of left cosets of C_h. Write $[h]$ for the conjugacy class of h. (Recall from Exercise 13.60 that this is the set of all elements of the form ghg^{-1} as g varies in G.) Define

$$\begin{aligned} f : S &\to [h], \\ gC_h &\mapsto ghg^{-1}. \end{aligned}$$

(a) Show that f is well-defined. (Recall what this means: f is defined in terms of a specific representative g of the coset gC_h. But gC_h could also be represented as $g'C_h$ for any $g' \in gC_h$ (see Proposition 13.34). In that case, $f(gC_h)$, which ought to be the same as $f(g'C_h)$, would have read $g'hg'^{-1}$. You need to show that $g'hg'^{-1}$ and ghg^{-1} are one and the same.)
(b) Show that f is a bijection between the set of left cosets of C_h and the conjugacy class of h.
(c) When G is finite, conclude that $[h]$ has $[G : C_h]$ elements in it. (Recall from Remark 13.31 that $[G : C_h]$ denotes the index of C_h in G.)

Exercise 13.64. In the exercise we will derive a relation known as the *class equation* of a finite group. Let G be a finite group.

(1) Recall from Exercise 13.63 that for an element $h \in G$, $[h]$ denotes the conjugacy class of G and C_h denotes the centralizer of h. Show that the following are equivalent:

(a) $|[h]| = 1$.
(b) $G = C_h$.
(c) h is in the center, $Z(G)$, of G. (See Exercise 13.59 for the center of a group.)

(Recall from Remark 12.56 in Chapter 12 that to say that the statements A, B, and C are equivalent is to say that A, B, and C are either simultaneously true or simultaneously false. One way of proving this is to prove it cyclically: show that $A \implies B \implies C \implies A$!)

(2) By Exercise 13.60 and general results about equivalence relations, G is a disjoint union of conjugacy classes. Dividing these conjugacy classes into those that have only one element and into those that have more than one element, and invoking the results of Exercise 13.63, derive the following *class equation* of G:

$$|G| = |Z(G)| + \sum_{\substack{h \text{ from distinct} \\ \text{conjugacy classes,} \\ \text{and } |[h]|>1}} [G : C_h].$$

Exercise 13.65. A finite group G is said to be a p-group if $|G| = p^n$ for some prime p and $n \geq 1$. We will use the class equation (see Exercise 13.64) to show that the center of a p-group is *nontrivial*, that is, it contains more than just the identity element. So let G be a p group for some prime p.

(1) In the class equation for G from Exercise 13.64, show that each summand on the right side of the equation other than the first summand $|Z(G)|$ must necessarily be divisible by p.

(2) Conclude that $|Z(G)|$ must also be divisible by p, and therefore must contain an element other than just the identity.

14

Graphs: An Introduction

In this chapter we will dip our feet into a branch of mathematics that started with something that can be described as a puzzle, a fun piece of curiosity, but which quickly morphed into a subject of immense applicability, particularly in computer science. This is the subject of *graph theory*, and it has a very precise origin, namely the *Königsberg Bridge Problem*. We will anchor our study of graph theory on this problem, developing enough material to give a complete solution to this problem.

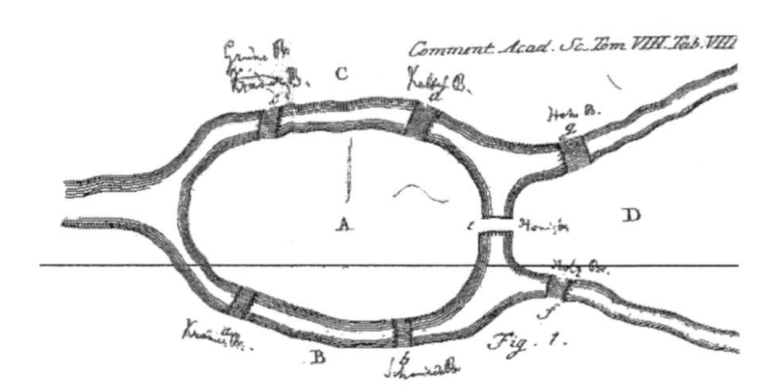

Figure 14.1. Layout of Königsberg Bridges from Euler's Paper "Solutio problematis ad geometriam situs pertinentis," available at `https://scholarlycommons.pacific.edu/euler-works/53/`

The problem dates to the first half of the18th century, and is centered on the town of Königsberg, then part of Prussia, but now called Kaliningrad and part of Russia. The river Pregel (now called by its Russian name Pregolya) flows as two channels through this town, circles an island then known as Kneiphof (now known as Knaypkhof) then comes together and flows eventually into the Baltic via the Vistula Lagoon. In the early 18th century, there were seven bridges on this river, configured as in the picture above (the region marked A is the island Kneiphof). The problem is the following: is it possible to stroll through the town crossing each bridge once and exactly once?

Euler, a prolific Swiss mathematician, then living in St. Petersburg, solved this problem: he showed that it is impossible to traverse the bridges as described! He did so in an ingenious manner, stripping away all inessentials from the problem, including length measurements and angles (which is the stuff of geometry!). He reduced the problem to one of simply *configuration:* the question of which landmass (one of the regions marked A, B, C, or D in Figure 14.1) is connected to which other landmass by bridges. He studied the number of bridges that lead to each landmass (5 for A, 3 each for B, C, and D), and went on to devise a technique that allows one to check if *any* configuration of landmasses and bridges between them admits such a stroll. His technique is based *solely* on how many landmasses have an odd number of bridges leading to them!

14.1 Königsberg Bridge Problem and Graphs

The current way to represent Euler's ideas is to imagine each landmass as a point, and each bridge between landmasses as a line (or curve) connecting the corresponding points. Thus, the configuration of landmasses and bridges from Figure 14.1 leads to the configuration of points and curves shown overlaid in Figure 14.2. Since the exact positioning of the points or shape of curves does not change the configuration, we may, for aesthetic purposes, draw the configuration in a more symmetric fashion: this is shown in Figure 14.3.

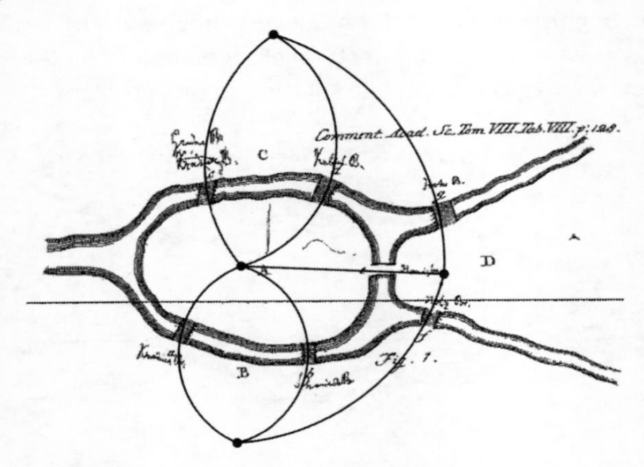

Figure 14.2. Layout of Königsberg Bridges from Euler's Paper (from `https://scholarlycommons.pacific.edu/euler-works/53/`) with bridge schematic overlaid

If you look closely at Euler's picture of the bridges in Figure 14.1, you will note that in addition to labeling the landmasses, Euler has also labeled the bridges using small letters a through g. Accordingly, as bridges correspond to our curves in our configuration diagram, we have labeled the edges using the small letters a through g, exactly as Euler did in his picture.

Figure 14.3 is the prototypical graph! A *graph* (note that the word appears by itself; it is not to be confused with the *graph of a function* with which you are very familiar

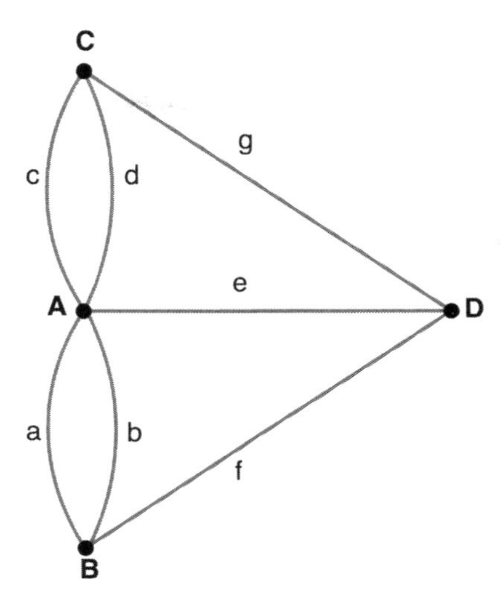

Figure 14.3. Schematic of bridges and landmarks

from high school!) is simply a configuration of points and lines or curves between these points, such as the one in Figure 14.3! We will now give a formal definition, but see Figure 14.4 for further examples of graphs.

Recall from Example 9.21 in Chapter 9 that if V is a set, and R is the equivalence relation on $V^2 = V \times V$ defined by subsets $\{(x, y), (s, t)\}$ of V^2 such that either $x = s$ and $y = t$ or else $x = t$ and $y = s$, then V^2/R is the set whose members are unordered pairs of elements of S. We will write $V \times_{\text{sym}} V$ for V^2/R below.

Here is the formal definition of a graph:

Definition 14.1. A graph G is a triple $G = (V, E, \phi)$, where V is a nonempty set whose members are called *vertices* (or *nodes*), E is a set whose members are called *edges*, and $\phi : E \to V \times_{\text{sym}} V$ is a function, so ϕ maps each edge $e \in E$ to an unordered pair of vertices in V. If $\phi(e) = (X, Y)$, the vertices X and Y are referred to as the *endpoints* of the edge e, and e is said to be *incident* on X and Y, or to *run between* X and Y. Given two vertices X and Y, if there exists an edge e such that $\phi(e) = (X, Y)$, we say that X and Y are *adjacent* or are *neighbors*.

Remark 14.2. We will unpack this definition below, but we first remark that the definition we gave is slightly restricted: it is actually that of a particular class of graphs known as *undirected graphs*, a class to which the graph in Figure 14.3 arising from the Königsberg Bridge problem belongs. We will focus on undirected graphs in this first exposure to graph theory.

Remark 14.3. Note that we have required the set of vertices V to be nonempty. We will not consider a graph on an empty set of vertices to have mathematical meaning!

Definition 14.1 merely says that a graph consists of a set of vertices (which we informally referred to as points in the context of the Königsberg Bridge problem above)

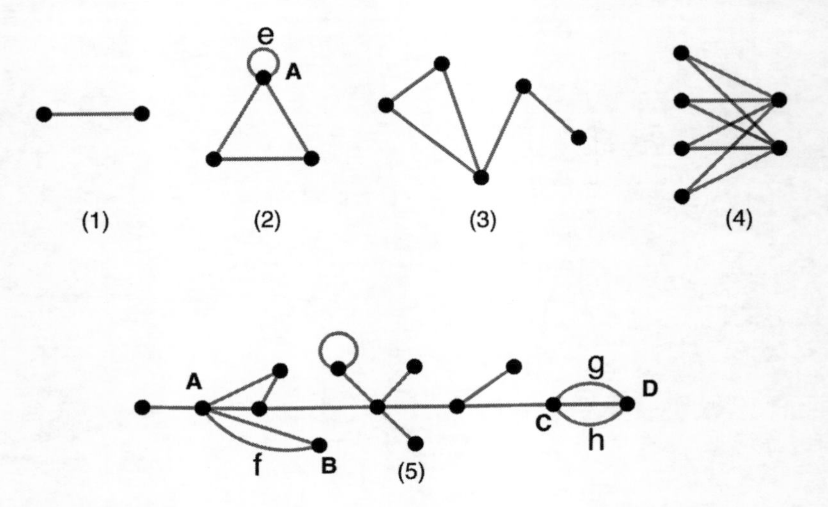

Figure 14.4. Some Examples of Graphs

and edges between them (which we informally referred to as lines or curves). The fact that the edges run *between* the vertices is captured by the function ϕ: it specifies which two vertices the given edge runs between.

Let us study Figure 14.3 in light of this definition. The set of vertices V for the graph in this figure is $\{A, B, C, D\}$. The set of edges E is $\{a, b, c, d, e, f, g\}$. The function $\phi : E \to V \times_{\text{sym}} V$ that sends edges to their endpoints is defined as follows: $\phi(a) = (A, B)$, $\phi(b) = (A, B)$, $\phi(c) = (A, C)$, $\phi(d) = (A, C)$, $\phi(e) = (A, D)$, $\phi(f) = (B, D)$, and $\phi(g) = (C, D)$. Any two vertices in this graph are adjacent.

Figure 14.4 shows more examples of graphs. Notice, for instance, that in the graph labeled (5) in that figure, the vertices labeled A and C are not adjacent. Observe that in both Figures 14.3 and 14.4 we have used capital letters to label vertices and small letters to label edges. This is a convention we will follow in this chapter when labeling graphs. But as Figure 14.4 shows, we do not always label all edges or all vertices of graphs if we do not need to explicitly refer to them.

Definition 14.4. A graph is said to be *finite* if it contains only finitely many vertices and finitely many edges. A graph that is not finite is said to be *infinite*.

Remark 14.5. We will consider only finite graphs in this chapter.

We observe something about Definition 14.1: it does not rule out two distinct edges running between the same endpoints. In other words, it does not insist that the function ϕ be injective. For example, in the graph labeled (5) in Figure 14.4, there are two edges with endpoints A and B: the edge marked f, and the line segment between A and B (which we have left unmarked so as to not clutter the diagram). Similarly, there are three edges in the same graph with endpoints C and D: the edges marked g and h, as well as the (unmarked) line segment between C and D. By contrast, the graphs labeled (1) through (4) in Figure 14.4 do not exhibit this phenomenon: between any two vertices, there is at most one edge that is incident upon these vertices.

The graph labeled (2) in Figure 14.4 shows a different peculiarity: the edge labeled e in that graph has the property that it runs between the vertex A and the vertex A! Put differently, $\phi(e) = (A, A)$. We make this a definition:

Definition 14.6. A *loop* in a graph $G = (V, E, \phi)$ is an edge $e \in E$ such that $\phi(e)$ is of the form (A, A) for some vertex $A \in V$. In such a situation we say that the loop e is *centered* on A.

We refer to graphs without either of these two peculiarities as *simple* graphs. The formal definition is as follows:

Definition 14.7. A graph (V, E, ϕ) as in Definition 14.1 is said to be *simple* if it contains no loops and if the function ϕ is injective. A graph is said to be a *multigraph* if it contains no loops and if ϕ is not injective.

We may refer informally to the graph (2) in Figure 14.4 as a simple graph with loops, and the graph (5) of Figure 14.4 as a multigraph with loops.

Note that Definition 14.1 does not insist that between any two pairs of distinct vertices there must run an edge between them. Thus, the function ϕ in the definition of the graph need not have as its image the entire set of unordered pairs (X, Y) with $X \neq Y$. Graphs labeled (3) and (5) in Figure 14.4 are examples where not every pair of distinct vertices has an edge between them. (Graphs for which any two pairs of distinct vertices have an edge running between them are known as *complete* graphs; see Exercise 14.49 ahead.)

Since graphs are tools for capturing configurations of points and lines or curves between them, and since such configurations can be quite complex and come in different categories, it is not surprising that graph theory contains a lot of definitions! But this is not something to be discouraged by; instead, it should be taken as a measure of the richness of the configurations that we are trying to study!

We need one more definition before we can present Euler's solution of the Königsberg Bridge problem; it captures something quite essential to the structure of a graph:

Definition 14.8. Let $G = (V, E, \phi)$ be a graph and A a vertex of G. The *degree* of A, denoted $\deg(A)$, is defined as the cardinality of the set of edges that are incident upon A, with one special case: when a vertex admits a loop centered on it, we count this edge *twice*. Thus, $\deg(A) = |\{e \in E \mid \phi(e) = (A, X) \text{ for some } X \in E,\ X \neq A\}| + 2 * |\{e \in E \mid \phi(e) = (A, A)\}|$.

Thus, in Figure 14.3, $\deg(A) = 5$, while $\deg(B) = \deg(C) = \deg(D) = 3$. In the graph labeled (2) in Figure 14.4, the degree of A is 4, with the loop e counted twice.

Note that it is possible for a graph to have vertices of degree zero! These are called *isolated* vertices, and there is nothing in Definition 14.1 that precludes their existence! The graph in Figure 14.5 exemplifies this: the vertices A and D have degree 0, while B and C have degree 1 each!

Euler's proof that it is impossible to walk through the town so that each bridge is crossed once and only once is extremely simple, once the basic features of a graph are

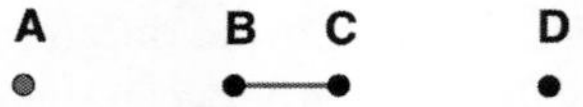

A B C D

Figure 14.5. A graph with isolated vertices.

identified and laid out! His proof, in the language of graphs, goes as follows: to start at one of the four landmasses and walk through the town crossing each bridge once and exactly once is to start at some vertex of the graph of Figure 14.3 and walk along the edges of the graph so that each edge (which corresponds to a bridge) is traversed once and exactly once. Let us focus on any one vertex, say A. We have three scenarios: During our walk,

Scenario 1: We may have started at A, or

Scenario 2: Ended at A, or else,

Scenario 3: Have started and ended at some other vertex or vertices but would have walked into and out of A, possibly multiple times, along different edges.

Note that when we walk into A and then walk out of A along a different edge, we would have accounted for two of the edges incident upon A. So, in Scenario 3, since we are walking into and out of A possibly several times along different edges, we would have used up an even number of edges incident upon A. But A has degree 5, which means one or more of the edges incident upon A would not have been traversed—this is not allowed! Thus Scenario 3 does not apply at A—we need either to have started or ended at A.

Notice that the exact degree of A did not matter; all that we needed to rule out Scenario 3 occurring at A was that the degree of A is odd. But then, exactly this same reasoning applies to vertices B, C, and D, as they all also have odd degree. Hence Scenario 3 does not apply to them as well. Thus, we would have either started or ended at B, and likewise for C and D. But this is now absurd. There is only one point we can start from, and at most one other point we can end at; there cannot be four points from which we either start or end! Hence, the Königsberg Bridges cannot be traversed as desired!

We can capture this very simple yet elegant set of ideas in the following:

Theorem 14.9. *If a graph contains more than two vertices with odd degree, then it is impossible to traverse the graph so that each edge is traversed once and exactly once.*

We can sharpen these discussions further. First, let us note the contrapositive to Theorem 14.9: If it is possible to traverse a graph so that each edge is traversed once and exactly once, then the graph must have at most two vertices of odd degree. Now let us push these same arguments above some more.

Suppose we are able to traverse our graph so that each edge is traversed once and exactly once, and suppose that our traversal is such that we start at some vertex X and end at a different vertex Y. Then, every vertex Z on the traversal different from X and Y falls under Scenario 3 above. The arguments above show that Z must have even degree: whenever we walk in to Z, we necessarily walk out since we neither start nor end at Z, and then each time we do this walking in and out, we use up two edges. As for the start point X: we start our traversal along one edge, but thereafter, if we return to X we necessarily leave X since X is not the endpoint of our traversal. Just like with Z, every time we enter and leave X we use up two edges, and if we add all these to the edge we began our traversal on, we find that X has odd degree. By exactly the same arguments, Y also has odd degree. We thus find that if we are able to traverse our graph so that each edge is traversed once and exactly once and so that our start and end vertices are different, then our graph must have exactly two vertices of odd degree (and the traversal must start at one of these vertices of odd degree and end at the other vertex of odd degree).

Now suppose our traversal above is such that it starts and ends at the same vertex X (so the traversal is "closed"). Then, as our arguments above showed, every vertex other than Z falls under Scenario 3, and therefore has even degree. But so does X! This is because, to start with, the edge along which we started our traversal and the edge along which we ended our traversal count for two edges incident on X. Next, between leaving X at the beginning of our traversal and landing at X at the end of our traversal we may have entered and exited X one or more times, but each of those times would have contributed to two more edges incident on X. Adding all these edges together, we find that X also has even degree!

We summarize these discussions:

Theorem 14.10. *Assume it is possible to traverse a graph so that each edge is traversed once and exactly once. If the traversal starts and ends on different vertices, then the graph must have exactly two vertices of odd degree (and then one of these vertices must be the start vertex and the other the end vertex). If on the other hand the traversal starts and ends on the same vertex, then every vertex in the graph must have even degree.*

Theorem 14.9 and its stronger form Theorem 14.10 provide necessary conditions for being able to traverse a graph so that each edge is traversed once and exactly once. To formulate a condition that is both necessary and sufficient, we will consider something known as the *degree-sum formula*. It is another elementary but pretty result proved by Euler:

Proposition 14.11. (*Degree-sum formula.*) *In a finite graph, the sum of the degrees of all the vertices equals twice the number of edges.*

Proof. Let $G = (V, E, \phi)$ be the finite graph. We wish to show the following:

$$\sum_{X \in V} (\deg(X)) = 2|E|.$$

The proof is simply a matter of counting in two different ways! The left-hand side adds up, across all vertices, the number of edges incident on each vertex. Now view this process from the viewpoint of the edges. Each edge appears in the sum on the left because it is counted when computing the degree of its endpoints, and further, each edge is actually counted twice, once for each end point. Thus, each edge contributes twice to the sum on the left. (Notice that a loop also contributes twice to the sum on the left. This is because, by definition, while computing the degree of the vertex it is incident upon, the loop counts as two; see Definition 14.8.) The formula immediately follows. □

Corollary 14.12. *The number of vertices of odd degree in a graph is even.*

Proof. Since the sum of degrees of all vertices is even by the proposition above, the number of odd summands on the left side of the degree-sum formula must necessarily be even. □

14.2 Walks, Paths, Trails, Connectedness

To prove a converse to Theorem 14.9, that is, to provide sufficient conditions on a graph to be able to traverse the graph so that each edge is traversed once and exactly once, we need to first consider the notion of *connectedness* of a graph, and its related notion of *connected components.* To study these notions, we will begin by making mathematically precise the concept of traversing (or walking or strolling through) a graph so that each edge is traversed (or walked, or strolled through) once and exactly once — a concept that we have considered to be self-explanatory so far!

We have the following:

Definition 14.13. Let $G = (V, E, \phi)$ be a graph. A *walk* W through the graph is a sequence

$$(X_0, e_1, X_1, e_2, X_2, \ldots, X_{k-1}, e_k, X_k)$$

where $k \geq 0$ is an integer, $X_i \in V$, $e_j \in E$, and each e_j runs between X_{j-1} and X_j. (When $k = 0$, the walk consists just of the vertex X_0.) The length of W is defined to be k, the number of edges. W is said to be a *closed walk* if $X_k = X_0$. If W is not closed then it is known as an *open walk.*The walk is said to be *from* X_0 *to* X_k, and X_0 and X_k are called the start and end points of the walk respectively. We also say informally that the various X_i and e_j are *in* W or *a part of* W.

Thus, a walk captures the notion of strolling through the graph, starting at some vertex X_0, proceeding along some edge e_1 that is incident on X_0 till we hit the vertex X_1 that is the other endpoint of e_1 (of course, if e_1 is a loop, X_1 will equal X_0), then proceeding along some edge e_2 that is incident on X_1 till we hit the other endpoint X_2, and so on, always proceeding from one vertex to one of its neighbors by walking along an edge connecting the two, till we end at the vertex X_k.

Note that in the definition of a walk, we do not insist that the edges or the vertices be different. Thus, in a walk, it is possible that $X_i = X_j$ for some distinct i and j, or for that matter, $e_i = e_j$ for some distinct i and j. For instance, for the graph in Figure 14.6 ahead, the walk $(A, p, B, q, C, x, F, t, B, q, C)$ has the vertices B and C repeated, as also the edge q. (If $e_i = e_j$ for distinct i and j, what can you conclude about X_{i-1}, X_{j-1}, X_i and X_j?)

Definition 14.14. Let $G = (V, E, \phi)$ be a graph. A *walk*

$$W = (X_0, e_1, X_1, e_2, X_2, \ldots, X_{k-1}, e_k, X_k)$$

in G is said to be a *trail* if $e_i \neq e_j$ for all i, j with $i \neq j$. A trail

$$(X_0, e_1, X_1, e_2, X_2, \ldots, X_{k-1}, e_k, X_k)$$

with $X_0 = X_k$ is called a *closed trail.* A closed trail is also known as a *circuit.* A trail that is not closed is known as an *open trail.*

The link between trails and the Königsberg Bridge problem is given by the following:

Definition 14.15. Let G be a graph. A trail

$$T = (X_0, e_1, X_1, e_2, X_2, \dots, X_{k-1}, e_k, X_k)$$

in G is said to be an *Eulerian trail* if every edge of G appears in T. An Eulerian trail

$$T = (X_0, e_1, X_1, e_2, X_2, \dots, X_{k-1}, e_k, X_k)$$

is said to be a *closed Eulerian trail* or an *Eulerian circuit* if $X_0 = X_k$. An Eulerian trail that is not closed is known as an *open Eulerian trail.*

Thus, an Eulerian trail (or Eulerian circuit) is precisely what we were seeking in our Königsberg Bridge problem. (Note that in an Eulerian trail, it is permissible that $X_i = X_j$ for some distinct pair i, j; all that matters is that every edge in G appears once and exactly once.)

Definition 14.16. Let G be a graph. A walk

$$W = (X_0, e_1, X_1, e_2, X_2, \dots, e_{k-1}, X_{k-1}, e_k, X_k)$$

in G is said to be a *path* if $X_i \neq X_j$ for all i, j with $i \neq j$. A walk

$$(X_0, e_1, X_1, e_2, X_2, \dots, e_k, X_k, e_{k+1}, X_{k+1})$$

is said to be a *closed path* or a *cycle* if

$$(X_0, e_1, X_1, e_2, X_2, \dots, e_k, X_k)$$

is a path, $X_0 = X_{k+1}$ (and in the case where $k = 1$, $e_{k+1} (= e_2) \neq e_1$). A path that is not closed is known as an *open path.*

Remark 14.17. Thus, in an open path, no vertex is repeated, while in a closed path, only the first and last are repeated. Note that a path is necessarily a trail: if it is not a trail, then some two edges e_i and e_j $(i \neq j)$ must be equal, so both their endpoints will be equal and hence repeat in the walk. But as we just noted, no vertex can repeat in an open path, while exactly one vertex, namely the start vertex, repeats in a closed path.

Definition 14.18. Let G be a graph, and let

$$W = (X_0, e_1, X_1, e_2, X_2, \dots, X_{k-1}, e_k, X_k)$$

be a walk in G. A *subwalk* W' of W is a walk obtained either as the subsequence

$$(X_i, e_{i+1}, X_{i+1}, \dots, e_j, X_j)$$

of W for some $i < j$, or by considering such a subsequence

$$(X_i, e_{i+1}, X_{i+1}, \dots, X_{j-1}, e_j, X_j)$$

and for two vertices X_s and X_t $(s < t)$ in this subsequence such that $X_s = X_t$ (if such a pair exists), further deleting the subsequence

$$(e_{s+1}, X_{s+1}, \dots, X_{t-1}, e_t, X_t (= X_s)\ ).$$

(Note that if such a pair exists, then the subwalk $(X_s, e_{s+1}, X_{s+1}, \dots, X_{t-1}, e_t, X_t (= X_s))$ is just a closed walk in G.)

If W' is a subwalk of W, W is said to *contain* W'.

Here are some examples to illustrate all this:

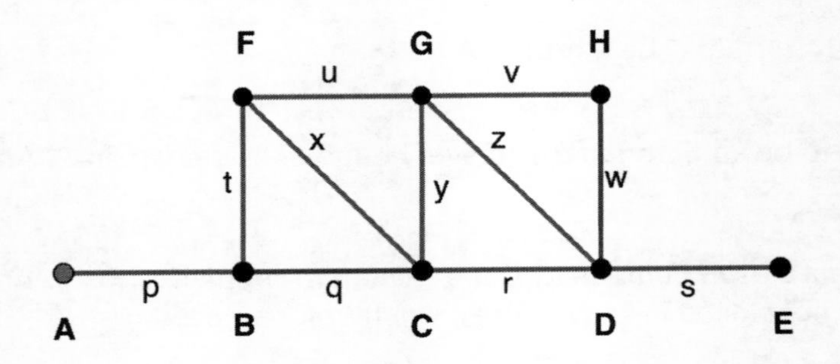

Figure 14.6. Walks in Graphs

Example 14.19. Our graph G will be as depicted in Figure 14.6. We have the following:

(1) The following are all walks in G:

$$W_1 = (A, p, B, t, F, x, C, q, B, t, F, u, G, z, D, s, E),$$
$$W_2 = (B, q, C, y, G, u, F, x, C, r, D),$$
$$W_3 = (A, p, B, t, F, u, G, v, H, w, D, s, E),$$
$$W_4 = (C, y, G, u, F, x, C, r, D),$$
$$W_5 = (C, r, D),$$
$$W_6 = (B, q, C, r, D, w, H, v, G, u, F, t, B).$$

(2) W_2 is a trail from B to D, as each edge in W_2 appears just once. It is not an Eulerian trail as all edges of G do not appear in W_2. (In fact, G cannot have any Eulerian trails by Theorem 14.9, as vertices A, B, E, and F all have odd degree. Note that the vertex C appears twice, so W_2 cannot be a path.)

(3) W_3 is a path in G from A to E. (It is also a trail, by Remark 14.17.)

(4) W_4 is a subwalk of W_2 obtained by considering the subsequence $(C, y, G, u, F, x, C, r, D)$ of W_3.

(5) W_5 is a subwalk of W_4 (and hence of W_2) obtained by noting that the vertex C appears twice in W_4 and deleting the sequence (y, G, u, F, x, C) from W_4. Observe that the subwalk (C, y, G, u, F, x, C) is a closed trail in G; it is even a closed path.

(6) W_6 is a closed path in G.

Exercise 14.20. The graph in Figure 14.6 did not admit any Eulerian trails. By contrast, the graph in Figure 14.7 has an Eulerian trail: find it!

As our examples show, a walk through a graph need not be a path; vertices could be repeated in the walk. Our first result, however, is the following:

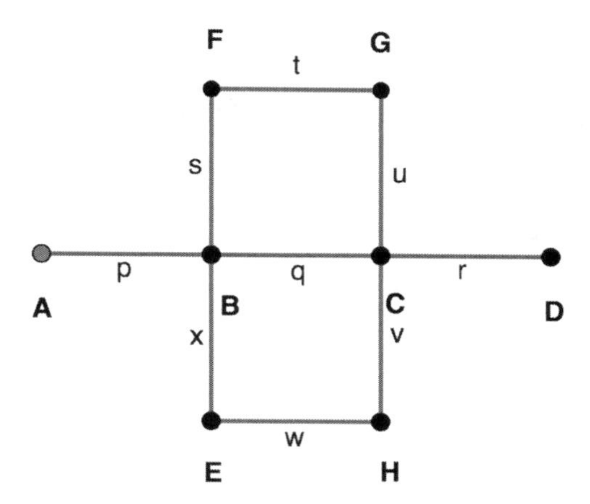

Figure 14.7. Find an Eulerian trail in this graph!

Lemma 14.21. *In a graph, any walk contains a subwalk that is a path from the start point to the end point of the walk.*

Proof. Let G be the graph, and

$$W = (X_0, e_1, X_1, e_2, X_2, \ldots, X_{k-1}, e_k, X_k)$$

be the walk. The proof of the lemma is simple. If any two vertices X_i and X_j that appear in W are the same for some $i < j$, this means that the subsequence

$$(X_i e_{i+1}, X_{i+1}, \ldots, X_{j-1}, e_j, X_j (= X_i))$$

forms a closed subwalk of W. We simply remove this closed walk, retaining X_i, and repeat for any other closed subwalk still present.

Thus, we start by deleting the subsequence

$$(e_{i+1}, X_{i+1}, \ldots, X_{j-1}, e_j, X_j (= X_i))$$

of W. The sequence W' that is left will still be a walk. Repeat for W': if any two vertices $X_{i'}$ and $X_{j'}$ are the same in W' for some $i' < j'$, delete the subsequence

$$(e_{i'}, X_{i'}, \ldots, X_{j'-1}, e_{j'}, X_{j'} (= X_{i'}))$$

of W'. What is left will be a walk W''. Proceed thus, identifying all repeated vertices and deleting the subsequence of edges and vertices between them. This procedure must stop, as there are only finitely many vertices in the sequence that defines W (and hence only finitely many pairs of vertices that could be equal). What will be left will be, by definition, a path from X_0 to X_k. □

We will now look at the notions of a subgraph of a graph and the connectedness of a graph:

Definition 14.22. Let $G = (V, E, \phi)$ be a graph. A subgraph of G is a graph $H = (V', E', \phi')$ such that $V' \subseteq V$, $E' \subseteq E$ and $\phi'(e) = \phi(e)$ for all $e \in E'$.

Thus, a subgraph of G is a graph H obtained from G by choosing some subset V' of the set of vertices V of G, and choosing some subset of those edges of G that join the various vertices in V'. So, given vertices X_1 and X_2 in V', if e is an edge in E' between X_1 and X_2, then e is also an edge in G between X_1 and X_2 (but it is not necessary that every edge between X_1 and X_2 in G exist as an edge in H).

Figure 14.8 gives an example of a subgraph.

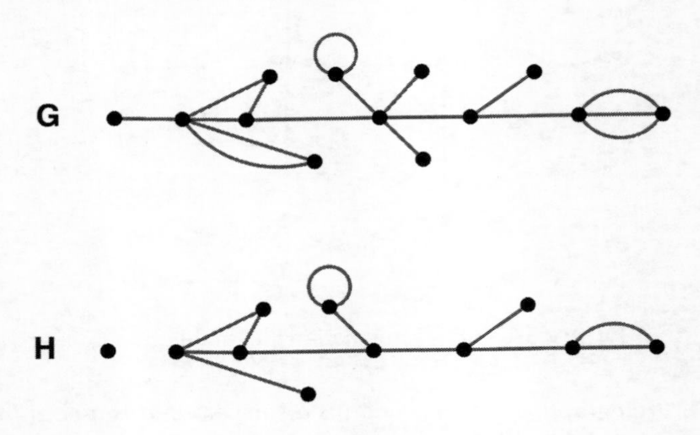

Figure 14.8. H is a subgraph of G

Example 14.23. Here are further examples of subgraphs; see Figure 14.9 for an illustration.

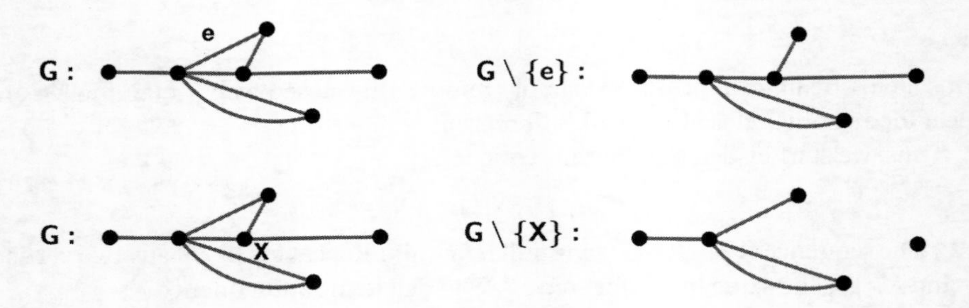

Figure 14.9. Subgraphs obtained by deleting an edge and deleting a vertex.

(1) If $G = (V, E, \phi)$ is a graph and e an edge in G, the graph denoted $G \setminus e$ is the subgraph of G obtained by deleting the edge e, and retaining everything else in G. More generally, if $E' \subseteq E$, then $G \setminus E$ is the graph obtained by deleting all edges $e \in E$.

(2) If $G = (V, E, \phi)$ is a graph and X a vertex in G, the graph denoted $G \setminus X$ is the subgraph of G obtained by deleting the vertex X and all edges incident on X, and retaining everything else in G. More generally, if $V' \subseteq V$, then $G \setminus V$ is the graph obtained by deleting all vertices $X_i \in V$, along with all edges incident on any vertex in V. (Note that if V contains only one element, $G \setminus X$ becomes meaningless; see Remark 14.3.)

Definition 14.24. Let G be a graph. Two distinct vertices X and Y are said to be *connected* if there is a path from X to Y (or vice versa). The graph itself is said to be connected if any two distinct vertices of G are connected. A graph that is not connected is also called *disconnected.*

For example, all the graphs in Figure 14.4 are connected, but the graph in Figure 14.5 is not connected.

Remark 14.25. Note that if G has only one vertex, then G is *vacuously* connected. For, the third sentence in Definition 14.24 is really saying the following: G is connected if the statement "If G contains distinct vertices then any two distinct vertices are connected" is true. By our discussions on the truth-table of the "implies" statement (see Table 3.6 in Chapter 3), the statement within the quotation marks is true when the hypothesis "If G contains distinct vertices" is false!

We will now impose an equivalence relation on the vertices of a graph that will lead to the notion of the connected components of a graph: Let $G = (V, E, \phi)$ be a graph. Define a relation $\sim$ on V as follows: For vertices X and Y, $X \sim Y$ if X and Y are connected. We will refer to this as the *connectedness relation on the vertices.*

Lemma 14.26. *The connectedness relation on the vertices of a graph is an equivalence relation.*

Proof. Let $G = (V, E, \phi)$ be the graph, and $\sim$ the connectedness relation on the vertices. Any $X \in V$ is related to itself since the path (X) with zero edges connects X to itself. If $X \sim Y$, there is a path

$$(X = X_0, e_1, X_1, \dots, X_{k-1}, e_k, Y)$$

so the reverse path

$$(Y, e_k, X_{k-1}, \dots, X_1, e_1, X_0 = X)$$

is a path from Y to X. Finally, if $X \sim Y$ and $Y \sim Z$, there are paths

$$(X = X_0, e_1, X_1, \dots, X_{k-1}, e_k, Y)$$

and

$$(Y = Y_0, f_1, Y_1, \dots, Y_{l-1}, f_l, Z),$$

so the concatenation

$$(X, e_1, X_1, \dots, X_{k-1}, e_k, Y, f_1, Y_2, \dots, Y_{l-1}, f_l, Z)$$

is a *walk* from X to Z. (Convince yourself with examples that such a concatenation need not be a path.) But by Lemma 14.21, such a walk contains a path from X to Z, so $X \sim Z$ as well. □

Now let $V_1, \dots, V_r$ be the equivalence classes of V under $\sim$. Thus, we have a disjoint union

$$V = V_1 \sqcup V_2 \sqcup \cdots \sqcup V_r.$$

Let $E_i \subset E$ consist of all edges e such that both endpoints of e belong to V_i (or *the* endpoint of e belongs to V_i if e is a loop). We claim that

$$E = E_1 \sqcup E_2 \sqcup \cdots \sqcup E_r.$$

To see this, first note that $E_i \cap E_j = \phi$ whenever $i \neq j$. For, if $e \in E_i \cap E_j$, then e has one endpoint, say X, in E_i and another endpoint, say Y, in E_j. Then (X, e, Y) is a path, so $X \sim Y$. But this is impossible, as X and Y are in different equivalence classes. It follows that the union $E_1 \cup E_2 \cup \cdots \cup E_r$ is actually a disjoint union $E_1 \sqcup E_2 \sqcup \cdots \sqcup E_r$.

To show that any $e \in E$ is in one of the E_i, note first that if e is a loop, then its (single) endpoint must obviously lie in some V_i, so $e \in E_i$. If e has two distinct endpoints X and Y, then again X must lie in some V_i and Y must lie in some V_j. If $i \neq j$, then exactly as in the arguments above for why the E_i are disjoint, we would find $X \sim Y$, a contradiction. Hence $i = j$, and e lies in $E_i (= E_j)$.

To summarize, the vertices of G break up into disjoint equivalence classes V_i under $\sim$, and every edge in G runs between vertices that are in the same V_i for some i. (In particular, there are no edges that connect a vertex in V_i to a vertex in V_j if $i \neq j$.) The graph now visibly decomposes into a "disjoint union"

$$G = (V_1, E_1, \phi_1) \sqcup (V_2, E_2, \phi_2) \sqcup \cdots \sqcup (V_r, E_r, \phi_r),$$

where $\phi_i : E_i \to V_i \times_{\text{sym}} V_i$ is obtained by the rule $\phi_i = \phi|_{E_i}$, and by the "disjoint union" of graphs we simply mean a graph as above, with the vertex set written as a disjoint union of subsets V_i, and with all edges running only between vertices in the same V_i.

Definition 14.27. The graphs (V_i, E_i, ϕ_i) above are called the connected components of the graph G.

Thus, in the language of Definition 14.27, a graph is connected precisely when it has just one connected component.

Figure 14.10 shows an example of the connected components of a graph.

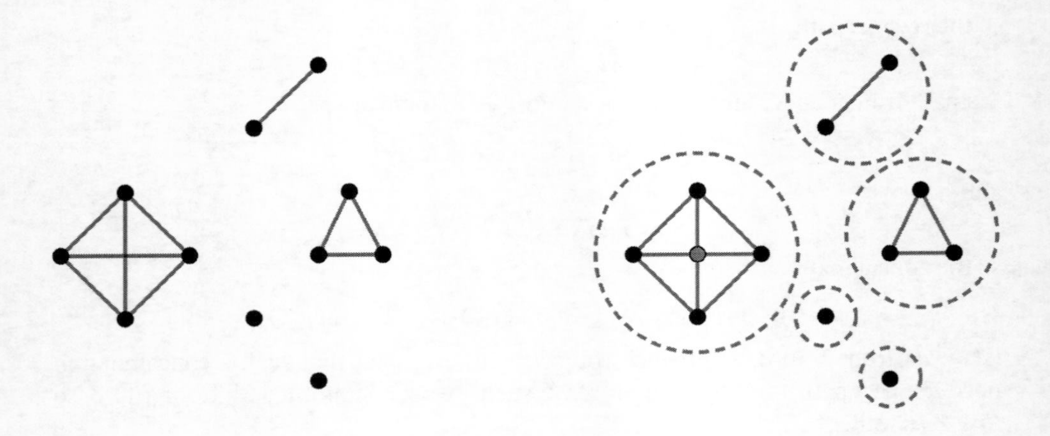

Figure 14.10. A graph on the left, with its connected components demarcated on the right

One result about the existence of Königsberg Bridge type walks (now called Eulerian trails) falls out very easily from our discussion of connected components:

Lemma 14.28. *If a graph G possesses an Eulerian trail, then it must contain at most one connected component with a positive number of edges (with the remaining components therefore consisting only of isolated vertices).*

Proof. Assume the statement is false. Then there is a graph G with an Eulerian trail that has at least two connected components G_1 and G_2 each containing edges e_1 and e_2 respectively. Therefore any Eulerian trail must look like

$$(Z_0, f_1, Z_1, \dots, X_1, e_1, Y_1, f_{k+1}, Z_{k+1}, \dots, X_2, e_2, Y_2, f_{l+1}, Z_{l+1}, \dots)$$

(or a similar one where the edge e_2 is traversed first and then the edge e_1). Here, X_1 and Y_1 are endpoints of e_1, and therefore belong to the component G_1, while X_2 and Y_2 are the endpoints of e_2 and belong to G_2. But the subwalk

$$(Y_1, f_{k+1}, Z_{k+1}, \dots, X_2)$$

contains (by Lemma 14.21) a path from Y_1 to X_2, which violates the fact that Y_1 and X_2 are in different connected components of G. □

Remark 14.29. The proof above works just as well if the Eulerian trail in the graph is actually a circuit; all that we used is that the trail must contain both e_1 and e_2. It follows that if G possesses an Eulerian circuit, then too, it must contain at most one connected component with a positive number of edges (with the remaining components therefore consisting only of isolated vertices).

We also need to consider the notion of a cut edge of a graph:

Definition 14.30. Let G be a graph. An edge e, lying in some connected component G_i of G, is called a *cut edge* of G if the graph $G_i \setminus \{e\}$ is disconnected.

(See Figure 14.11 for an example.)

Figure 14.11. A cut-edge. G on the left has two connected components; on removing e, $G \setminus \{e\}$ now has three connected components.

Exercise 14.31. Show that a loop can never be a cut edge.

Here is an easy result regarding cut edges that we will need:

Lemma 14.32. *If e is a cut edge in a connected graph G and if S and T are its endpoints, then S and T are in different components of $G \setminus \{e\}$. Moreover, $G \setminus \{e\}$ has precisely two connected components, the component of S and the component of T.*

Proof. We first show that S and T lie in different components of $G \setminus \{e\}$. Assume not. This means there is a path P from S to T in $G\setminus\{e\}$. Since $G\setminus\{e\}$ is a subgraph of G, P can be thought of as a path in G that avoids e. Since G is connected, any two points X and Y are connected by a path Q. If this path does not contain e, then X and Y will clearly remain connected via this same path Q in $G \setminus \{e\}$. If on the other hand Q contains e, then Q must look like

$$Q = (X, e_1, X_1, \dots, e_k, S, e, T, e_{k+1}, \dots, e_n, Y),$$

or else, if T appeared first in the path, it would look like the above but instead contain the subsequence $(\dots, e_k, T, e, S, e_{k+1}, \dots)$. By switching the labeling of S and T, we may assume it is as in the first displayed form. Note that because Q is a path and hence a trail (see Remark 14.17), e would not appear in the subpath, $(X, e_1, X_1, \dots, e_k, S)$ or in the subpath $(T, e_{k+1}, \dots, e_n, Y)$. By removing the edge e from Q and inserting the path P between S and T instead, we get a walk from X to Y that does not contain e, and is therefore in $G \setminus \{e\}$. By Lemma 14.21, this walk contains a path, so we find once again that X and Y are connected in $G \setminus \{e\}$. Since X and Y are arbitrary, we find that $G \setminus \{e\}$ is connected, a contradiction. Hence, S and T lie in different components of $G \setminus \{e\}$.

Now consider any vertex X of $G \setminus \{e\}$. Since G is connected, X is connected to S in G by a path P. If this path does not contain e, then P is a path in $G\setminus\{e\}$, and therefore, X and S are connected in $G \setminus \{e\}$. If P contains e, then S and T must appear immediately on either side of e in P. Since S can appear just once in P because P is a path, P must look like

$$P = (X, e_1, X_1, \dots, e_k, T, e, S).$$

Since P is also a trail (Remark 14.17), the subwalk $(X, e_1, X_1, \dots, e_k, T)$ is then a path that does not contain e, and therefore X and T are connected in $G \setminus \{e\}$. It follows that any vertex of $G \setminus \{e\}$ is in the connected component of either S or T. Hence there are precisely two components. □

14.3 Existence of Eulerian Trails and Circuits: Sufficiency

Armed with our consideration of walks and paths in the previous section, we will now describe sufficient conditions for the existence of Eulerian trails and circuits in a graph G.

We have seen in Theorem 14.10 necessary conditions for the existence of an Eulerian trail that starts at one vertex X and ends at a different vertex Y (X and Y should be the only two vertices of odd degree in the graph) and for the existence of an Eulerian circuit (all vertices should be of even degree).

We have also seen in Lemma 14.28 and Remark 14.29 another necessary condition for the existence of an Eulerian trail or circuit: the graph must either be connected (with a positive number of edges) or have at most one connected component that has a positive number of edges (with the remaining components therefore having only isolated vertices). If G_1 is this connected component of G with a positive number of edges, then it is easy to see that an Eulerian trail (respectively circuit) for G is the same as an Eulerian trail (respectively circuit) for G_1. We may thus restrict our discussion of sufficient conditions for Eulerian trails to graphs with the property that they are connected.

We will show that the conditions described in Theorem 14.10 are not only necessary but sufficient, once the connectedness hypothesis is included:

Theorem 14.33. *A connected graph $G = (V, E, \phi)$ has an open Eulerian trail if and only if it has exactly two vertices of odd degree (in which case the trail starts at one of the vertices of odd degree and ends at the other). G has an Eulerian circuit if and only if all of its vertices have even degree.*

Proof. The "only if" portion of the theorem constitutes Theorem 14.10 which we have already proven. We concentrate here on the "if" part.

We will first prove that if all vertices have even degree, then G has an Eulerian circuit.

We will do induction on $|E|$. If $|E| = 0$, then G consists of a single isolated vertex X (so of degree 0), and (X) is an Eulerian circuit. If $|E| = 1$, say $E = \{e\}$, then we have two possibilities for G (see Figure14.12):

(1) e is a loop and G consists of a single vertex X with the loop e, in which case (X, e, X) is an Eulerian circuit.

(2) e is not a loop, so it has distinct endpoints X and Y, and G consists of just X and Y with the edge e running in between. This case is not possible under the hypothesis that vertices of G have even degree, since X and Y have degree 1.

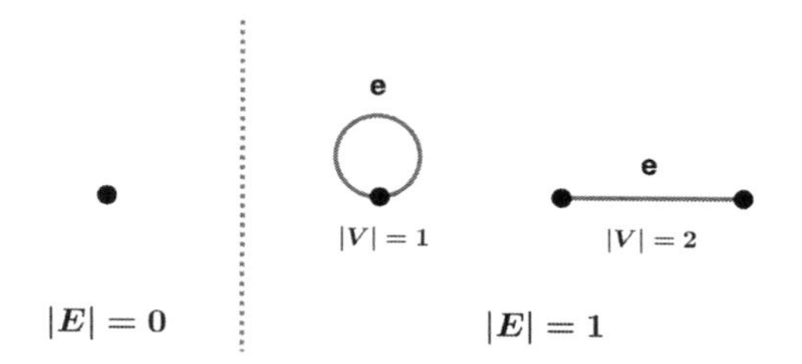

Figure 14.12. Graphs with no vertices of odd degree, cases where $|E| = 0, 1$.

We have thus established the result when $|E| = 0, 1$. We will also take care of another special case: assume $|E| \geq 2$, and that G consists of a single vertex X. Then all edges in G must be loops centered on X, and it is clear how to construct an Eulerian circuit: just traverse the loops one by one. So we can assume in what follows that G has more than one vertex.

Now assume that $|E| = m \geq 2$ in our graph G (and that G has more than one vertex), and that we have established the result for all graphs $G' = (V', E', \phi')$ with $|E'| < m$ and whose vertices are all of even degree. If any edge in G is a loop e, centered at say X, we form the graph $G \setminus \{e\}$. This graph is also connected (see Exercise 14.31, a loop cannot be a cut edge), and has one less edge than G. Note that since X had even degree in G, it continues to have even degree in $G \setminus \{e\}$ since e, being a loop, contributed 2 to the degree of X in G. The degrees of all other vertices of G are unchanged in $G \setminus \{e\}$, so they continue to be even. We may thus apply induction to $G \setminus \{e\}$ and conclude that it has an Eulerian circuit.

Now note that there are edges in G other than e that are incident on X. For, there exists another vertex Y in G by assumption, so any path from Y to X, which exists by the connectedness of G, yields such an edge. Moreover, the fact that the degree of X in $G \setminus \{e\}$ is even shows that there must be more than just one such edge. The Eulerian circuit we have found in $G \setminus \{e\}$ must include these other edges of G incident on X.

Hence, such a circuit would be of the form

$$T = (Q_0, \dots, f_k, X, f_{k+1}, \dots, Q_0).$$

Then the trail

$$T = (Q_0, \dots, f_k, X, e, X, f_{k+1}, \dots, Q_0)$$

formed by pausing T at X, traversing the loop e to return to X, and then resuming T is now a circuit in G, and we are done in this case.

So assume that no edge in E is a loop. Pick an edge e that runs between two vertices of G, say X and Y. If all other edges also run between X and Y, then the graph consists of vertices X and Y with edges $e_1 = e, e_2, \dots, e_{2k-1}, e_{2k}$. It is clear how to construct an Eulerian circuit:

$$T = (X, e_1, Y, e_2, X, \dots, X, e_{2k-1}, Y, e_{2k}, X)$$

will do the job.

Now assume that not every edge in E runs between X and Y. This implies that there exists at least one other vertex, say Z', in V. By the connectedness of G, there is a path T from X to Z'. If this path does not contain e (Case (a)), then it reads (see Figure 14.13(a))

$$T = (X, f, Z, \dots, e_{k-1}, Z'),$$

where Z is the next vertex after X in the path, and f is an edge connecting X and Z. If T contains e (Case (b)), then it reads (see Figure 14.13(b))

$$T = (X, e, Y, f, Z, \dots, e_{l-1}, Z'),$$

where Z is the next vertex after Y in the path, and f is an edge connecting Y and Z. Thus, there exists a vertex Z that is a neighbor of X in Case (a), and neighbor of Y in Case (b). We may assume by switching the labels X and Y that we are in Case (a), and that f joins X and Z.

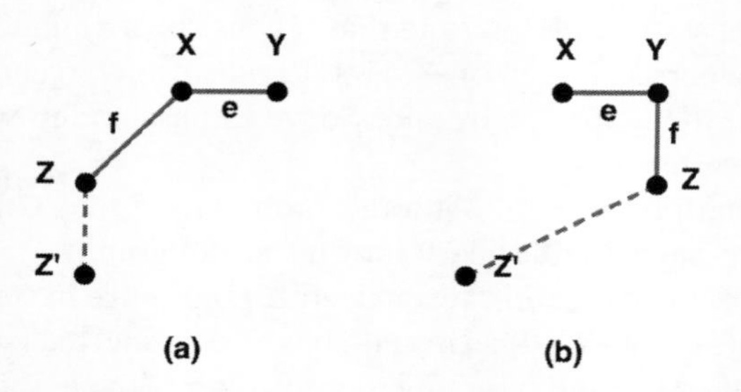

Figure 14.13. Path from X to Z' yields vertex Z and edge f

Consider the graph $G' = G \setminus \{e, f\}$. From this, form the new graph G'' by joining Y and Z with an edge g. See Figure 14.14. (The edge g will be in addition to any other edges between Y and Z already present.) Then in relation to G, we find that the degree of X in G'' has dropped by 2, so is still even, while the degree of Y and Z remain the same. Moreover, the number of edges in G'' is now $m - 1$. We now divide the proof into two cases.

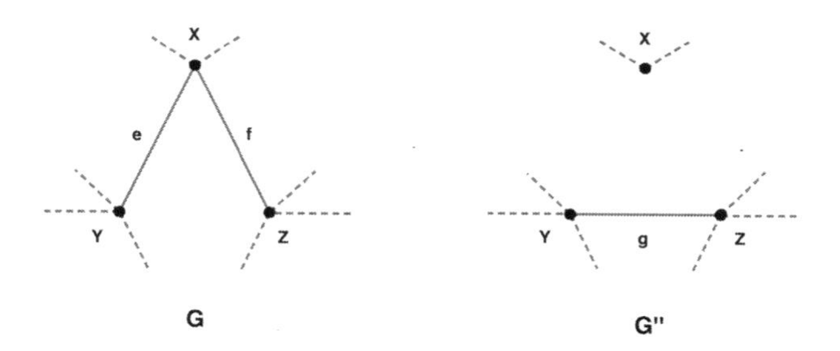

Figure 14.14. The Graphs G and G''

We assume in the first case that G'' is connected. Then, by induction, G'' contains an Eulerian circuit

$$T = (W, \ldots, h_{k-1}, \underbrace{Y, g, Z,} h_k, \ldots, W)$$

where W is some vertex in G'. We replace the portion of T that is a traversal from Y to Z along g in G'' as follows:

$$T' = (W, \ldots, h_{k-1}, \underbrace{Y, e, X, f, Z,} h_k, \ldots, W).$$

Then T' is an Eulerian circuit in G.

In our next case, we assume that G'' is not connected. We will show that in this case, G'' has precisely two components: one containing X, and the other containing Y and Z.

If G'' is not connected, then $G' = G \setminus \{e, f\}$ already must not be connected. This is because G'' is obtained by adding a new edge g to G', and addition of edges does not cause the number of connected components to increase (see Exercise 14.44 ahead). It follows that either e or f must be cut edges of G. By switching the labels of Y and Z we may suppose that e is a cut edge. By Lemma 14.32, $G \setminus \{e\}$ contains exactly two connected components, the equivalence class of X and that of Y. See Figure 14.15. At this point Z is in the connected component of X since the edge f has not been severed.

Now note that f must be a cut edge in $G \setminus \{e\}$! For, if not, X and Z would still stay connected in $G' = G \setminus \{e, f\}$, which means that G' has two components: one containing Y, and the other containing X and Z. On adding the edge g between Y and Z, we would connect these two components, so G'' would be connected. But this violates our assumption that G'' is not connected!

Next, in $G' = G \setminus \{e, f\}$, another application of Lemma 14.32 shows that the connected component of X in $G \setminus \{e\}$ further breaks up into two connected subcomponents: the connected component of X and the connected component of Z. See Figure 14.15

again. Finally, in G'', where we join Z and Y, this connected component of Z from G' becomes part of the connected component of Y from G' (which originally was the connected component of Y in $G\setminus\{e\}$), and we find that G'' has two connected components: the connected component of X and the connected component of Y (or Z); once again see Figure 14.15.

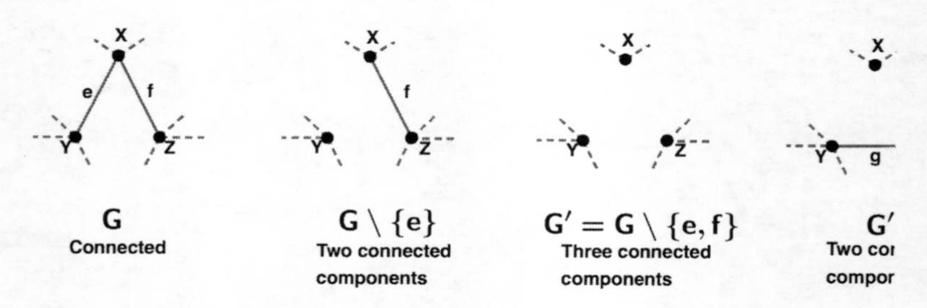

Figure 14.15. The Graphs G, $G\setminus\{e\}$, $G\setminus\{e,f\}$, and G'' in the case where G'' is not connected.

Each of these components still has the property that the degrees of vertices are all even (as already observed, the degree of X drops by 2 while the degrees of Y and Z stay the same on going from G to G''). Moreover, the number of vertices in each component is less than m, since the component of X is missing Y and Z, and the component of Y is missing X.

By induction, each of these components has an Eulerian circuit. If X is not the only vertex in its component, then this component has a circuit that looks like:

$$T_X = (P_0, \ldots, a_{k-1}, X, a_k, \ldots, P_0),$$

where P_0 is some vertex in the connected component containing X. This circuit may also be written by a cyclic shift as the circuit

$$T_X = (X, a_k, \ldots, P_0, \ldots, a_{k-1}, X).$$

If X is the only vertex in its connected component, then T_X looks like (X) if there are no edges incident on X, else, if there are loops $a_1, \ldots, a_k$ incident on X, then T_X looks like $(X, a_1, X, \ldots, a_k, X)$. Note that these special forms of T_X are already written as paths starting and ending at X.

The component containing Y has an Eulerian circuit that contains the edge g. If Z comes before Y in this circuit, we can simply traverse the circuit in the reverse direction and assume Y comes before Z. Therefore, this circuit looks like

$$T_Y = (Q_0, \ldots, b_{l-1}, Y, g, Z, b_l, \ldots, Q_0).$$

If we now replace the (Y, g, Z) portion of T_y with the trail T_X, then in the case where X is not the only vertex in its component, we get the trail

$$T = (Q_0, \ldots, b_{l-1}, Y, e, \underbrace{X, a_k, \ldots, P_0, \ldots, a_{k-1}, X}_{T_X}, f, Z, b_l, \ldots, Q_0),$$

which is our desired Eulerian circuit in G!

Of course, in the special case where X is the only vertex in its component, the newly created trail looks like

$$T = (Q_0, \dots, b_{l-1}, Y, e, \underbrace{X}_{T_X}, f, Z, b_l, \dots, Q_0),$$

or

$$T = (Q_0, \dots, b_{l-1}, Y, e, \underbrace{X, a_1, \dots, X, \dots, a_k, X}_{T_X}, f, Z, b_l, \dots, Q_0),$$

which once again is our desired Eulerian circuit in G.

To finish the proof of the theorem, we consider the case where G has exactly two vertices of odd degree, say X and Y. We add a new vertex Z and add an edge e between X and Z and an edge f between Y and Z. In the new graph G', X and Y have even degree (as we have added a new edge incident on each), Z has degree 2, and all other vertices have the same degree as in G, and are hence of even degree. By what we have proved above, G' must have an Eulerian circuit

$$T' = (Z, e, X, a_1, \dots, a_{k-1}, Y, f, Z)$$

which may also written by a cyclic shift as the circuit

$$T' = (X, a_1, \dots, a_{k-1}, Y, \underbrace{f, Z, e, X}).$$

We now form the trail obtained from T' by removing the tail portion (f, Z, e, X), i.e.,

$$T' = (X, a_1, \dots, a_{k-1}, Y).$$

Since T' visited every edge in G' exactly once, T will visit every edge in G exactly once as we have only removed that portion of T' that was outside G anyway. T is thus an Eulerian trail in G that starts at X and ends at Y. □

14.4 Further Exercises

Exercise 14.34. Practice Exercises:

Exercise 14.34.1. For each of the graphs depicted in Figure 14.4, determine the degree of each vertex. Classify the graphs into those that possess an Eulerian trail and those that do not. Determine Eulerian trails for those graphs that possess them.

Exercise 14.34.2. Repeat Exercise 14.34.1 for the graphs depicted in Figure 14.22 ahead.

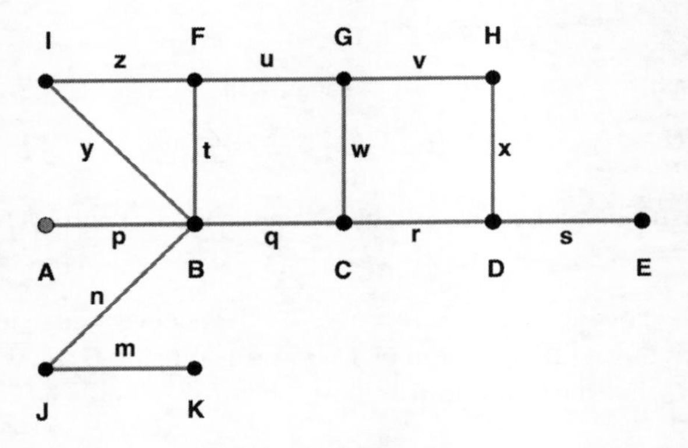

Figure 14.16. Exercise 14.34.3.

Exercise 14.34.3. The following are all walks in the graph depicted in Figure 14.16. Determine which of these are trails, which are closed trails, which are open trails, which are paths, and which are closed paths, which are open paths. Determine all pairs (i, j) such that W_i is a subwalk of W_j.

$$W_1 = (A, p, B, y, I, z, F, t, B, p, A),$$
$$W_2 = (A, p, B, y, I),$$
$$W_3 = (B, y, I, z, F, t, B),$$
$$W_4 = (K, m, J, n, B, y, I, z, F, u, G, w, C, q, B, n, J),$$
$$W_5 = (K, m, J, n, B, q, C, w, G, v, H, x, D, s, E),$$
$$W_6 = (K, m, J, n, B, q, C, r, D, s, E).$$

Exercise 14.34.4. For each of the graphs depicted in Figure 14.4, determine the cut edges, and identify the connected components formed after deleting the cut edges.

Exercise 14.34.5. Repeat Exercise 14.34.4 for the graphs depicted in Figure 14.19.

Exercise 14.35. Let $G = (V, E, \phi)$ be a simple graph (Definition 14.7), and let $|V| = n$. Fix some labeling of the vertices $\{V_1, V_2, \ldots, V_n\}$. With respect to this labeling, we define an $n \times n$ matrix $A = (a_{i,j})$ known as the *adjacency matrix* of G as follows:

$$a_{i,j} = \begin{cases} 1 \text{ if there is an edge between } V_i \text{ and } V_j, \\ 0 \text{ otherwise.} \end{cases}$$

(Thus, $a_{i,i} = 0$ for all i, because of the assumption of simplicity.)

(1) Show that the (i, j)-th entry of A^t is the number of walks of length t between V_i and V_j. (Hint: Use induction on t. For $t > 1$ consider the last vertex before V_j in walks between V_i and V_j.)

(2) If V_i and V_j are in the same component of G, show that the length of the shortest path from V_i to V_j $(i \neq j)$ is the smallest positive integer t such that the (i, j)-th entry of A^t is nonzero.

(Note that if V_i and V_j are in different components, then the (i, j)-th entry of A^t is zero for all t by part (1), as there are no paths between V_i and V_j, and hence by Lemma 14.21 no walks.)

Exercise 14.36. If G is a graph with vertices $\{V_1, V_2, \ldots, V_n\}$, the sequence

$$\{d_1, d_2, \ldots, d_n\},$$

where $d_i = \deg(V_i)$, is known as the *degree sequence* of the graph. We already know from Proposition 14.11 that $\sum_{i=1}^{n} d_i$ must be even. Show that if G is simple, then the degree sequence must satisfy the further property that for each k with $1 \leq k \leq n$,

$$\sum_{i=1}^{k} d_k \leq k(k-1) + \sum_{j=k+1}^{n} \min(k, d_j).$$

(Hint: Partition the vertex set into two sets $X = \{V_1, V_2, \ldots, V_k\}$ and $Y = V \setminus X$. Relate the sum on the left to the number of edges between pairs of vertices in X and the number of edges between vertices in X and vertices in Y. Now find an upper bound for each of these numbers.) As it turns out, the inequalities above, along with the easier fact that the sum of the degree sequence must be even, are also sufficient conditions for there to exist a simple graph with the given degree sequence (provided the sequence is ordered, by relabeling the vertices if necessary, so that $d_1 \geq d_2 \geq \ldots \geq d_n$). This is known as the Erdös-Gallai theorem.

Exercise 14.37. Let G be a connected graph, and X a vertex of degree 1. Show that $G \setminus \{X\}$ is still connected. (Hint: Note that G must have at least two vertices for us to be able to delete X; see Example 14.23(2) or directly see Remark 14.3. Invoke Remark 14.25 and assume that G has at least three vertices. Let P and Q be distinct vertices of G, neither equalling X. There is a path from P to Q in G because G is connected: what can you say about this path?)

Exercise 14.38. In this exercise we will derive a lower bound on the number of edges of a connected graph G. We will show that $|E| \geq |V| - 1$.

(1) Establish the result in the case where G contains a vertex of degree 0. (Hint: What should G look like in this case?)

(2) Assume that all vertices of G have degree 1. Establish the result in this case as well. (Hint: Again, what should G look like in this case?)

(3) Assume that all vertices of G have degree at least 2. Apply Proposition 14.11 to establish the result in this case.

(4) Now assume that we are not in the three cases considered above in parts (1), (2), and (3). Thus G has at least one vertex of degree 2 or higher, and at least one vertex of degree 1 (and of course, none of degree 0). Let $W = \{W_1, W_2, \ldots, W_k\}$ be the set of vertices of degree 1. Establish the result for $G \setminus W$ and then extend the result to G. (Hint: Exercise 14.37 may be helpful.)

Exercise 14.39. Let G be a graph, and assume that all vertices have degree at least 2. We will show that G must contain a cycle.

(1) Show that if G is not simple, it has to contain a cycle, independent of any degree condition on its vertices.

(2) Let P be the longest path in G (which at this point we may assume to be simple, although the arguments will go through even without this assumption), and assume $P = (X_0, e_1, Y_1, \ldots, e_k, Y_k)$. Let X be any vertex adjacent to X_0 (recall this means that there is an edge between X_0 and X). Show that X already appears in P.

(3) From part (2) above, we find that all vertices that are adjacent to X_0 appear in P. Let $\{X_1, X_2, \ldots, X_l\}$ be all the neighbors of X_0, arranged in the order in which they appear in P. (Thus, each X_i appears as some Y_{j_i} in the description of P in part (2), and $j_1 < j_2 < \cdots < j_l$. Moreover, Y_1, is obviously a neighbor as it is connected to X_0 by e_1, and it is the first neighbor that appears in P, so Y_1 is just X_1.) Show that $l \geq 2$ and that the length of P is therefore at least 2.

(4) For any $m \geq 2$, consider the neighbor X_m (which appears as some Y_{j_m} in P). This vertex is connected to X_0 by some edge f, by virtue of being a neighbor. Show that f does not appear in the subwalk

$$(X_0, e_1, X_1 (= Y_1), e_2, Y_2, \ldots, e_m, X_m (= Y_{j_m})).$$

(5) Now use the previous results to exhibit a cycle in G!

Exercise 14.40. In this exercise we will build on the result in Exercise 14.39 to show if a graph $G = (V, E, \phi)$ is such that $|E| \geq |V|$, then G must necessarily have a cycle. (Recall from Exercise 14.38 that in general we have the bound $|E| \geq |V| - 1$, so the current exercise shows that if $|E|$ exceeds its lower bound by just 1, we are guaranteed a cycle!)

As in Exercise 14.39, we may assume if we wish that G is simple, although the arguments will work even without this assumption. We will use induction on $|V|$:

(1) Show that if $|V| \leq 3$ then V must have a cycle. (Hint: Enumerate all possible graphs with $|E| \geq |V|$ and $|V| \leq 3$.)

(2) Now assume $|V| > 3$ and that the result has been established for all simple graphs which have fewer vertices than G. Use induction to show that G must contain a cycle. (Hint: By Exercise 14.39 we can assume that G has a vertex X of degree 0 or 1.)

Exercise 14.41. Show that e is a cut edge in a connected graph G if and only if there exist distinct vertices X and Y in G such that every path between X and Y contains e.

Exercise 14.42. In analogy with Definition 14.30, we may define a *cut vertex* of a graph G as a non-isolated vertex X lying in a connected component G_i of G such that the subgraph $G_i \setminus \{X\}$ is disconnected.

(1) Show by example that for each $n \geq 2$, there exists a connected graph G with a vertex X such that $G \setminus \{X\}$ has n connected components.

(2) Show that a vertex X of G lying in a connected component G_i is a cut vertex if and only if there exist distinct vertices Y and Z in G_i (neither equal to X) such that every path from Y to Z passes through Z.

Exercise 14.43. Show that if G is connected, and e is an edge that is part of a cycle, then $G \setminus \{e\}$ is still connected.

Exercise 14.44. Show that if G is a graph and a new graph G' is formed by adding an edge between two vertices of G, then the number of connected components of G' is at most that of G. What can you say about the two vertices that have now been connected if the number of connected components of G' is less than that of G?

Exercise 14.45. A graph $G = (V, E, \phi)$ is said to be *bipartite* if $V = V_1 \cup V_2$, where V_1 and V_2 are proper subsets of V, and such that every edge $e \in E$ has one endpoint in V_1 and the other in V_2. Some examples of bipartite graphs are shown in Figure 14.17.

The purpose of this exercise is to show that a connected graph is bipartite if and only if it has no cycles of odd length.

(1) Show that if G is bipartite, then G does not have any cycles of odd length. (Hint: If V_1 and V_2 are as in the definition, then any path in G has to go back and forth between V_1 and V_2.)

(2) Now we will show that if G has no cycles of odd length then it is bipartite.

   (a) Show that no connected component of G can have a cycle of odd length. (Hint: this is indeed as easy as it seems!)

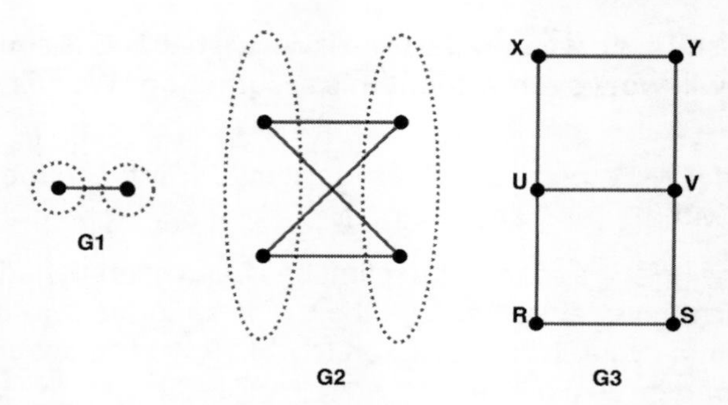

Figure 14.17. Some bipartite graphs. Can you see why G3 is bipartite?

(b) Show that if each connected component of G is bipartite then G is bipartite.

(c) Thanks to parts (2a) and (2b), it is enough to assume that G is a connected graph with no cycles of odd length and show that G must be bipartite. Pick an arbitrary vertex X in G, and let

$$V_1 = \{Y \in V \mid \text{length of shortest path from } X \text{ to } Y \text{ is even}\},$$

and let

$$V_2 = \{Z \in V \mid \text{length of shortest path from } X \text{ to } Z \text{ is odd}\}.$$

Note that $V_1 \cup V_2 = V$ (why?) and $V_1 \cap V_2 = \Phi$. We will show that every edge in G has one end point in V_1 and the other in V_2; this will prove that G is bipartite.

(i) Consider Y and Z in V_1, and let P_Y be a path of shortest length between X and Y, and P_Z a path of shortest length between X and Z. Let W be the last vertex where P_Y and P_Z intersect (see Figure 14.18).

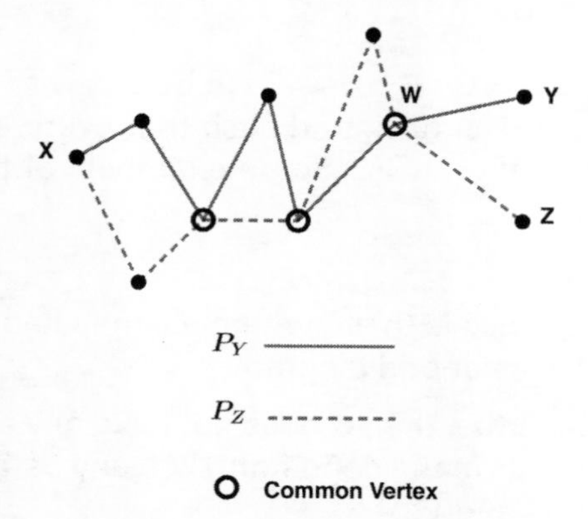

Figure 14.18. W is the last vertex where P_Y and P_Z intersect.

Let P'_Y be the portion of P_Y that runs from X to W, and let P'_Z be the corresponding portion of P_Z. Use the fact that P_Y and P_Z are of shortest length to show that P'_Y and P'_Z must be of equal length.

(ii) We will show that there is no edge between Y and Z. Assume to the contrary that e is an edge between Y and Z. Show that the walk along G that starts at W, proceeds to Y along P_Y, walks along e to Z, and then returns to W along the reverse of P_Z is a cycle of odd length. Conclude that no two vertices in V_1 are joined by an edge.

(iii) Show that no two vertices in V_2 are joined by an edge.

(iv) Conclude that G is bipartite.

Exercise 14.46. We consider a special class of graphs in this exercise. A *tree* is a connected graph with no cycles. (A graph whose connected components are trees is called a *forest*.) Figure 14.19 shows examples of trees and a forest. Show that a graph is a tree if and only if any two points are connected by a *unique* path.

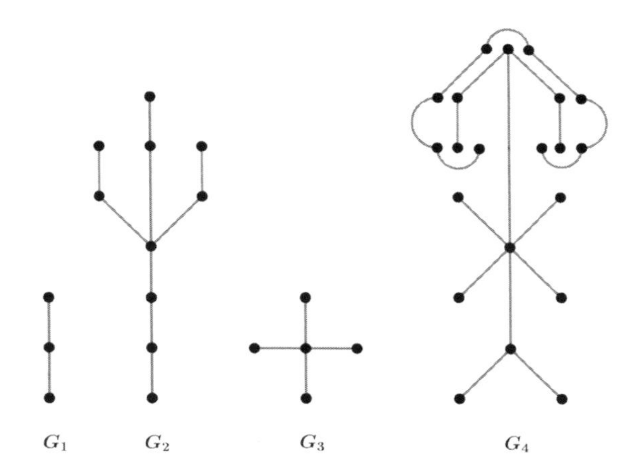

Figure 14.19. Graphs G_1, G_2 and G_3 are trees, G_4 is a forest with two components.

Exercise 14.47. Show that a graph $G = (V, E, \phi)$ is a tree if and only if G is connected and $|E| = |V| - 1$. (Hint: For one direction of the implication, "only if," use Exercise 14.40. For the other direction, assume that G has a cycle. Let e be an edge of this cycle, and count the number of edges and vertices of $G \setminus \{e\}$ after applying Exercise 14.43 to obtain a contradiction.)

Exercise 14.48. A *spanning tree* of a graph $G = (V, E, \phi)$ is a subgraph $T = (V', E', \phi' = \phi|_{E'})$ of G that is a tree, and whose vertex set V' equals V. (See Figure 14.20.)

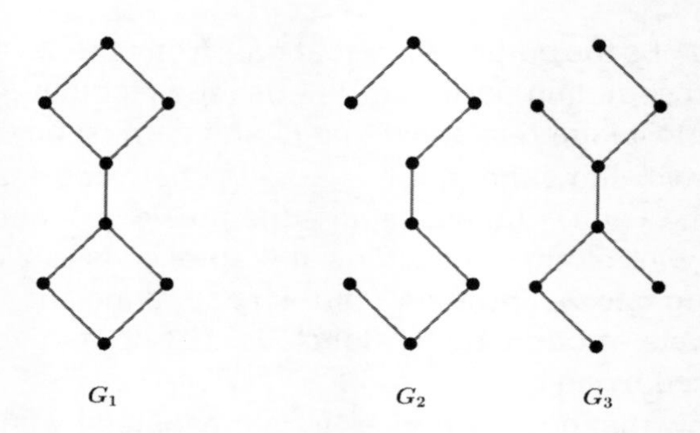

Figure 14.20. A Graph G_1, along with two of its spanning trees G_2 and G_3.

Show that a graph with several components cannot have a spanning tree, but every connected graph has a spanning tree. (Hint: for the second part, successively delete edges from cycles, applying Exercise 14.43 at each stage.)

Exercise 14.49. A graph $G = (E, V, \phi)$ is said to be *complete* if it is simple and any two vertices are adjacent. Show that if G is complete and $|V| = n$, then $|E| = \dfrac{n(n-1)}{2}$ and each vertex has degree $n-1$. Conclude that G has an Eulerian cycle if and only if n is odd, and an (open, i.e., not closed) Eulerian trail if and only if $n = 2$.

Complete graphs with $|V| = n$ are denoted K_n. (There is only one graph for each n.) Some complete graphs for small values of n are shown in Figure 14.21.

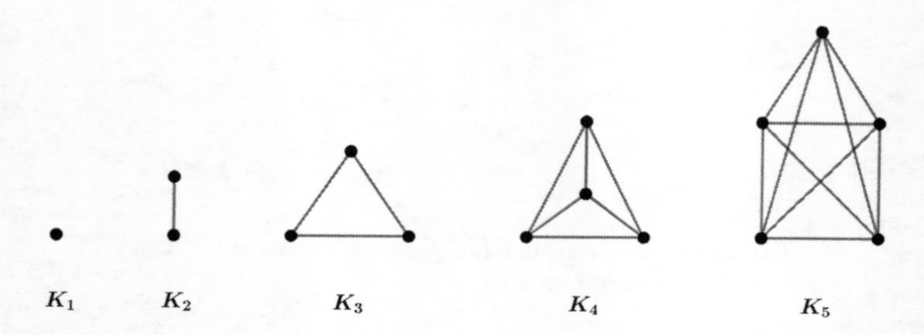

Figure 14.21. The complete graphs K_n for $n = 1, 2, 3, 4, 5$.

Remark 14.49.1. Complete graphs are examples of *regular* graphs, i.e., graphs in which each vertex has the same degree. Some regular graphs that are not complete are shown in Figure 14.22.

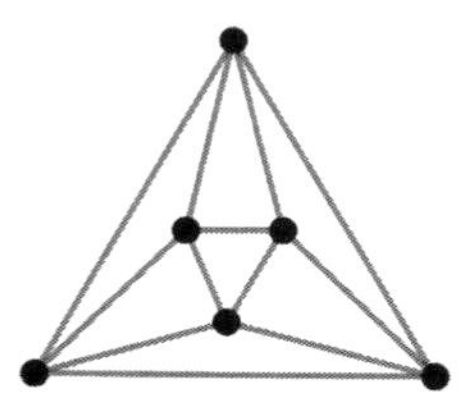
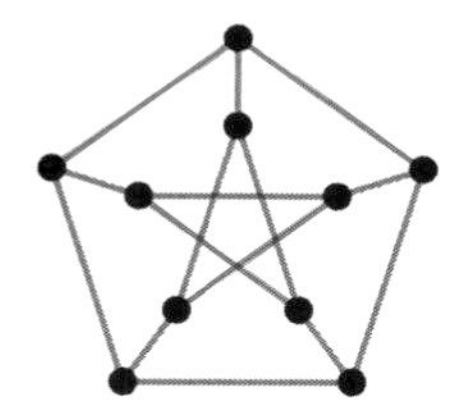
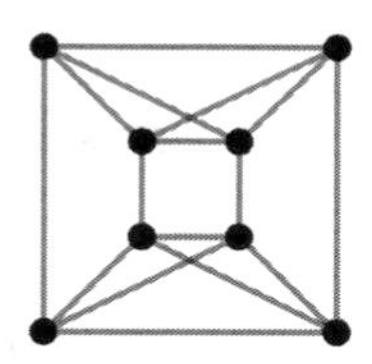

Figure 14.22. Some examples of regular graphs (that are not complete).

Index